AF868573

QUANTUM ELECTRONICS

Volume 1: Basic Theory

QUANTUM ELECTRONICS

Volume 1: Basic Theory

V. M. FAIN and YA. I. KHANIN

TRANSLATED BY

H. S. H. MASSEY

EDITED BY

J. H. SANDERS

THE M.I.T. PRESS
MASSACHUSETTS INSTITUTE OF TECHNOLOGY
CAMBRIDGE, MASSACHUSETTS

Copyright © 1969
Pergamon Press Ltd.

First English edition 1969

This book is Volume 13 of the International Series of Monographs in Natural Philosophy published by Pergamon Press Ltd., Oxford, England.

Distributed in the United States and Canada
by M. I. T. Press, Cambridge, Massachusetts

This book is a translation of Part I of Квантовая Радиофизика by V. M. Fain and Ya. I. Khanin published in 1965 by Sovetskoje Radio, Moscow, and includes corrections and revisions supplied by the authors.

Library of Congress Catalog Card No. 67–22832

PRINTED IN GERMANY

08 011820 8

Contents of Volume 1

Contents of Volume 1

Contents of Volume 2

Foreword

THE BIRTH of a new independent field of physics, now known by the name of quantum electronics, was heralded about ten years ago by the creation of the molecular oscillator. This field at once attracted the attention of a large number of research workers, and rapid progress took place. Extensive experimental and theoretical material has now been accumulated. The present book attempts to give a résumé of this material and, to a certain extent, to generalize it.

We have tried to arrange the material so that, as far as is possible, the reader need not continually refer elsewhere. The references to literature of a theoretical nature make no pretence of completeness, but when citing experimental work we have tried to give as full a list as possible since readers may be interested in details for which there is no space in the book.

The theoretical sections of the books are by no means a survey of present work. We have tried to highlight the basic principles and their results. It is natural that slightly more attention has been paid to fields in which the authors themselves have been involved. The experimental material is given in the form of a survey, with only a brief description of the technical details of devices.

The book as a whole is designed for the reader with a knowledge of theoretical physics (quantum mechanics in particular) at university level. It should be pointed out that the material in the various sections of the book is of differing degrees of complexity. Readers will need less preparation for Volume 2. The most difficult paragraphs are marked with an asterisk.

In conclusion we should mention that sections 1–20, 22–40 and Appendix I were written by V. M. Fain; sections 41–49 and 51–58 by Ya. I. Khanin. At the authors' request section 21 was written by V. N. Genkin, section 50 by E. G. Yashchin, Appendix II by V. I. Talanov and Appendix III by Ye. L. Rozenberg.

We are grateful to Professor A. V. Gaponov and Professor V. L. Ginzburg for reading the book in manuscript and making a number of useful suggestions. We are also grateful to A. P. Aleksandrov, V. N. Genkin, G. M. Genkin, N. G. Golubeva, G. L. Gurevich, G. K. Ivanova, M. I. Kheifets, Yu. G. Khronopulo, Ye. E. Yakubovich and E. G. Yashchin for their great help in reading the proofs.

Radiophysics Scientific Research Institute,
Gor'kii

V. M. FAIN
YA. I. KHANIN

Preface to the English Edition

WE WERE very pleased to learn that our book was to be translated into English and would thus become available to a wide circle of English-speaking readers.

Research in the field of quantum electronics has continued since our manuscript was handed over to the Soviet publishers, but there has been no essential change in the basic theory or understanding of the physical processes in quantum devices. Among the most interesting questions on which work has been done of late, mention should be made of the development of coherence theory and holography, which is based on it. Non-linear optics and its application are developing rapidly. Although the book treats the fundamentals of non-linear optics, one of the important non-linear optical effects—the phenomenon of self-trapping—is not discussed in the book.

Among the other important problems recently worked on and not reflected in the book are the electrodynamics of gas lasers, the theory of the natural width of lasers (gas lasers in particular) and the important work on semiconductor lasers.

Despite the great importance of these questions we did not include them in the present edition, not merely for lack of time but also because one is here dealing with subjects under development, not all of whose facets are completely clear and whose discussion would be previous.

We are preparing material on the majority of these questions, and also a detailed treatment of the theory of the non-linear properties of materials (mainly solids) which is of great importance for non-linear optics and quantum electronics; this will be included in the second Soviet edition, which is planned for 1969.

In the present edition we have confined ourselves to correcting any errors that have been found and making some slight additions.

February 1967

V. M. FAIN — Institute of Solid-State Physics, Moscow, Academy of Sciences U.S.S.R.

YA. I. KHANIN — Radiophysics Scientific Research Institute, Gor'kii.

Introduction

QUANTUM electronics as an independent field of physics came into prominence in the middle fifties when the first quantum oscillators and amplifiers were made. The immediate precursor of quantum electronics was radiofrequency spectroscopy, which is now one of its branches. An enormous quantity of experimental material concerned with the resonant properties of substances had been accumulated by radiofrequency spectroscopy. Such research had made it possible to establish the structure of levels, the frequencies and intensities of transitions, and the relaxation characteristics of different substances. Investigations of paramagnetic resonance spectra in solids and the inversion spectrum of ammonia have been of particular importance to quantum electronics.

During radiofrequency investigations the state of a substance is not, as a rule, subject to significant changes and remains close to thermodynamic equilibrium. But besides the investigation of substances under undisturbed conditions, other methods began to appear which were connected with the action of strong resonant fields on a substance. These methods, which we can call active, were first applied in nuclear magnetic resonance. They include nuclear magnetic induction, spin echo and the Overhauser effect. The main outcome of these methods was the possibility of producing strongly non-equilibrium states in quantum systems which could emit coherently. Therefore the actual material was accumulated through radiofrequency spectroscopy, and resulted in the birth of experimental ideas which were then used as the basis of quantum oscillators and amplifiers.

The concept of stimulated emission, which is important for quantum electronics, was first formulated by Einstein as early as 1917. Ginzburg (1947) pointed out the importance of this phenomenon in radiofrequency spectroscopy.

The idea of amplifying electromagnetic waves by non-equilibrium quantum systems was first mooted by Fabrikant, Vudynskii and Butaeva. The patent (Fabrikant, Vudynskii and Butaeva, 1951) obtained by this team in 1951 contains a description of the principle of molecular amplification. Slightly later, in 1953, Weber made a suggestion about a quantum amplifier. Basov and Prokhorov (1954) discussed an actual design for a molecular oscillator and amplifier operating with a beam of active molecules and developed their theory. Gordon, Zeiger and Townes independently had the same idea and, in the same year, 1954, published a report on the construction of an oscillator that operated with a beam of ammonia molecules. Gordon, Zeiger and Townes introduced the now well-known term "maser"†.

The successful operation of a beam molecular oscillator stimulated the search for new methods and results were not long in coming. Basov and Prokhorov (1954) suggested the principle of a three-level gas-beam oscillator. In 1956 Bloembergen discussed the possibility of making a quantum amplifier with a solid paramagnetic working medium. The estimates he made confirmed that the idea was feasible and in 1957 such an instrument was made by Scovil, Feher and Seidel. After this, reports appeared on the production of a whole series of similar instruments based on different paramagnetic crystals.

Instruments based on quantum principles have a number of exceptional properties when compared with ordinary amplifiers and oscillators. The molecular beam maser oscillator is not particularly powerful but its stability is far better than the stability of the best quartz oscillators. This has brought about the use of the maser as a frequency standard. The paramagnetic maser amplifier has an extremely low noise level and satisfactory gain and bandwidth characteristics.

The next stage in the development of quantum electronics was the extension of its methods into the optical range. In 1958 Schawlow and Townes discussed the question theoretically and came to the conclusion that it was perfectly possible to make an optical maser oscillator. They suggested gases and metal vapour as the working substances. The question of possible working substances and the methods of producing the necessary non-equilibrium states in them was also discussed in a survey by Basov, Krokhin and Popov (1960). These authors discussed paramagnetic crystals and semiconductors as well as gases.

In 1960 Maiman made the first pulsed ruby quantum optical generator which is called a "laser"‡. For the first time science and technology had available a coherent source of light waves. The future prospects of devices of this kind were obvious and in a very short time a large number of teams had come onto the scene of laser research. The list of crystals suitable for use in lasers quickly grew. Then certain luminescent glasses and liquids were used for the same purpose. In 1961 Javan, Bennett and Herriott made the first continuous laser operating with a mixture of the inert gases neon and helium.

† Maser is an acronym formed from *M*icrowave *A*mplification by *S*timulated *E*mission of *R*adiation.

‡ The term laser is an acronym from *L*ight *A*mplification by *S*timulated *E*mission of *R*adiation. It must be pointed out that there is not yet any firmly established terminology in quantum electronics. Besides "laser" the name "optical maser" is frequently used.

Introduction

Quantum electronics is very young; its basic trends are still far from clear. A whole series of problems is still unsolved. Under these conditions the writing of a monograph discussing the theoretical and experimental basis of quantum electronics is a rather complex affair. It must be understood that the book reflects to only a limited extent the position as it is today.

Quantum electronics as a theoretical science possesses a number of characteristic features which separate it both from quantum physics and from electronics. Unlike ordinary "classical" electronics, quantum electronics is characterized by the extensive application of the methods of quantum theory. However, the application of quantum field theory to quantum electronics has a specific feature which distinguishes it from ordinary quantum electrodynamics (see, e.g., Heitler, 1954; Akhiezer and Berestetskii, 1959).

Quantum electronics makes wide use of the resonant properties of matter both for the study of matter itself (radiofrequency spectroscopy, paramagnetic resonance), and for its use in quantum amplifiers and oscillators. It is obvious that resonances with a high Q-factor in a substance are essential for both purposes. To obtain sharp resonances discrete energy levels must exist in the substance. The presence of discrete electron levels means that these electrons cannot be free but must be in bound states in the atoms, molecules or solid. We notice that the characteristic feature of ordinary "classical" electrodynamics is the interaction of the radiation field with free electrons. It is true that quasiclassical systems (harmonic oscillator, electron in a magnetic field, etc.) may also have a discrete spectrum but the essential feature of classical and quasi-classical systems is that the energy levels of such systems are quasi-equidistant. For example, the harmonic oscillator has equidistant levels with no upper limit. The energy spectrum of quantum systems is much more diverse than the spectrum of quasi-classical systems. In particular the energy levels may be so arranged that there are two levels whose spacing is not the same as the spacing of any other levels in the same system. Under certain conditions no attention need be paid (during an interaction with radiation of the corresponding frequency) to any other levels of the system and we can use the idealization of a two-level system. The idealizations of a three-level system, etc., are introduced likewise.

As we have already pointed out, wide use is made of the resonant properties of matter in quantum electronics. As may be easily understood, during the resonant interaction of matter with a field it is particularly important to allow for different kinds of dissipative relaxation processes. Unlike ordinary quantum electrodynamics in which, as a rule, we are not interested in relaxation processes in matter, in quantum electronics the concept, and thus the description, of the different relaxation processes plays a major part.

The concept of stimulated emission plays an important and even predominant part in quantum electronics. All the active quantum-electronic instruments—maser amplifiers and oscillators—use the phenomenon of stimulated emission. The phenomenon of stimulated emission is closely linked (as will become clear from the appropriate sections of the book) with the non-linear properties of quantum systems used in quantum electronics. The non-linear properties, in their turn, are caused by the non-equidistant nature of the energy levels.

When describing the processes of the interaction of matter with radiation we must, strictly speaking, use quantum theory, i.e. quantum theory is used to treat the matter and the field. For many problems, however, the classical description of an electromagnetic field is a fully justified approximation. This is because the fields discussed in quantum electronics are large, and because the mean quantum values of the electric and magnetic fields are accurately described by the classical Maxwell equations. It is essential to allow for the quantum properties of the field when investigating the quantum fluctuations of the field, in particular when studying the noise properties of amplifiers and oscillators.

The arrangement and selection of the material in the present book have been made with these features of quantum electronics in mind. The book is composed of two parts: Volume 1 "Basic Theory" and Volume 2 "Maser Amplifiers and Oscillators". A large amount of material has been kept for the Appendixes.

In Volume 1 an attempt is made to give the basic theory of quantum electronics. In this part we have tried to show how the concepts and equations used in quantum electronics follow from the basic principles of theoretical physics. When doing this we make frequent use of very simple models so as not to complicate the treatment. Such models are necessary for the understanding of a particular process, but the models can often not be used for direct comparison with experiment.

The first chapter of the book deals with general questions of the interaction of radiation with matter. The basic concepts of quantum theory are briefly treated in this chapter. The reader's attention is particularly drawn to the density matrix description of the quantum state. This is because in its various applications quantum electronics deals with mixed states and not with pure states. Quantum theory allows us, by the use of the density matrix, to give a unified description of both pure and mixed states. In the first chapter we discuss in sequence the quantum theory of fields in resonators, in waveguides and in free space and also the concept of phase in quantum field theory, of the indeterminacy relation between the phase and the number of particles, the question of the transition to classical physics, etc. Section 4 discusses in more detail than usual the question of the different forms of interaction energy between a field and charged particles.

The second chapter deals with the general question of relaxation. When there are relaxation processes present the behaviour of quantum systems is governed by the cause of the dissipation—a dissipative system which possesses a continuous spectrum and an infinite number of degrees of freedom. In this case we must derive approximate equations which will take into account the relaxation processes (the transport equations). Therefore Chapter II deals essentially with the applicability of the different equations used in quantum electronics. In particular, by proceeding from basic principles, we derive the conditions for applicability of the frequently used population balance equations. The same chapter discusses the questions of the irreversibility of real systems and the principle of the increase of entropy. We also show how the transport equations can be used to describe fluctuations. Some long calculations are given in this chapter but they may be omitted on the first reading without making it difficult to understand the other parts of the book. The results which are necessary for reading subsequent chapters are given in the introduction to this chapter.

In Chapter III we have gathered together the possible quantum effects in ordinary electronics which may appear at very high frequencies and at low temperatures. These effects, as a rule, are small. In Chapter III an account is also given of the quantum theory of real resonators with finite Q.

In Chapters IV and V we discuss the behaviour of quantum systems in fields, which are here described classically. Particular attention is paid to the response of a system to such fields. This response, for example in the form of the mean magnetization of the system, is described in terms of the susceptibility. A number of general susceptibility properties are discussed, particularly the dispersion relations, the fluctuation-dissipation theorem, and the symmetry properties. In these same chapters we treat the idealizations of two- and three-level systems and find the equations of motion for these systems. In § 20 of Chapter IV we show how it is possible to give a rigorous description of systems which are not subject to the equations derived in Chapter II. The method of moments is used; the rigorous basis of this method is given in § 20. In § 21 it is used to examine cross-relaxation processes.

In Chapters VI, VII and VIII we deal with a number of questions concerning the theory of spontaneous and stimulated emission. In particular we discuss the connection with classical theory, the part played by non-linearity, the phase relations, etc. Chapter VII treats the theory of coherent spontaneous emission in free space and the theory of the natural line width. In Chapter VIII we deal with the physical nature of the processes of spontaneous and stimulated emission in a resonator.

Recently a new branch of quantum electronics—non-linear optics—has appeared. The development of non-linear optics, connected with success in the field of optical quantum light generators (lasers), is only just beginning. However, in our view a number of the essential features of the interaction of matter with optical waves can already be stated. The ninth and last chapter of the first volume is devoted to relating these features to the general scheme of quantum electronics.

In Chapters X and XI of Volume 2 we discuss the elements of the theory of quantum oscillators and amplifiers working in the microwave region and review the practical achievements in this field. A relatively large amount of attention is paid to two-level paramagnetic masers although they have not been put to practical use. This is done because two-level systems are simpler, and their theoretical analysis can be carried out in detail; in addition, this material is not contained in other books on the physics of quantum electronics (Singer, 1959; Troup, 1959; Vuylsteke, 1960).

The quantum paramagnetic amplifier theory discussed in Chapter X is of a general nature and its results are also fully applicable to the case of multi-level amplifiers. In the section dealing with quantum oscillators most attention is paid to the dependence of the form of the emitted signal on the different parameters. Unfortunately there is at present no satisfactory theory of transient modes in quantum oscillators. Material is therefore presented which, although it reflects the present level of the theory, is more illustrative in nature. Questions connected with methods of exciting the working substance are discussed in a fair amount of detail in Chapters X and XI. In all cases when approximate methods of calculation are used we have tried to explain the limits of their applicability, since this is generally not discussed in published works.

The maser oscillator operating with a beam of active molecules is described somewhat briefly in Chapter XI. We considered that we could limit ourselves to a short description by assuming that this material is known to the reader from the books of Singer (1959), Troup (1959) and Vuylsteke (1960).

Chapter XII is devoted to optical masers. Most of the space here is occupied by a survey of experimental achievements and a description of the features of laser operation. Theoretical questions are touched upon in so far as the present state of the theory permits. It should be pointed out that the stream of original papers about lasers is so thick and fast at present that the material in Chapter XII will probably be out of date by the time this book sees the light of day.

The book contains three Appendixes. To them is relegated material which lies a little outside the general plan of the book. Nevertheless the importance of this material is quite clear.

In Appendix II we give the elements of the theory of optical resonators. It should be pointed out that interest in problems concerning the electrodynamics of the optical waveband has been aroused only quite recently because of the development of laser technology. As far as we know the attempt made in Appendix II to give a systematic treatment of the material is one of the first.

Appendix III discusses the spectra of the paramagnetic crystals used in maser amplifiers and oscillators. This chapter is by way of a short review and assumes the reader's acquaintance with the problems discussed. In this chapter there is a detailed bibliography and list of sources which must be used as an introduction to the subject.

In conclusion we would remark that the reading of the book requires the reader's acquaintance with quantum theory at the level of a university course in theoretical physics.

We make frequent reference to the excellent course of theoretical physics by Landau and Lifshitz. This does not mean, of course, that the reader has to know the whole of Landau and Lifschitz. It is sufficient merely to understand those parts of the book mentioned here. Among other books we can recommend *The Quantum Theory of Radiation* by Heitler, to which we make frequent reference.

It must be pointed out that the present book does not discuss a number of problems in quantum electronics such as the radio-spectroscopy of gases, nuclear magnetic resonance and paramagnetic resonance. These are dealt with in a number of monographs (Townes and Schawlow, 1955; Andrew, 1955; Ingram, 1955; Al'tshuler and Kozyrev, 1961; Gordy, Smith and Trambarulo, 1953) to which we refer the interested reader.

In conclusion we should like to say that a number of chapters in the book can to a certain extent be read independently of the others. For example, the reader chiefly interested in masers and lasers can concentrate his attention on the second Volume of the book. From the first Volume he may need § 24 and an acquaintance with Chapters I and VIII.

VOLUME 1

Basic Theory

CHAPTER I

The Quantum Theory of the Interaction of Radiation with Matter

THE PRESENT chapter gives a short account of the basic concepts of quantum theory and its application to the interaction of an electromagnetic field with matter. We have limited ourselves to non-relativistic quantum mechanics when describing the behaviour of matter. Since quantum electronics generally has to deal with mixed states matter and of field this chapter introduces the density matrix which can be used for a unified description of pure and mixed states.

1. The Basic Concepts of the Quantum Theory

1.1. Basic Postulates

The nature of the phenomena occurring at the atomic level is very different from the nature of the phenomena of the macrocosm. For this reason the basic concepts of the classical theory proved to be invalid in describing the microcosm. The concept of the state of a physical system underwent a most radical re-examination. In classical physics it is assumed that the physical quantities (or properties of a system) found from various measurements made on a system are characteristics of the particular state of the system, that they are always present in a given system in a definite amount and that this does not depend on the observational methods and equipment. In quantum physics they are at the same time characteristics of the methods and equipment used for the observations. In the microcosm we cannot ignore the effect of the measuring apparatus on the measured object. Therefore the concept of the quantum state takes into account both the object which is in this state and the possible experimental devices used to make the measurement. Accordingly, the quantum theoretical description of quantum objects differs essentially from the classical description. Quantum theory, unlike the classical theory, is a statistical theory in principle. The laws of quantum theory do not govern the actual behaviour of a particular object, but give the probabilities of the various ways in which the object may behave as a result of an interaction with its surroundings.

The following three postulates form the basis of the quantum description of physical phenomena.

1. *Each physical quantity has corresponding to it a linear Hermitian operator or matrix.* For example the radius vector of a particle $\boldsymbol{r}$ is associated with the multiplication operator $\hat{\boldsymbol{r}}$, the momentum of the particle with the operator $\hat{\boldsymbol{p}} = -i\hbar\nabla$, the angular momentum with the operator $\hbar\hat{\boldsymbol{L}} = [\hat{\boldsymbol{r}} \wedge \hat{\boldsymbol{p}}] = -i\hbar[\hat{\boldsymbol{r}} \wedge \nabla]$. The operators corresponding to the physical quantities are, generally speaking, not commutative. There are commutation relations between the coordinate and the momentum operators:

$$\hat{x}\hat{p}_x - \hat{p}_x\hat{x} = i\hbar, \quad \hat{y}\hat{p}_y - \hat{p}_y\hat{y} = i\hbar, \quad \hat{z}\hat{p}_z - \hat{p}_z\hat{z} = i\hbar, \tag{1.1}$$

and there are also commutation relations between the operators of the components of the angular momentum:

$$\hat{L}_x\hat{L}_y - \hat{L}_y\hat{L}_x = i\hat{L}_z, \quad \hat{L}_y\hat{L}_z - \hat{L}_z\hat{L}_y = i\hat{L}_x, \quad \hat{L}_z\hat{L}_x - \hat{L}_x\hat{L}_z = i\hat{L}_y, \tag{1.2}$$

where $\hbar$ is Planck's constant divided by 2π.

Commutation relations such as (1.1) and (1.2) are basic characteristics of operators.

2. *Only the eigenvalues of the operator $\hat{A}$ can be the result of a precise measurement of a physical quantity represented by this operator.* The characteristic difference from classical theory is the fact that physical quantities may take up a discrete, as well as a continuous, series of values. It is well known, for example, that the energy spectrum of atoms is discrete in nature.

3. *When the state of a system can be described by a wave function Ψ the mean value of the physical quantity represented by the operator $\hat{A}$ is given by*

$$\langle A \rangle = \int \Psi^* \hat{A} \Psi \, dq, \tag{1.3}$$

where q stands for all the arguments (with the exception of the time t) of the function Ψ.

Let us examine the last postulate in greater detail. It is first necessary to define more closely the concept of state in quantum theory and to define clearly the statistical ensemble to which the mean value of a given quantity refers.

1.2. States and Statistical Ensembles

States in quantum theory are subdivided into "pure" states and "mixed" states, or "mixtures". Let us first examine pure states, to which the third postulate relates. A pure state can be defined by a wave function Ψ. The arguments of this function are the eigenvalues of a certain complete set of quantities $\hat{A}, \hat{B}, \hat{C} \ldots$†. We denote this set of arguments by the symbol q. In the course of time the wave function alters and the state of the system alters accordingly. It is assumed in this case that if the wave function is given at some fixed point in time its future behaviour is also determined. The wave function satisfies the principle of superposition, the essence of which is that if $\Psi_1(q, t)$ and $\Psi_2(q, t)$ are two possible wave functions describing two different states, then the functions $c_1\Psi_1(q, t) + c_2\Psi_2(q, t)$ also describes a state of the system.

As has already been pointed out, quantum theory is a statistical theory, unlike classical theory. The statements of quantum theory are generally statistical in nature. A wave function fully defines the statistical properties of the system's state. A statistical set (ensemble) in quantum theory is a set of identical measurements (experiments) made on an object in a given quantum state. A measurement or experiment, generally speaking, changes the state of the object. It is therefore necessary (in order not to go outside the framework of the given ensemble) to return the object after each measurement to the original quantum state or to deal with a set of objects in one and the same quantum state. In this case the measurement is made once on each object. In an ensemble produced in this way we can introduce the probability distribution for obtaining a particular result of a measurement. Therefore to define a statistical ensemble in quantum theory we must first state the type of measurement to be made on the object, and secondly give the state of the object.

The question arises of how to find the probability distribution in the different ensembles formed during the measurement of one quantity or another in a given state with the wave function $\Psi(q, t)$.

It can be shown, proceeding from (1.3), that the probability distribution of the quantity q in the ensemble as a result of measuring this quantity is given by the square of the modulus of the function $\Psi(q, t)$ in the q-representation‡. For example, the probability distribution of the coordinate of an electron in a state with the wave function $\Psi(\mathbf{r})$ (in the ensemble appearing when measuring the coordinate $\mathbf{r}$) is given by $|\Psi(\mathbf{r})|^2$. In order to obtain the probability distribution of the momenta p (in the ensemble appearing when

† We should mention that a complete set of values is defined as that set of physical quantities that can be measured simultaneously. Here, in accordance with the uncertainty principle, it is assumed that if these quantities have simultaneously determined values, then no other quantity (which is not a function of them) can have a definite value.

‡ The representation of the wave function is determined by all its arguments. The mean value of the quantity $\hat{A}$, which is a function only of q, in accordance with (1.3) is

$$\langle A \rangle = \int |\Psi(q)|^2 A(q)\, dq.$$

It follows from this and the normalization condition $\int |\Psi(q)|^2\, dq = 1$ that $|\Psi(q)|^2$ gives the probability distribution.

measuring the momentum of an electron in the same state) we must change to the p-representation; this is done by expanding $\Psi(\boldsymbol{r})$ into a series (or integral) in the eigenfunctions of the momentum operator $\hat{\boldsymbol{p}}$

$$\Psi(\boldsymbol{r}) = \sum_{p} \Phi(\boldsymbol{p})\, \Psi_p(\boldsymbol{r}).$$

The set of coefficients $\Phi(\boldsymbol{p})$ is the wave function in the $\boldsymbol{p}$-representation and $|\Phi(\boldsymbol{p})|^2$ gives the probability distribution of the momenta. During an individual act of measurement, for example of the coordinate of an electron, its state changes in such a way that its wave function becomes $\delta(\boldsymbol{r} - \boldsymbol{r}_0)$ (i.e. the electron transfers into a state with a definite coordinate). Likewise, measurement of the momentum transfers the electron into the state Ψ_p with a definite momentum value. Therefore in order to investigate an ensemble in a given state it is necessary to make the measurements on a series of identical objects in one and the same state, or in some way to return the object after each measurement to the original state.

It must be stressed that all representations of the wave function (which are analogous to the use of different frames of reference in relativity theory) can be used with equal validity to describe the state. Stating the representation, in the same way as stating the quantum state, in no way assumes a unique statement of the statistical ensemble. (In order to give the ensemble it is necessary to state which measurements are being made on the system.)

In the general case we can change from one representation to another by using the appropriate unitary transformation†

$$\Psi(q) = \hat{U}\Phi(p).$$

Therefore the presence of a definite wave function leads to the probability distributions for different measurements in a given state being connected by the unitary transformation

$$|\Psi(q)|^2 = |\hat{U}\Phi(p)|^2, \quad |\Phi(p)|^2 = |\hat{U}^{-1}\Psi(q)|^2. \tag{1.4}$$

This connection between the probability distributions is applicable to pure states which have a definite wave function. In the general case of a mixed state relation (1.4) does not hold.

Before moving on to deal with mixed states let us compare the quantum

† We should mention that the unitary transformation changes the wave function Ψ into $\hat{U}\Psi$, the operator $\hat{A}$ changing into $\hat{U}^{-1}\hat{A}\hat{U}$. The characteristic unitary property of $\hat{U}\hat{U}^{+} = 1$ or $\hat{U}^{+} = \hat{U}^{-1}$ (where the $+$ sign denotes Hermitian conjugation) ensures invariance of all the observed quantities, i.e. the unitary transformation does not alter the spectrum, the mean values, the transition probabilities, etc.

and classical descriptions of the state of a system. In classical mechanics the state is given by all the coordinates and momenta of the system. It is assumed that as the result of a measurement made on the system these values of the coordinates and momenta are known. It is therefore sufficient to make one measurement to check that the system is in the given state. It is assumed here that the measurement process introduces negligibly small changes in the state of the system. In quantum theory this is not the case. Here the state of the system is given by the wave function, which is a function only of the coordinates or only of the momenta. The coordinates and momenta cannot be stated simultaneously. They do not exist simultaneously. This follows from the first postulate, which relates the operators to the physical quantities. Unlike the classical description, the wave function description of the state provides information not on what really happens in a system but on what *may* happen to the system (with a certain probability) as the result of interaction with a measuring instrument. Here the measurement may introduce a considerable change in the state of the system. It must be stressed that the statistical description in quantum theory is not the result of lack of information about a property of the system. Let us say, for example, that a particle is in a state described by the wave function $\Psi(\boldsymbol{r}, t)$. Then if $\Psi(\boldsymbol{r}, t)$ is not the same as $\delta(\boldsymbol{r} - \boldsymbol{r}_0)$ the question of what is the actual coordinate of the system has no meaning. This question is just as meaningless as the question of the wavelength of a finite wave packet (or of its coordinate). A quantum particle is a more complex object than a classical object. A quantum particle has no coordinate or momentum as such, as in Newtonian mechanics. As the result of measuring the coordinate the particle is localized—it changes into a state with a definite coordinate. Knowing the wave function we can predict the relative frequency of the result of a given measurement. Therefore in order to check that the system is in a given quantum state we must make not one measurement, but a series consisting of a large enough number of measurements. In quantum theory, just as in classical theory, defining the state at one point in time predetermines it at any subsequent point in time. In the second section of the present chapter we shall examine the equations that define the change of the wave function with time.

1.3. Mixed States. The Density Matrix

We now move on to an examination of mixed states, which cannot be described by a wave function. The fact that a state cannot always be described by a wave function can be understood if we examine the sub-system A of a certain system $A + B$. Let the system $A + B$ be described by the wave function $\Psi = \Psi(x_A, x_B)$, where x_A and x_B are the coordinates of the sub-systems A and B respectively. This function, generally speaking, does not break down into the product of the functions Ψ_A and Ψ_B

$$\Psi \neq \Psi_A \cdot \Psi_B$$

even in the case when the systems A and B do not interact†.

This means that neither A nor B can be described by wave functions (for greater detail see Mandel'shtam, 1950). The question now arises of how to describe the state of a quantum system in the case when there is no wave function. Neumann (1932) found the following method. Mixed states (just as pure ones) can be described in a unified way by a *density matrix*. Let us first take the density matrix for a pure state. To do this we change from the q-representation of the wave function to another representation which is characterized here by the integer n which denotes the eigenvalues of the Hermitian operator $\hat{N}$ (which describes a certain physical quantity)

$$\Psi(q, t) = \sum_n b_n \Psi_n(q). \tag{1.5}$$

Here $\Psi_n(q)$ are the eigenfunctions of the operator $\hat{N}$;
b_n is the wave function in the N-representation.

Substituting (1.5) in (1.3), which defines the mean value of A, we find the expression for the mean value of A in the N-representation‡

$$\langle A \rangle = \sum_{n,n'} b_n^* b_{n'} A_{nn'}, \tag{1.6}$$

where

$$A_{nn'} = \int \Psi^*(q)\, \hat{A} \Psi_{n'}(q)\, dq \tag{1.7}$$

are the matrix elements of the operator $\hat{A}$. Since we may consider the N-representation to be arbitrary, (1.6) is the expression for the mean value in an arbitrary representation. Denoting $b_n^* b_n$ by $\sigma_{n'n}$ we can rewrite the expression for the mean (1.6) in a more symmetrical form:

$$\langle A \rangle = \sum_{n'n} \sigma_{n'n} A_{nn'} = \mathrm{Tr}(\hat{\sigma}\hat{A}) = \mathrm{Tr}(\hat{A}\hat{\sigma}). \tag{1.8}$$

The expression Tr (Trace) denotes the sum of the diagonal elements of the matrix

$$(\sigma \hat{A})_{n'n''} = \sum_n \sigma_{n'n} A_{nn''}.$$

† It can easily be shown that the wave function of the system $A + B$ can differ from $\Psi_A \cdot \Psi_B$ if an interaction has occurred between these sub-systems (a collision), even if they are not interacting at the present time t.

‡ The question may arise of the ensemble in which we have taken the mean value of the quantity $\langle A \rangle$, which may in particular be a function of the noncommuting operators $\hat{q}$ and $\hat{p}$. It is obvious that this cannot be the ensemble in which measurements of $\hat{p}$ (or $\hat{q}$) are made. In fact this ensemble is determined by the measurements of the quantity A, i.e. $\langle A \rangle = \Sigma w(a)\, a$, where $w(a)$ is the probability that the system changes into a state with a definite value a as the result of an individual act of measurement.

When writing the last equation we made use of the matrix multiplication rule. It is easy to see that the operators within the trace sign can change places: $\mathrm{Tr}\, \hat{A}\hat{B} = \mathrm{Tr}\, \hat{B}\hat{A}$. The matrix $\hat{\sigma}$ is called the *density matrix* or the *statistical operator*. It is obvious that the density matrix can be used to obtain the same statistical information as the wave function (when the latter exists). Formula (1.8) can be used to find various means, whilst the diagonal elements of the matrix, as can easily be seen, give the probability distribution of the quantity N. From this follows the normalization condition†

$$\sum_n \sigma_{nn} = \mathrm{Tr}\, \hat{\sigma} = 1. \tag{1.9}$$

The density matrix can also be used to describe mixed states. Let $\hat{F}$ be an operator relating to the sub-system A of the system $A+B$. Then the matrix elements of $\hat{F}$ taken by means of the eigenfunctions $\Psi_{nu} = \Psi_n(x_A)\, \Psi_u(x_B)$ are of the form

$$F_{nu;u'u'} = F_{nn'}\delta_{uu'}, \tag{1.10}$$

where $\Psi_n(x_A)$ are the eigenfunctions of a certain operator $\hat{A}$ relating to the sub-system A;
$\Psi_u(x_B)$ are the eigenfunctions of the operator $\hat{B}$ of the sub-system B;
$\delta_{uu'}$ is Kronecker's symbol, which is equal to 1 when $u = u'$ but is otherwise zero.

We notice that using the eigenfunctions of Ψ_{nu} in the form of products in no way implies statistical independence of the sub-systems A and B. In fact the arbitrary wave function of the system $A+B$ can be expanded into a series in Ψ_{nu}, and the latter, generally speaking, cannot be stated in the form of the product $\Psi_A \cdot \Psi_B$. Using (1.10) and (1.8) (this can be done since in accordance with our assumption the whole system $A+B$ is in a pure state) we obtain

$$\langle F \rangle = \sum_{nn'u} \sigma_{n'u;nu} F_{nn'} = \sum_{nn'} \varrho_{n'n} F_{nn'} = \mathrm{Tr}(\hat{\varrho}\hat{F}), \tag{1.8'}$$

where

$$\varrho_{n'n} = \sum_u \sigma_{n'u;nu} \quad \text{or} \quad \hat{\varrho} = \mathrm{Tr}_B\, \hat{\sigma} \tag{1.11}$$

(Tr_B denotes the sum over the indices for the sub-system B). The matrix $\hat{\varrho}$ is by definition the density matrix of the sub-system A. It is easy to see that this matrix can be used to find all the mean values and probability distributions in the sub-system A. Therefore mixed states can be described by the

† If the density matrix is not normalized then from (1.8) and (1.9) follows the expression for the mean value in the form

$$\langle A \rangle = \frac{Tr(\hat{\sigma}\hat{A})}{\mathrm{Tr}\, \hat{\sigma}}.$$

density matrix $\varrho_{nn'}$. This density matrix possesses the following properties (see, e.g., Fano, 1957).

(a) The density matrix is Hermitian

$$\varrho_{n'n} = \varrho^*_{nn'}.$$

(b) It is normalized to unity

$$\operatorname{Tr} \hat{\varrho} = 1.$$

(c) The diagonal elements of the density matrix which give the probability of the state n are

$$\varrho_{nn} \geqq 0.$$

(d)

$$\operatorname{Tr} \hat{\varrho}^2 \leqq 1,$$

where the equals sign applies to a pure state.

Criteria can also be given by which we can distinguish the pure from the mixed state. In the pure state

$$\varrho_{n'n} = b_n^* b_{n'}$$

and

$$(\varrho^2)_{nm} = \sum_l \varrho_{nl}\varrho_{lm} = \sum_l b_l^* b_n b_m^* b_l = b_m^* b_n,$$

i.e. in this case

$$\hat{\varrho}^2 = \hat{\varrho}. \tag{1.12}$$

It follows from the invariance of the mean value of F in (1.8′) under the unitary transformation $\hat{U}$ that with this kind of transformation ϱ becomes

$$\hat{\varrho}' = \hat{U}^{-1}\hat{\varrho}\hat{U} \tag{1.13}$$

(whilst $\hat{F}$ becomes $\hat{F}' = \hat{U}^{-1}\hat{F}\hat{U}$).

If we use an analogy with tensors, $\hat{\varrho}$ and $\hat{F}$ can be called second-rank tensors and the wave function a vector†. The different representations play the part of the different coordinate systems. The mean values are transformation invariants or scalars. It is clear from this that in the general case the state should be described by the second-rank tensor $\hat{\varrho}$, since the operators $\hat{F}$ are second-rank tensors, so the invariants can be obtained by contracting these tensors with tensors of the same dimensionality. Only in special cases can $\hat{\varrho}$ be stated in the form of the product of two vectors $\varrho_{n'n} = b_{n'}b_n^*$.

† In a space of an infinite number of measurements.

It follows from (1.13) that the probability distributions in the various ensembles appearing when measuring the different quantities are given by

$$\varrho'_{kk} = \sum_{nn'} (U^{-1})_{kn}\varrho_{nn'}U_{n'k}, \quad \varrho_{nn} = \sum_{kk'} U_{nk}\varrho'_{kk'}(U^{-1})_{k'n}, \tag{1.14}$$

which replace the relations (1.4), which are valid only for pure states.

If the system is in a pure state the measurement of all the quantities which give the wave function of the state† leads with certainty to the original state, i.e. the probability distribution in the ensemble appearing when there is a complete measurement consists of the two terms 0 and 1. Moreover, in the mixed state we can introduce as the complete set of quantities the commuting operators $\hat{L}$, $\hat{M}$, $\hat{N}$, ... which have the feature that the density matrix is diagonal in the representation that is diagonal with respect to these quantities. The measurement of these quantities leads to an ensemble in which the probability distribution fully defines the density matrix (since there are elements that are not diagonal)‡. We shall also call this measurement the full measurement and the corresponding ensemble the full ensemble. It can easily be seen that this full measurement in the special case of a pure state becomes the measurement of the full set of quantities. In this case the density matrix satisfies (1.12) and

$$\varrho^2_{nn} = \varrho_{nn}$$

is diagonal. This equation has two solutions: 0 and 1. It may be said that the pure state differs from the mixed state in that in the pure state there is always an ensemble in which the probability distribution consists of the two terms 0 and 1.

Therefore the density matrix of the pure state after reduction to the diagonal form by means of the unity transformation becomes

$$\sigma_{kl} = \delta_{nk}\delta_{nl} \tag{1.15}$$

or

$$\hat{\sigma}_n = \left| \begin{matrix} 0 & 0 & 0 & 0 & 0 & 0 & 0 \\ 0 & . & . & . & . & . & 0 \\ 0 & . & 0 & . & . & . & 0 \\ 0 & . & . & 1 & . & . & 0 \\ 0 & . & . & . & . & 0 & 0 \\ 0 & . & . & . & . & . & 0 \\ 0 & . & . & . & . & . & 0 \end{matrix} \right| . \tag{1.15'}$$

† It is not a question of the arguments of the wave function, which also make up the complete set of quantities, but of the indices characterizing the actual function. For example atomic electrons are characterized by the following set of values: n, l, m, s (the principal quantum number, the orbital quantum number, the magnetic quantum number and the electron spin respectively). The wave function is of the form $\Psi_{nlms}(q)$.

‡ In the general case the distribution of the probabilities in a single ensemble does not make it possible to define the density matrix (just as is the case with the wave function).

Therefore one of the eigenvalues of $\hat{\sigma}_n$ is equal to unity and the others are equal to zero. It is easily shown that this condition is also a sufficient condition for the existence of a pure state.

It is obvious that the density matrix of the most general mixed state is equal to the linear superposition of the possible matrices of $\hat{\sigma}_n$ of the pure states

$$\hat{\varrho} = \sum_n p_n \hat{\sigma}_n, \tag{1.16}$$

where

$$\sum_n p_n = 1; \quad p_n \geqq 0. \tag{1.17}$$

Condition (1.17) ensures the normalization of $\hat{\varrho}$ to 1; carrying out the unitary transformation of all the matrices $\hat{\sigma}_n$ in (1.16) we come to the most general mixed state density matrix that satisfies all the conditions listed above.

If the object is in a pure state with a given wave function this wave function describes the behaviour of this individual object. It characterizes those properties of the object which are found when a measurement is made on the object when it is in definite ambient conditions. If, however, the object is in a mixed state and has no wave function it is described by a density matrix which is also a characteristic of this individual object. At the same time, however, another situation is possible when the density matrix describes an ensemble of objects each of which is in a definite state (pure or mixed)†, i.e. the density matrix can also describe the statistical features connected with our *lack of knowledge* of certain properties of the object‡ (although these properties are peculiar to the object in a definite form). In fact let us examine a set of objects each of which has a probability p_n of being in one of the pure states with a density matrix $\hat{\sigma}_n$ from (1.15′). Then all the mean values in this ensemble are defined by the density matrix (1.16) in accordance with (1.8′)

$$\langle A \rangle = \mathrm{Tr}(\hat{A}\hat{\varrho}) = \sum p_n \, \mathrm{Tr}(\hat{A}\hat{\sigma}_n), \tag{1.18}$$

i.e. the mean over the ensemble is equal to the sum of the mean values in the state σ_n multiplied by the probability of this state in the ensemble.

To conclude this sub-section let us take a simple example which illustrates the use of the density matrix. Let a system consisting of two spins of $\frac{1}{2}$ be in a pure state with a total spin of zero (i.e. all the spin components and its absolute value are equal to zero).

As is well known§, the wave function of a state of this kind is of the form

$$\psi_{0,0} = \frac{1}{\sqrt{2}} \{\psi_{1/2}(1)\,\psi_{-1/2}(2) - \psi_{1/2}(1)\,\psi_{-1/2}(2)\}, \tag{1.19}$$

† Or it is sometimes said that the density matrix in this case describes the ensemble of states. These ensembles should be distinguished from the ensembles that appear when measuring in a given state.

‡ As occurs in classical statistics.

§ See any course of quantum mechanics, e.g. Landau and Lifshitz (1963).

where $\psi_{\pm\frac{1}{2}}$ are the eigenfunctions of the operator of the projection of an individual spin onto the z-axis.

We shall now examine the state of the first spin only, without taking any notice of the state of the second. The first spin as a separate system has no wave function and is in a mixed state. The density matrix of the state can be obtained from the density matrix of the whole system by using (1.11). The matrix elements of the density matrix of the whole system are of the form

$$\sigma_{1/2,-1/2;1/2;-1/2} = \tfrac{1}{2}, \qquad \sigma_{-1/2,1/2;-1/2,1/2} = \tfrac{1}{2},$$

$$\sigma_{1/2,-1/2,-1/2,1/2} = -\tfrac{1}{2}, \quad \sigma_{-1/2,1/2;1/2,-1/2} = -\tfrac{1}{2},$$

where the first suffix denotes the state of the first spin and the second the state of the second spin. Retaining only the diagonals for the second spin we can use (1.11) to find the density matrix of the first spin in the form

$$\hat{\varrho} = \frac{1}{2}\begin{pmatrix}1 & 0\\ 0 & 1\end{pmatrix}. \tag{1.20}$$

It follows from this that measurement of the projection of the spin on to the z-axis with a probability of $\frac{1}{2}$ leads to a value of $\frac{1}{2}$, and with a probability of $\frac{1}{2}$ to a value of $-\frac{1}{2}$. Before measurement, however, there was no definite value for the projection of the first spin on to the z-axis. This can be seen from (1.19): the spin was in a superposition of the states $\frac{1}{2}$ and $-\frac{1}{2}$. Besides this situation another is possible when (1.20) describes an ensemble of pure states $\frac{1}{2}$ and $-\frac{1}{2}$ and reflects the property not of an individual spin but of the ensemble of spins.

It should be stressed that, in the case where (1.20) describes the state of an individual spin, it cannot in the general case be interpreted by saying that the projection of the spin on to the z-axis has a definite value, but that *we do not know it.* Let us assume that in the state (1.20) the z-component of the spin has in fact a definite, but unknown, value. We now change to another representation in which the x-component of the spin is diagonal. In this representation the density matrix will have the same form† (1.20) but its elements are the probabilities for the x-components of the spin. Just as for the case of the z-component, we should have assumed that the x-component of the spin has a definite value but we do not know it. A similar argument can be adduced for the y-component. Therefore the assumption that (1.20) in the general case describes a definite, but unknown, value of the component of an individual spin leads to the result that is absurd from the point of view of quantum theory that all three components of the spin have definite values. This contradicts the uncertainty relation for the components of the spin.

† The matrix (1.20) is proportional to the unitary matrix and the latter, as is well known, is of exactly the same form in all representations.

If in fact the density matrix (1.20) describes the state of an individual spin (and not an ensemble of spins), then this means that before measurement the spin had no definite z-component (and the same applies to the x- and y-components).

2. The Change of Quantum State with Time

2.1. The Schrödinger Equation

The basic equation of quantum theory is the Schrödinger equation. This equation describes the change of the quantum state with time. Since the characteristic of the (pure) quantum state is the wave function Ψ it is natural that the basic equation of quantum theory should describe the change with time of the wave function of the system.

The Schrödinger equation takes the form

$$i\hbar \frac{\partial \Psi}{\partial t} = \hat{H}\Psi, \tag{2.1}$$

where $\hat{H}$ is the Hamiltonian operator of the system.

The Schrödinger equation expresses the principle of causality in quantum theory. Therefore causality in quantum theory cannot be understood in a mechanical sense (as in Laplace determinism). According to quantum theory there are no physical laws in existence which would give a unique answer to the results of a subsequent experiment if the results of a preceding experiment are known. All we can do is to indicate the probability of one result or another.

2.2. The Equation for the Density Matrix

Equation (2.1) describes the change with time of a pure state. In order to establish how a mixed state changes with time we must find an equation to describe the change in the density matrix. Strictly speaking this equation cannot be derived from the Schrödinger equation but must be postulated, since the density matrix cannot generally be reduced to the wave function. However, by making certain additional assumptions we can derive the equation for the density matrix from the equation (2.1).

Let us first examine the change with time of the density matrix of a pure state. Using the expansion (1.5) we can rewrite (2.1) in the form

$$i\hbar \sum_n \dot{b}_n \Psi_n(q) = \sum_n b_n \hat{H} \Psi_n(q).$$

Multiplying the right- and left-hand sides of this equation by $\Psi_{n'}^*(q)$, integrating for all q, and remembering the property of orthonormalization of the eigenfunctions†

$$\int \Psi_n^*(q)\,\Psi_m(q)\,dq = \delta_{nm},$$

we obtain the Schrödinger equation in the N-representation

$$i\hbar \dot{b}_{n'} = \sum_n H_{n'n} b_n. \tag{2.2a}$$

The equation for the complex-conjugate quantity $b_{n''}^*$ is of the form

$$i\hbar \dot{b}_{n''}^* = -\sum_n H_{nn''} b_n^*. \tag{2.2b}$$

Multiplying equation (2.2a) by $b_{n''}^*$ and equation (2.2b) by $b_{n'}$ and adding them we obtain

$$i\hbar \frac{\partial}{\partial t}(b_{n''}^* b_{n'}) = \sum_n (H_{n'n} b_n b_{n''}^* - b_n^* b_{n'} H_{nn''}). \tag{2.2bA}$$

Introducing the pure state density matrix we reach the required relation

$$i\hbar \frac{\partial \hat{\sigma}}{\partial t} = \hat{H}\hat{\sigma} - \hat{\sigma}\hat{H}. \tag{2.3a}$$

We can further establish that the mixed state density matrix $\hat{\varrho}$ is subject to the same equation. To do this it is sufficient to assume that there is a more general system $A+B$ which includes the sub-system A (the sub-systems A and B do not interact) and that this system $A+B$ is in a pure state. Then

$$i\hbar \frac{\partial \hat{\sigma}_{A+B}}{\partial t} = [\hat{H}_A + \hat{H}_B, \hat{\sigma}_{A+B}],$$

where $[\hat{A}\hat{B}] \equiv \hat{A}\hat{B} - \hat{B}\hat{A}$;

$\hat{H}_A$ and $\hat{H}_B$ are the Hamiltonians of the sub-systems A and B respectively;

$\hat{\sigma}_{A+B}$ is the density matrix of the combined system $A+B$.

Taking the trace of the B sub-system on the right- and left-hand sides of the equation we find without difficulty

$$i\hbar \frac{\partial \hat{\varrho}}{\partial t} = [\hat{H}, \hat{\varrho}], \tag{2.3b}$$

† Without limiting the generality eigenfunctions can always be chosen so that they satisfy this property.

where $\hat{\varrho} = \mathrm{Tr}_B \hat{\sigma}_{A+B}$ is the density matrix of the system A and the A suffix has been omitted from $\hat{H}_A$.

We notice that the Schrödinger equation is contained in equation (2.3b) as a special case since equation (2.3b) is also valid in the case when $\hat{\varrho}$ cannot be written as the product of the amplitudes of the probabilities b_n.

2.3. The Schrödinger Representation

When equations (2.1) or (2.3b) are used to describe a change of quantum state it is assumed that the operators of the physical quantities are not time-dependent†, and the whole time-dependence is contained in the wave function (or density matrix). This description or representation is named after Schrödinger. The time-dependence of the density matrix in this representation can be written in the form

$$\varrho(t) = \exp\left(-\frac{i}{\hbar}\hat{H}t\right)\hat{\varrho}(0)\exp\left(\frac{i}{\hbar}\hat{H}t\right). \tag{2.4}$$

It is easy to check that this solution satisfies equation (2.3b). Despite the fact that in the general case we can write down the solution of the equation for the density matrix this solution has no great practical value because of the difficulty in calculating the operators $\exp(\pm(i/\hbar)\hat{H}t)$. We notice that the operator $e^{\hat{A}}$ is defined by its expansion into a Taylor series

$$e^{\hat{A}} = 1 + \hat{A} + \frac{1}{2!}\hat{A}^2 + \frac{1}{3!}\hat{A}^3 + \cdots$$

2.4. The Heisenberg Representation

We now move on to another way of describing the change in a quantum state—the Heisenberg representation. In this representation all the time-dependence is transferred to the operators and the wave function or density matrix are not time-dependent. Both these representations are, of course, physically equivalent. For example when taking the mean

$$\langle A(t)\rangle = \mathrm{Tr}\,\hat{\varrho}(t)\,\hat{A} = \mathrm{Tr}\,\hat{\varrho}\hat{A}(t)$$

the same time function is obtained no matter which is time-dependent—the density matrix or the operator. We change from the Schrödinger to the Heisenberg representation, as always when changing from one representation to another, by means of a unitary transformation. Using equation (2.3b) we

† Here for the sake of simplicity we are discussing the case when the operators in the Schrödinger representation have no explicit time-dependence.

can find the unitary transformation that makes the change to the Heisenberg representation (in which the density matrix $\hat{\varrho}' = \hat{U}^{-1}\hat{\varrho}\hat{U}$ is not time-dependent)

$$i\hbar \frac{\partial \hat{\varrho}'}{\partial t} = i\hbar \frac{\partial \hat{U}^{-1}}{\partial t} \hat{\varrho}\hat{U} + \hat{U}^{-1}[\hat{H}, \hat{\varrho}]\, \hat{U} + i\hbar \hat{U}^{-1}\hat{\varrho} \frac{\partial \hat{U}}{\partial t}$$

$$= \left(i\hbar \frac{\partial \hat{U}^{-1}}{\partial t} + \hat{U}^{-1}\hat{H}\right)\hat{\varrho}\hat{U} + \hat{U}^{-1}\hat{\varrho}\left(i\hbar \frac{\partial \hat{U}}{\partial t} - \hat{H}\hat{U}\right) = 0.$$

From this (remembering that this equation holds for any $\hat{\varrho}$) we have

$$i\hbar \frac{\partial \hat{U}^{-1}}{\partial t} = -\hat{U}^{-1}\hat{H}, \tag{2.5}$$

$$i\hbar \frac{\partial \hat{U}}{\partial t} = \hat{H}U. \tag{2.5'}$$

Because of the unitary nature of $\hat{U}$ and the Hermitian nature of $\hat{H}$ both these equations are compatible and follow from one another.

Using equation (2.5) we can find the equations of motion for the operators in the Heisenberg representation

$$\frac{\partial \hat{A}_t}{\partial t} = \frac{i}{\hbar}\, [\hat{H}_t, \hat{A}_t], \tag{2.6}$$

where

$$\hat{H}_t = \hat{U}^{-1}\hat{H}\hat{U}, \quad \hat{A}_t = \hat{U}^{-1}\hat{A}\hat{U}. \tag{2.7}$$

Equation (2.5′) can be integrated and its solution is of the form

$$\hat{U} = \exp\left(-\frac{i}{\hbar}\, \hat{H}t\right), \tag{2.8}$$

where we make use of the initial condition

$$\hat{U} = 1 \qquad \text{when} \qquad t = 0.$$

Therefore the solution of the Heisenberg equations of motion is of the form

$$\hat{A}_t = \exp\left(\frac{i}{\hbar}\, \hat{H}t\right) \hat{A} \exp\left(-\frac{i}{\hbar}\, \hat{H}t\right), \tag{2.9}$$

where

$$\hat{H}_t = \hat{H}$$

(we should mention that here $\hat{A}$ and $\hat{H}$ are Schrödinger operators which are not explicitly time-dependent). It is not difficult to see that (2.4) makes the transition from the Heisenberg representation $\hat{\varrho}' = \hat{\varrho}(0)$ to the Schrödinger representation $\hat{\varrho}(t)$.

2.5. Stationary States

Let us examine the special but very important case of stationary states. By definition these are states in which all the probability distributions and all the means are independent of time. For this to be the case the density matrix must commute with the Hamiltonian of the system

$$[\hat{\varrho}, \hat{H}] = 0. \tag{2.10}$$

Here, according to (2.3b), the Schrödinger density matrix $\hat{\varrho}(t)$ is not time-dependent (and is the same as the Heisenberg density matrix). It follows that all the means (and the probability distribution) are independent of time. In the special case of a pure state, when $\hat{\varrho}^2 = \hat{\varrho}$, it follows that the wave function of a stationary state is the eigenfunction of the Hamiltonian operator and its time dependence in the Schrödinger representation is of the form

$$\Psi(t) = \Psi_n \exp\left(-\frac{i}{\hbar} E_n t\right), \tag{2.11}$$

where E_n is the eigenvalue of the energy operator.

This kind of state always remains an eigenstate of the Hamiltonian operator and can be distinguished from the original state only by the phase factor.

2.6. Interaction Representation

Apart from the Heisenberg and Schrödinger representation, use is often made of the *interaction representation.* Let us assume that the Hamiltonian of a system can be split into two parts: $\hat{H}_0$ is the unperturbed Hamiltonian and $\hat{V}$ is the perturbation (or interaction) Hamiltonian

$$\hat{H} = \hat{H}_0 + \hat{V}.$$

This breakdown is, of course, somewhat arbitrary and is determined in each case by the nature of the problem. $\hat{V}$ is generally chosen so that it may be considered small. The interaction representation is intermediate between the Heisenberg and Schrödinger representations. The operators in this representation are time-dependent like the Heisenberg operators for an unperturbed system (with an operator $\hat{H}_0$) and the change with time of the wave function (density matrix) is entirely determined by the perturbation $\hat{V}$.

Let us make the transition to the interaction representation in the equation for the density matrix (2.3). The transformation of the operators $\hat{A}$ from the Schrödinger representation to the interaction representation $\hat{A}_{int}$ can be expressed by

$$\hat{A}_{int} = \hat{U}_0^{-1}\hat{A}\hat{U}_0, \tag{2.12a}$$

where

$$\hat{U}_0 = \exp\left(-\frac{i}{\hbar}\hat{H}_0 t\right). \tag{2.12b}$$

$$\hat{V}_{int} = \hat{U}_0^{-1}\hat{V}\hat{U}_0, \quad \hat{\varrho}_{int} = \hat{U}_0^{-1}\hat{\varrho}\hat{U}_0 \quad \text{and} \quad \hat{H}_{0int} = \hat{H}_0. \tag{2.13}$$

In order to find the equation for $\hat{\varrho}_{int}$ we differentiate $\hat{\varrho}_{int}$ from (2.13) and use (2.3b)

$$\frac{\partial\hat{\varrho}_{int}}{\partial t} = -\frac{i}{\hbar}\hat{U}_0^{-1}[\hat{H}_0 + \hat{V}, \hat{\varrho}]\,\hat{U}_0 + \frac{i}{\hbar}[\hat{H}_0, \hat{\varrho}_{int}]$$

$$= -\frac{i}{\hbar}\{\hat{U}_0^{-1}\hat{V}\hat{U}_0\hat{U}_0^{-1}\hat{\varrho}\hat{U}_0 - \hat{U}_0^{-1}\hat{\varrho}\hat{U}_0\hat{U}_0^{-1}\hat{V}\hat{U}_0\}$$

$$= -\frac{i}{\hbar}[\hat{V}_{int}, \hat{\varrho}_{int}].$$

Thus we finally obtain

$$\frac{\partial\hat{\varrho}_{int}}{\partial t} = -\frac{i}{\hbar}[\hat{V}_{int}, \hat{\varrho}_{int}]. \tag{2.14}$$

The mean value of A, as usual, can be found by the rule

$$\langle A\rangle = \mathrm{Tr}\,(\hat{\varrho}_{int}\hat{A}_{int}). \tag{2.14a}$$

If we make $\hat{V} = 0$ we obtain the Heisenberg representation of the density matrix. The time-dependence of the operator is expressed by (2.12) (or (2.6)) and the density matrix is not time-dependent. Making $\hat{H}_0 = 0$ ($\hat{V} \neq 0$) we obtain the ordinary Schrödinger representation ($\hat{U} = 1$).

To conclude this section we point out that the question of which representation to use is to be decided in each case from the standpoint of convenience and simplicity only. In every other way all three representations are completely equivalent.

3. The Quantum Theory of Fields in Ideal Resonators, Waveguides and Free Space

3.1. The Canonical Form of the Field Equations

In quantum theory the strengths of electric and magnetic fields are described by appropriate operators and the field state by a wave function or a density matrix.

Our task will now be to find the field operators, the quantum equations of motion and the Schrödinger equations from the well-known classical field equations (Maxwell's equations). We shall use the classical analogue method which is treated in detail in Dirac's book (Dirac, 1958). Canonical coordinates and momenta must be introduced into the classical theory of an electromagnetic field in order to make use of this method. Unlike an ordinary mechanical system with a finite number of degrees of freedom an electromagnetic field must, of course, be described by an infinite number of canonical coordinates and momenta.

The classical equations of a free electromagnetic field are of the form†:

$$\text{curl}\,\boldsymbol{E} + \frac{1}{c}\frac{\partial \boldsymbol{H}}{\partial t} = 0, \tag{3.1}$$

$$\text{div}\,\boldsymbol{H} = 0, \tag{3.2}$$

$$\text{curl}\,\boldsymbol{H} - \frac{1}{c}\frac{\partial \boldsymbol{E}}{\partial t} = 0, \tag{3.3}$$

$$\text{div}\,\boldsymbol{E} = 0. \tag{3.4}$$

In order to change from the variables $\boldsymbol{E}$ and $\boldsymbol{H}$ to the canonical variables q_ν and p_ν we must expand the fields $\boldsymbol{E}$ and $\boldsymbol{H}$ in the natural oscillations (modes) of the electromagnetic field. We shall first derive the equations which describe the modes of an electromagnetic field in a volume bounded by an ideal conductor (resonator) and then we shall examine the case of waveguides and free space. Taking the curl of both sides of (3.1), using the relation

$$\text{curl curl}\,\boldsymbol{E} = \nabla\,\text{div}\,\boldsymbol{E} - \nabla^2\boldsymbol{E}$$

and equations (3.4) and (3.3) we easily obtain

$$\nabla^2\boldsymbol{E} - \frac{1}{c^2}\frac{\partial^2 \boldsymbol{E}}{\partial t^2} = 0 \tag{3.5a}$$

† Gaussian units are used—Ed.

and likewise

$$\nabla^2 \boldsymbol{H} - \frac{1}{c^2}\frac{\partial^2 \boldsymbol{H}}{\partial t^2} = 0. \tag{3.5b}$$

These fields must satisfy the boundary conditions on a closed surface

$$[\boldsymbol{E} \wedge \boldsymbol{n}_0] = 0 \quad \text{and} \quad (\boldsymbol{H} \cdot \boldsymbol{n}_0) = 0, \tag{3.6}$$

where $\boldsymbol{n}_0$ is the normal to this surface.

We find the required solution of (3.5a) in the form

$$\boldsymbol{E} = a(t)\, \boldsymbol{E}_\nu(\boldsymbol{r}). \tag{3.7}$$

Substituting this expression in (3.5a) we obtain

$$\frac{\nabla^2 \boldsymbol{E}_\nu(\boldsymbol{r})}{\boldsymbol{E}_\nu(\boldsymbol{r})} = \frac{1}{c^2}\frac{\ddot{a}(t)}{a(t)} = -k_\nu^2,$$

where k_ν^2 is a constant†.

Therefore the equations which describe the resonant modes of the cavity are of the form

$$\nabla^2 \boldsymbol{E}_\nu + k_\nu^2 \boldsymbol{E}_\nu = 0 \quad \text{and} \quad \nabla^2 \boldsymbol{H}_\nu + k_\nu^2 \boldsymbol{H}_\nu = 0 \tag{3.8}$$

with the boundary conditions (3.6). $\boldsymbol{E}_\nu$ and $\boldsymbol{H}_\nu$, which satisfy (3.8) and (3.6), form the complete system of the eigenfunctions of the problem in question. One of the properties of these functions is their orthogonality. We normalize these functions as follows:

$$\int (\boldsymbol{E}_\nu \cdot \boldsymbol{E}_{\nu'})\, dV = 4\pi\delta_{\nu\nu'}, \qquad \int (\boldsymbol{H}_\nu \cdot \boldsymbol{H}_{\nu'})\, dV = 4\pi\delta_{\nu\nu'}, \tag{3.9}$$

where $\delta_{\nu\nu'} = \begin{cases} 1 & \nu = \nu' \\ 0 & \nu \neq \nu' \end{cases}$ is the ordinary Kronecker symbol and the integration is carried out over the whole volume of the resonator. Now any electromagnetic field in a resonator can be written in the form

$$\boldsymbol{E}(\boldsymbol{r}, t) = \sum_\nu p_\nu(t)\, \boldsymbol{E}_\nu(\boldsymbol{r}) \quad \text{and} \quad \boldsymbol{H}(\boldsymbol{r}, t) = -\sum_\nu \omega_\nu q_\nu(t)\, \boldsymbol{H}_\nu(\boldsymbol{r}), \tag{3.10}$$

where $\omega_\nu = k_\nu c$.

The variables p_ν and q_ν are the canonical variables describing the electromagnetic field in a resonator.

† It is not difficult to show that k_ν^2 must be positive. In actual fact the k_ν^2 are the eigennumbers of the square of the Hermitian operator—$i\nabla$. From this it follows in particular that the E_ν which satisfy the boundary condition (3.6) form the complete system of eigenfunctions.

A characteristic property of the canonical variables is that they satisfy Hamilton's canonical equations

$$\dot{q}_\nu = \frac{\partial H}{\partial p_\nu}, \quad \dot{p}_\nu = -\frac{\partial H}{\partial q_\nu}. \tag{3.11}$$

It is easy to check that these are equivalent to the field equations (3.1)–(3.4). To do this we must find the Hamiltonian function—the field energy expressed in terms of the canonical variables q_ν and p_ν. By means of (3.9) and (3.10), using the well-known expression for the energy of a radiation field, we find

$$H = \frac{1}{8\pi} \int (\boldsymbol{H}^2 + \boldsymbol{E}^2)\, dV = \frac{1}{2} \sum_\nu (p_\nu^2 + \omega_\nu^2 q_\nu^2). \tag{3.12}$$

Substituting (3.10) in (3.1) and (3.3) and using the relations

$$k_\nu \boldsymbol{E}_\nu = \operatorname{curl} \boldsymbol{H}_\nu \quad \text{and} \quad k_\nu \boldsymbol{H}_\nu = \operatorname{curl} \boldsymbol{E}_\nu, \tag{3.13}$$

we find for p_ν and q_ν†

$$p_\nu = \dot{q}_\nu, \quad \dot{p}_\nu = -\omega_\nu^2 q_\nu. \tag{3.14}$$

Exactly the same equations can be obtained from the Hamiltonian function (3.12). Here we should point out that equations (3.14) are the equations of a system of independent harmonic oscillators, whilst the Hamiltonian function is equal to the sum of the Hamiltonian functions

$$H_\nu = \tfrac{1}{2}(p_\nu^2 + \omega_\nu^2 q_\nu^2) \tag{3.15}$$

of the harmonic oscillators. Therefore each mode of the resonator is harmonically time-dependent and can be described, as can a mechanical system, by a Hamiltonian function (3.15). We shall call each such field oscillator a radiation oscillator. The field excitation can, as a result, be described as excitation of the corresponding radiation oscillators. Since $\boldsymbol{E}_\nu$ and $\boldsymbol{H}_\nu$ are given functions of the coordinates, finding the field in the resonator is reduced to finding the variables p_ν and q_ν as functions of the time.

3.2. Quantum Field Theory

Up to now we have been discussing classical field theory. We can now proceed to the construction of the quantum theory. To do this we must replace p_ν and q_ν by the appropriate operators. A basic feature of these operators, which is independent of the choice of representation, is the commutation relation to which these operators are subject. According to the

† It is easy to check that the relations (3.13) satisfy (3.8) and (3.9).

classical analogue method (Dirac, 1958) the commutation relation for the operators $\hat{u}$ and $\hat{v}$ can be found from

$$\hat{u}\hat{v} - \hat{v}\hat{u} = -i\hbar\{u, v\}, \tag{3.16}$$

where

$$\{u, v\} = \sum_\nu \left(\frac{\partial u}{\partial p_\nu} \frac{\partial v}{\partial q_\nu} - \frac{\partial u}{\partial q_\nu} \frac{\partial v}{\partial p_\nu} \right)$$

is the classical Poisson bracket.

For the actual variables q_ν and p_ν the Poisson brackets are of the form

$$\{q_\nu, p_{\nu'}\} = -\delta_{\nu\nu'}, \quad \{q_\nu, q_{\nu'}\} = 0, \quad \{p_\nu, p_{\nu'}\} = 0.$$

From this we can find the commutation relations between the operators $\hat{p}_\nu$ and $\hat{q}_\nu$†

$$[\hat{q}_\nu, \hat{p}_{\nu'}] = i\hbar\delta_{\nu\nu'}, \quad [\hat{q}_\nu, \hat{q}_{\nu'}] = 0, \quad [\hat{p}_\nu, \hat{p}_{\nu'}] = 0. \tag{3.17}$$

The Hamiltonian operator of the electromagnetic field now becomes

$$\hat{H} = \sum_\nu \tfrac{1}{2}(\hat{p}_\nu^2 + \omega_\nu^2 \hat{q}_\nu^2) \equiv \sum_\nu \hat{H}_\nu. \tag{3.18}$$

The relations (3.17) and (3.18) are a complete definition of the Hamiltonian operator of the electromagnetic field (it is equal to the sum of the Hamiltonian operators of the quantum harmonic oscillators). In the q_ν-representation the q_ν operators are multiplication operators and

$$\hat{p}_\nu = -i\hbar \frac{\partial}{\partial q_\nu}.$$

Therefore in the same representation

$$\hat{H} = \sum_\nu \frac{1}{2}\left(-\hbar^2 \frac{\partial^2}{\partial q_\nu^2} + \omega_\nu^2 q_\nu^2 \right).$$

The time-dependent Schrödinger equation is of the form

$$i\hbar \frac{\partial \Psi}{\partial t} = \left\{ \sum_\nu \frac{1}{2}(\hat{p}_\nu^2 + \omega_\nu^2 \hat{q}_\nu^2) \right\} \Psi. \tag{3.19}$$

The quantum equations of motion can be found by means of relations (2.6), (3.18) and (3.17)

$$\hat{p}_\nu = \dot{\hat{q}}_\nu, \quad \dot{\hat{p}}_\nu = -\omega_\nu^2 \hat{q}_\nu. \tag{3.20}$$

† We have used the notation $\hat{A}\hat{B} - \hat{B}\hat{A} \equiv [\hat{A}\,\hat{B}]$.

In appearance they are the same as the classical equations (3.14) of a harmonic oscillator. By eliminating $\hat{p}_\nu$ we come to the ordinary form of the harmonic oscillator equation

$$\ddot{\hat{q}}_\nu + \omega_\nu^2 \hat{q}_\nu = 0. \tag{3.21}$$

The operator of the derivative $\dot{\hat{A}}$ by definition satisfies

$$\left\langle \frac{d\hat{A}}{dt} \right\rangle = \frac{d}{dt} \langle \hat{A} \rangle. \tag{3.22}$$

From this we can immediately obtain the equation for the mean coordinate $\langle \hat{q}_\nu \rangle \equiv \bar{q}_\nu$, which is the same as the classical equation

$$\ddot{\bar{q}}_\nu + \omega_\nu^2 \bar{q}_\nu = 0. \tag{3.21 a}$$

Therefore the classical Maxwell equations for a free field are precisely the same as the equations which govern the quantum averages derived from the electric and magnetic field operators. These operators themselves are of the form

$$\hat{\boldsymbol{E}}(\boldsymbol{r}, t) = \sum_\nu \hat{p}_\nu \boldsymbol{E}_\nu(\boldsymbol{r}), \quad \hat{\boldsymbol{H}}(\boldsymbol{r}, t) = -\sum_\nu \omega_\nu \hat{q}_\nu \boldsymbol{H}_\nu(\boldsymbol{r}). \tag{3.23}$$

Here we should point out that the conclusion that the quantum averages are the same as the classical quantities is not, in general, true, and in the case of Maxwell's equations is a consequence of the linearity of these equations. This conclusion is invalid in particular for the quadratic field quantities since $\langle \hat{A}^2 \rangle \neq \langle A \rangle^2$.

3.3. Eigenfunctions

It is well known that an arbitrary solution (3.19) of the Schrödinger equation can be written as a superposition of stationary states. We shall now proceed to find the wave functions of the stationary states. These are the eigenfunctions of the Hamiltonian operator

$$\hat{H}\Psi_n = \sum_\nu \hat{H}_\nu \Psi_n = E_n \Psi_n, \tag{3.24}$$

where E_n are the eigenvalues of the field energy.

It is easy to confirm that the stationary wave functions which satisfy the equations (3.19) and (3.24) are of the form

$$\Psi_n = \Psi_{n_1, n_2, n_3, \dots n_\nu \dots n_{\nu'}} = \prod_\nu \Psi_{n_\nu} \mathrm{e}^{-\frac{iE_{n_\nu}}{\hbar} t} = \mathrm{e}^{-\frac{iE_n}{\hbar} t} \prod_\nu \Psi_{n_\nu}, \tag{3.25 a}$$

where the symbol Π denotes a product;

Ψ_{n_ν} is the eigenfunction of the Hamiltonian of the νth oscillator†;

$$\hat{H}_\nu \Psi_{n_\nu} = \tfrac{1}{2}(\hat{p}_\nu^2 + \omega_\nu^2 \hat{q}_\nu^2)\, \Psi_{n_\nu} = (n_\nu + \tfrac{1}{2})\, \hbar\omega_\nu \Psi_{n_\nu};$$

$E_{n_\nu} = (n_\nu + \frac{1}{2})\, \hbar\omega_\nu$ is the eigenvalue of the energy of the νth oscillator,

and the eigenvalue of the energy of the field is

$$E_n = \sum_\nu E_{n_\nu} = \sum_\nu (n_\nu + \tfrac{1}{2})\, \hbar\omega_\nu. \tag{3.26}$$

We have therefore found the complete system of wave functions and eigenvalues of the energy of a free electromagnetic field.

3.4. *The Properties of a Radiation Oscillator*

Let us examine in greater detail the properties of the νth radiation oscillator. It is well known (see, e.g., Landau and Lifshitz, 1963) that the non-zero elements of the canonical coordinates and momenta $\hat{q}_\nu$ and $\hat{p}_\nu$ are of the form (for the sake of simplicity we have omitted the suffix ν)

$$\begin{aligned} q_{n,n+1} &= q^*_{n+1,n} = \sqrt{\frac{\hbar(n+1)}{2\omega}}, \\ p_{n,n+1} &= p^*_{n+1,n} = -i\sqrt{\frac{\hbar\omega(n+1)}{2}}. \end{aligned} \tag{3.27}$$

These matrix elements, energy eigenvalues and harmonic oscillator wave functions can be derived by using only the commutation relations between $\hat{q}$ and $\hat{p}$ and the appropriate form of the Hamiltonian operator (see Landau and Lifshitz, 1963). We shall now derive the relations (3.27) and find the eigenfunctions and eigenvalues of the harmonic oscillator Hamiltonian function by a slightly different method from that given by Landau and Lifshitz (1963).

Instead of the operators (and their corresponding matrices) $\hat{q}$ and $\hat{p}$ we shall introduce the operators $\hat{a}$ and $\hat{a}^+$ which are called the "annihilation" and "creation" operators respectively‡

$$\hat{q} = \sqrt{\frac{\hbar}{2\omega}}\,(\hat{a} + \hat{a}^+), \qquad \hat{p} = i\sqrt{\frac{\hbar\omega}{2}}\,(\hat{a}^+ - \hat{a}). \tag{3.28}$$

† As is well known (see, e.g., Landau and Lifshitz, 1963) Ψ_{n_ν} is of the form

$$\Psi_n = N_n\, e^{-\alpha^2 q_\nu^2/2}\, H_n(aq_\nu), \tag{3.25b}$$

where $N_n = (\alpha/\pi^{1/2}\, 2^n\, n!)^{1/2}$; $\alpha = \sqrt{\omega_\nu/\hbar}$; $H_n(x)$ is the nth Hermite polynomial.

‡ The meaning of these names will be explained later.

By using the commutation relations (3.17) it is easy to find that the operators $\hat{a}$ and $\hat{a}^+$ are subject to the following commutation relation:

$$[\hat{a}, \hat{a}^+] = 1 \tag{3.29}$$

(the other commutations, in particular $[\hat{a}_\nu, \hat{a}_{\nu'}]$, are equal to zero). We notice that the operators $\hat{a}$ and $\hat{a}^+$ are non-Hermitian, unlike $\hat{q}$ and $\hat{p}$. The operators $\hat{a}$ and $\hat{a}^+$ can be used to write the energy of the oscillator in the form

$$\hat{H} = \tfrac{1}{2}\,\hbar\omega(\hat{a}^+\hat{a} + \hat{a}\hat{a}^+) = \hbar\omega(\hat{n} + \tfrac{1}{2}), \tag{3.30}$$

where the symbol $\hat{n}$ denotes the particle-number operator

$$\hat{n} = \hat{a}^+\hat{a} \tag{3.31}$$

(and when changing to the latter equality (3.30) we used the commutation relation (3.29)). The name "particle-number operator" (number of photons in our case) is connected with the fact that $\hat{n}$ takes integral values, as will follow from the further derivation. From the Hamiltonian (3.30) and the commutation relations (3.29) follow the equations of motion of the quantities $\hat{a}$ and $\hat{a}^+$

$$\dot{\hat{a}} = -i\omega\hat{a}, \quad \dot{\hat{a}}^+ = i\omega\hat{a}^+. \tag{3.32}$$

Let us now move on to describing the harmonic oscillator in terms of an angular variable—the oscillator's phase φ (Fain, 1967). Here we first recall that in classical mechanics the value of the canonically conjugate phase φ is an action variable (Landau and Lifshitz, 1958; ter Haar, 1961)

$$J = \frac{1}{2\pi}\oint p\,dq. \tag{3.32 a}$$

By using the relation

$$q = \frac{\sqrt{2H}}{\omega}\cos\varphi, \tag{3.33}$$

to define φ and using the expression

$$\tfrac{1}{2}p^2 + \tfrac{1}{2}\omega^2 q^2 = H$$

for the energy, we find

$$p = \pm\sqrt{2H}\sin\varphi, \qquad H = \mp\omega J. \tag{3.34}$$

The variable J has the dimensionality of angular momentum and its sign is determined by the velocity of the mapping point in the q, p plane, i.e. by the sign of $d\varphi/dt$. Negative J corresponds to a minus sign and positive J to a plus sign. The quantum generalization of the relations (3.33), (3.34) is achieved by introducing the operator $\hat{I}$, which is the sign of rotation operator. The action of the operator $\hat{I}$ is to multiply the function $e^{im\varphi}$ by 1 if $m > 0$ and by

-1 if $m < 0$. This also defines its action on any arbitrary periodic function $\Psi(\varphi)$ expanded in a series in $e^{im\varphi}$ (with non-zero values of m). Using (3.28) and (3.30) it is not hard to see that the quantum generalization of the relations (3.34) is of the form

$$\hat{n} = -i\hat{I}\frac{\partial}{\partial\varphi} - \frac{1}{2}, \tag{3.35}$$

$$\hat{a} = \left[e^{-i\varphi}\frac{1+\hat{I}}{2} + e^{i\varphi}\frac{1-\hat{I}}{2}\right]\sqrt{\hat{n}}, \tag{3.36}$$

$$\hat{a}^+ = \sqrt{\hat{n}}\left[\frac{1+\hat{I}}{2}e^{i\varphi} + \frac{1-\hat{I}}{2}e^{-i\varphi}\right]. \tag{3.37}$$

The eigenfunctions of the particle-number operator are of the form

$$\Psi_{n,1} = \frac{1}{\sqrt{2\pi}}e^{i(n+\frac{1}{2})\varphi}, \qquad \Psi_{n,-1} = \frac{1}{\sqrt{2\pi}}e^{-i(n+\frac{1}{2})\varphi}, \tag{3.38}$$

(where $\Psi_{n,\pm 1}$ corresponds to $I = \pm 1$) and the eigen-values are $n = 0, 1, 2, \ldots$†. From (3.36) and (3.37) we obtain that the operators $\hat{a}_+$ and $\hat{a}$ act as follows on the wave functions

$$\hat{a} + \Psi_{n-1} = \sqrt{n}\,\Psi_n, \qquad \hat{a}\,\Psi_{n+1} = \sqrt{n+1}\,\Psi_n.$$

$$\hat{a}^+\Psi_{n-1} = \sqrt{n}\,\Psi_n, \quad \hat{a}\Psi_{n+1} = \sqrt{n+1}\,\Psi_n. \tag{3.39}$$

It follows from this that the non-zero matrix elements $\hat{a}$ and $\hat{a}^+$ are of the form

$$a_{n,n+1} = \sqrt{n+1}, \quad \overset{+}{a}_{n,n-1} = \sqrt{n}, \tag{3.40}$$

and all other matrix elements, in particular

$$a_{n+1,n} = \overset{+}{a}_{n-1,n} = 0.$$

The meaning of the names "annihilation" and "creation" operators can be understood from (3.39). The operator $\hat{a}^+$ creates photons—it converts the system from a state with $(n-1)$ photons to a state with n photons; the operator $\hat{a}$ likewise annihilates the photons. The relations (3.27) can be obtained from (3.40) and (3.28). We have thus found the eigenfunctions $\Psi_n(\varphi)$, the eigenvalues of the energy E_n and the matrix elements of the harmonic oscillator.

† Integral values of n correspont to half-odd-integral values of the action variable $J/\hbar$. As long ago as 1926 Dirac (Dirac, 1926) showed that for spinless systems the action takes up only half-odd-intedral values. This also connected with the fact that the wave functions of a harmonic oscillator satisfy the cyclic condition

$$\Psi(\varphi + 2\pi) = -\Psi(\varphi).$$

3.5. The Heisenberg Relations for the Phase and the Number of Particles

We notice that in a state (3.38) which a definite n (and I) the phases are completely uncertain; all the values of φ (from 0 to 2π) are equally probable

$$dw_\varphi = |\Psi_{n,+1}|^2\, d\varphi = \frac{d\varphi}{2\eta}. \tag{3.41}$$

In the general case the state of an oscillator is a superposition of wave functions. The Heisenberg relations are a consequence of the commutation relations. We know that if there is a commutation relation

$$[\hat{A}, \hat{B}] = -2i\hat{C},$$

between the operators $\hat{A}$ and $\hat{B}$, then there is a Heisenberg relation

$$\langle(\Delta A)^2\rangle\, \langle\Delta B)^2\rangle \geqslant \langle C\rangle^2 \tag{3.42}$$

for the physical quantities A and B. Therefore from the commutation relations (3.17) we have

$$\Delta q_\nu\, \Delta p_\nu \geqslant \tfrac{1}{2}\hbar,$$

where Δq_ν, Δp_ν are the mean square fluctuations of the quantities q_ν and p_ν.

The commutation relations

$$[\hat{J}, f(\varphi)] = -i\hbar\,\frac{\partial f}{\partial \varphi},$$

where $f(\varphi)$ is an arbitrary periodic function (with period 2π), follow from the canonically conjugate nature of the action and the phase. From this and from (3.34), (3.35) it is not hard to obtain (Fain, 1967) that in the general case the Heisenberg relation

$$\sqrt{(\Delta n)^2 + (\langle n\rangle + \tfrac{1}{2})^2}\, \Delta f \geqslant \frac{1}{2}\left|\left\langle\frac{\partial f}{\partial \varphi}\right\rangle\right|, \tag{3.43}$$

holds, whilst in the quasi-classical approximation we have the more rigorous inequality†

$$\Delta n\, \Delta\varphi \geqslant \tfrac{1}{2}. \tag{3.44}$$

3.6. The Comparison of Classical and Quantum Field Theory

Let us now compare the classical and quantum theories of an electromagnetic field. Since all the properties of a free electromagnetic field are determined by the properties of the radiation oscillators we need do no more than compare the behaviours of classical and quantum harmonic oscillators.

† Remembering the energy uncertainty of a radiation oscillator is $\Delta E = \Delta n\hbar\omega$, the relation (3.44) can be written in the form

$$\Delta\varphi\, \Delta E \geqslant \hbar\omega/2. \tag{3.44'}$$

We shall deal in particular with the question of the limiting transition to classical theory. When making a transition to the limit of classical theory (when $\hbar$ approaches zero, or at large quantum numbers) the motion described by the wave function (or density matrix) does not generally change into motion on a fixed trajectory (see, e.g., Landau and Lifshitz, 1963). When changing to classical physics the Schrödinger equation becomes the Hamilton-Jacobi equation and the latter describes the motion of an ensemble of particles moving along classical trajectories.

Therefore at the classical limit the wave function (or density matrix) corresponds, generally speaking, to an ensemble of classical states.

Let us illustrate these ideas by an example which is of interest to us, the harmonic oscillator. When n approaches infinity (the classical limit) a quantum harmonic oscillator does not generally acquire a definite energy or phase. We shall examine the case when the harmonic oscillator is in a state with a definite n_0 (stationary state). In this case, in accordance with the uncertainty relation (3.44), the phase is completely uncertain. This state (for large enough n_0) corresponds to the ensemble of classical oscillators whose energy is fixed and equal to

$$E_0 = n_0 \hbar\omega,$$

whilst the phases are completely random; their distribution is given by (3.41). In order to confirm that this is so we shall introduce the concept of a classical precise measurement (or simply a classical measurement). By a classical measurement we shall understand an extremely accurate measurement from the classical point of view which does not invoke the uncertainty relation. In other words, if we measure the quantities n and φ (or E and φ) connected by the uncertainty relations (3.44) and (3.44′) the accuracies of the measurement of Δn and $\Delta\varphi$ should satisfy the inequalities

$$\left.\begin{array}{ll} 1 \ll \Delta n \ll \bar{n}, & \hbar\omega \ll \Delta E \ll \bar{E} \\ \dfrac{1}{\Delta n} \ll \Delta\varphi \ll 2\pi, & \dfrac{\hbar\omega}{\Delta E} \ll \Delta\varphi \ll 2\pi. \end{array}\right\}. \tag{3.45}$$

In the transition to the classical limit $\hbar \to 0$ we can make ΔE and $\Delta\varphi$ equal to zero. (The quantity Δn does not, generally speaking, approach zero since n is not a classical quantity.) We notice that the inequalities (3.45) can be satisfied in the classical case only when the mean energy of the oscillator is large enough. Let a classical measurement of the phase φ be made on an oscillator in a state Ψ_{n_0} with a definite value $n = n_0$. As a result of this measurement the oscillator transfers into the state

$$\Psi_{n_0\varphi_0} = \sum_n c_n \Psi_n, \tag{3.46}$$

where the ensemble of different n taking part in the formation of this wave packet can be determined from the uncertainty relation

$$\Delta n \sim \frac{1}{\Delta\varphi}\,.$$

We shall now show that the wave function (3.46) describes the state in which n, E and φ, in the same way as p_ν and q_ν, are determined with the accuracy of classical measurement. For this purpose let us examine the action of the different operators on the wave function (3.46). Using the definition of the eigenfunction and the fact that, in accordance with (3.45), in the state (3.46)

$$\bar{n} \approx n_0 \gg \Delta n_0\,,$$

we obtain

$$\hat{n}\Psi_{n_0,\varphi_0} = \sum_n c_n n \Psi_n \approx n_0 \sum_n c_n \Psi_n = n_0 \Psi_{n_0,\varphi_0}\,,$$

i.e. with classical accuracy the operator $\hat{n}$ acts as a multiplication operator on the function Ψ_{n_0,φ_0}, and we may consider that in the state Ψ_{n_0,φ_0} n is sufficiently (classically) defined. Likewise we obtain

$$\hat{a}\Psi_{n_0,\varphi_0} = \sum_n c_n \sqrt{n}\,\Psi_{n-1} = \sum_n c_n \sqrt{n}\,\mathrm{e}^{i\varphi}\,\Psi_n \approx \sqrt{n_0}\,\mathrm{e}^{i\varphi}\,\Psi_{n_0,\varphi_0}\,,$$

$$\hat{a}^{+}\Psi_{n_0,\varphi_0} \approx \sqrt{n_0}\,\mathrm{e}^{-i\varphi}\,\Psi_{n_0,\varphi_0}\,,$$

from which it follows that p_ν and q_ν are also determined in the state Ψ_{n_0,φ_0}. Therefore the measurement of φ makes the oscillator change into a state with a classical well-defined phase and energy. It is now clear that a wave function description using Ψ_{n_0} (when $n_0 \gg 1$) can be replaced by the description of a statistical ensemble of classical states with a definite n_0 and a random phase. The mean values of all the quantities in the state Ψ_{n_0} are the same as the averages for the ensemble of classical oscillators with a random distribution of phases (3.41). Let us examine, for example, the coordinate q_ν and the momentum p_ν of a radiation oscillator. Using definition (3.28) we obtain in the classical case

$$q_\nu = \sqrt{\frac{2\hbar n_0}{\omega_\nu}}\cos\varphi_\nu = \sqrt{\frac{2E_0}{\omega_\nu^2}}\cos\varphi_\nu\,,$$

$$p_\nu = \sqrt{2E_0}\sin\varphi_\nu\,,\quad \varphi_\nu = -\omega_\nu t + \varphi_\nu(0)\,.$$

Averaging over the phases we find that

$$\bar{q}_\nu = \bar{p}_\nu = 0, \quad \bar{p}_\nu^2 = \hbar\omega_\nu n_0 = E_0, \quad \bar{q}_\nu^2 = \frac{E_0}{\omega_\nu^2},$$

$$\Delta p_\nu \Delta q_\nu = \frac{E_0}{\omega_\nu}, \quad \bar{E}_\nu = E_0.$$

The same result can be obtained by using Ψ_{n_0} (with $n_0 \gg 1$) when averaging. In the quantum case when n_0 is small a precise measurement of the quantity q_ν leads to a state in which the quantity p_ν completely loses its certainty. On the other hand, in the classical case when q_ν is measured the quantity p_ν remains classically determined. Therefore the ensembles of the measurements of p_ν and q_ν are compatible and not mutually exclusive (as occurs in the quantum case). The complete determination of the classical statistical ensemble presupposes that the distributions of the quantities p_ν and q_ν are found simultaneously. If a single oscillator and not an ensemble of oscillators is considered, in the classical case this oscillator behaves as an oscillator with definite E and φ (which may, however, remain unknown). After measurement we find that the oscillator has definite E_0 and φ_0 and repeated classical measurement will lead to the same values of E_0 and φ_0.

It is interesting to follow how the classical probability distribution is obtained from the wave function of the oscillator in the quasi-classical approximation. The stationary wave function of the oscillator in the q_ν-representation in the quasi-classical approximation is of the form (see, e.g., Landau and Lifshitz, 1963)

$$\Psi(q_\nu) = \sqrt{\frac{2\omega_\nu}{\pi p_\nu}} \sin\left[\frac{1}{\hbar}\int_0^{q_\nu} p_\nu \, dq_\nu + \frac{\pi}{4}\right].$$

From this follows the probability distribution of q_ν in the form

$$|\Psi(q_\nu^0)| \, \Delta q_\nu = \int_{q_{\nu 0}^0}^{q_\nu^0 + \Delta q_\nu} \frac{2\omega_\nu}{\pi p_\nu} \sin^2\left[\frac{1}{\hbar}\int_0^{q_\nu} p_\nu \, dq_\nu + \frac{\pi}{4}\right] dq_\nu.$$

Remembering that the expression following the sine symbol is (in the quasi-classical approximation) a rapidly changing function, and Δq_ν (the accuracy of the classical measurement) contains many sine periods (many wave function nodes), we can replace the square of the sine by its mean value, i.e. $\frac{1}{2}$.

Then we obtain

$$|\Psi(q_\nu^0)|^2 \, \Delta q_\nu = \frac{\omega}{\pi p_\nu} \Delta q_\nu.$$

This is just the distribution that should be obtained for a classical oscillator with a random phase since with classical motion the time spent by the oscillator in the range dq_ν is inversely proportional to the momentum p_ν of the particle ($p_\nu = dq_\nu/dt$) so that $dt/(T/2) = (\omega/\pi)\,(dq_\nu/p_\nu)$, where T is the period of the motion.

Besides states which, when $\hbar \to 0$, do not change into a state with definite phase and amplitude, we can make up ensembles of stationary states which, when $\hbar \to 0$, change into a classical state (and not into an ensemble of states).

3.7. A Wave Packet in the Classical Limit

We now move on to the construction of a field whose properties are nearly classical. The solution of the quantum equations of motion can be written in the form

$$\hat{p}_\nu = \hat{p}_\nu(0)\cos\omega_\nu t - \omega_\nu \hat{q}_\nu(0)\sin\omega_\nu t,$$

$$\hat{q}_\nu = \hat{q}_\nu(0)\cos\omega_\nu t + (\hat{p}_\nu(0)/\omega_\nu)\sin\omega_\nu t. \tag{3.47}$$

Here $\hat{p}_\nu(0)$ and $\hat{q}_\nu(0)$ are operators in the Schrödinger representation and $\hat{p}_\nu$ and $\hat{q}_\nu$ in the Heisenberg representation. We can find expressions for the averages that are the same as the classical solutions†. We then select an initial state (Heisenberg wave function) so that the product of the uncertainties $\Delta\varphi\,\Delta n$ and $\Delta p_\nu(0)\,\Delta q_\nu(0)$ respectively has its minimum value

$$\Delta p_\nu(0)\,\Delta q_\nu(0) = \tfrac{1}{2}\hbar. \tag{3.48}$$

It is not hard to construct such a state.

In accordance with (3.25b) the stationary wave function of the ground state (with $n = 0$) is of the form

$$\Phi_0 = \left(\frac{\omega_\nu}{\pi\hbar}\right)^{1/4} \mathrm{e}^{-\frac{\omega_\nu}{2\hbar}q_\nu^2}.$$

In this state (3.48) is satisfied. The wave function Φ_0 cannot be used directly since this function describes the state with $\langle q_\nu(0)\rangle = \langle p_\nu(0)\rangle = 0$. It is not difficult, however, to alter Φ_0 so that the averages $\langle p_\nu(0)\rangle$ and $\langle q_\nu(0)\rangle$ are finite

$$\Phi = \left(\frac{\omega_\nu}{\pi\hbar}\right)^{1/4} \exp\left\{-\frac{\omega_\nu}{2\hbar}(q_\nu - \overline{q_\nu(0)})^2 + \frac{i\overline{p_\nu(0)}}{\hbar}q_\nu\right\}. \tag{3.49}$$

It may readily be seen that the averages of $\langle p_\nu(0)\rangle$ and $\langle q_\nu(0)\rangle$ are just equal to $\overline{p_\nu(0)}$ and $\overline{q_\nu(0)}$ and that (3.48) is satisfied at the same time. Let us find the probability distribution for different numbers of photons n. To do this we must expand the function Φ into a series of stationary wave functions (3.25b)

$$\Phi = \sum_n A_n \Psi_n.$$

† This occurs for any of the states of a radiation oscillator, as has been pointed out above.

The squares of the moduli of A_n give the required probability distribution. It turns out to be a Poisson distribution (Schiff, 1955)

$$|A_n|^2 = \frac{\bar{n}^n \cdot e^{-\bar{n}}}{n!}, \tag{3.50}$$

where $\bar{n} = E_{cl}/\hbar\omega_\nu$;

$E_{cl} = \frac{1}{2}(\bar{p}_\nu^2(0) + \omega_\nu^2 \overline{q_\nu(0)^2})$ is the classical energy of the νth mode.

As is well known, the fluctuation of the number of photons in the case of the distribution (3.50) is

$$\Delta n = \sqrt{\overline{n^2} - \bar{n}^2} = \sqrt{\bar{n}}. \tag{3.51}$$

From this it is easy to find that the fluctuation of the energy of the field is of the form

$$\Delta E = \sqrt{\hbar\omega}\,\sqrt{E_{cl}}. \tag{3.52}$$

Therefore the fluctuation of the energy approaches zero, in just the same way as the product of the uncertainties (3.48), when Planck's constant $\hbar$ approaches zero.

3.8. *The Radiation Field in Free Space*

Up to now we have been considering fields in resonators. We have found that the energy in each mode of a resonator is quantized and takes up values differing by $\hbar\omega_\nu$, where ω_ν is the eigen frequency of the resonator.

We now come to the discussion of a field in free space. The change to free space can be made by examining the field in a very large resonator whose dimensions approach infinity. In this case the spectrum of the radiation field (or of the oscillations whose wavelengths are much less than the dimensions of the resonator) becomes very dense and almost continuous. A continuous spectrum is obtained at the limit when the resonator dimensions are infinite. This means that summation over the spectrum must be replaced by integration. However, we shall leave the summation sign until we need to calculate the appropriate sum. Since the actual form of the resonator and the selection of the boundary conditions are not important when discussing the oscillations whose wavelengths (or, strictly speaking, the characteristic dimensions of the field inhomogeneity) are much less than the dimensions of the resonator, we can for simplicity use the eigenfunctions of a rectangular (cubic) resonator with periodic boundary conditions (Heitler, 1954). Therefore as the eigenfunctions $\boldsymbol{E}_\nu$ and $\boldsymbol{H}_\nu$ for free space we can take

$$\boldsymbol{E}_\nu = -\frac{\sqrt{8\pi}}{\sqrt{L^3}}\,\boldsymbol{e}_\nu \cos(\boldsymbol{k}_\nu \cdot \boldsymbol{r}), \quad -\frac{\sqrt{8\pi}}{\sqrt{L^3}}\,\boldsymbol{e}_\nu \sin(\boldsymbol{k}_\nu \cdot \boldsymbol{r}),$$

$$\boldsymbol{H}_\nu = \frac{c}{\omega_\nu}\,\text{curl}\,\boldsymbol{E}_\nu. \tag{3.53}$$

Here $\boldsymbol{e}_\nu$ is the unit polarization vector at right angles to the wave vector $\boldsymbol{k}_\nu$; each $\boldsymbol{k}_\nu$ corresponds to two directions of polarization $\boldsymbol{e}_\nu$ and two forms of the cos and sin functions (we shall consider that all these are denoted by the choice of the suffix ν); L in the formula (3.53) is the linear dimension of the resonator (the edge of the cube). It is easy to check that the functions (3.53) satisfy (3.8), which describe the modes of a field, and the normalization conditions (3.9).

The boundary conditions lead to a series of discrete values of the wave vector $\boldsymbol{k}_\nu$:

$$k_{\nu x} = \frac{2\pi}{L} n_{\nu x}, \quad k_{\nu y} = \frac{2\pi}{L} n_{\nu y}, \quad k_{\nu z} = \frac{2\pi}{L} n_{\nu z}, \tag{3.54}$$

where $n_{\nu x}, n_{\nu y}, n_{\nu z}$ are integers filling a half-space in the space $n_{\nu x}, n_{\nu y}, n_{\nu z}$. For example we can select ranges of variation for the integers n_ν of the form

$$0 < n_{\nu x} < \infty, \quad -\infty < n_{\nu y} < \infty, \quad -\infty < n_{\nu z} < \infty.$$

Besides the functions (3.53) which describe standing waves, we can choose travelling waves as the free space modes (Heitler, 1954). In this case the integers $n_{\nu x,y,z}$ take positive and negative values. As can be seen from (3.54), when $L \to \infty$ the spectrum of $\boldsymbol{k}$ (and hence ω) becomes continuous. It is not difficult to find the number of modes in a range of frequencies $d\omega_\nu$ with a given frequency, direction of polarization and direction of propagation in the element $d\Omega$ of a solid angle

$$dZ = \eta_\nu \, d\omega_\nu \, d\Omega \, L^3 = \frac{\omega_\nu^2 \, d\omega_\nu \, d\Omega}{(2\pi c)^3} L^3. \tag{3.55}$$

3.9. Waveguides

We shall now move on to a discussion of waveguides. They are distinguished from free space in that the dimensions of the equivalent resonator must now be considered to approach infinity only in the z-direction. It is not difficult to construct the corresponding eigenfunctions. Leaving this for the reader to do, we shall now calculate only the number of modes in the range of frequencies $d\omega_\nu$. Let $k_\perp$ be the projection of the wave vector onto a surface at right angles to the axis of the waveguide. Then

$$k_\perp^2 + k_z^2 = \frac{\omega_\nu^2}{c^2}. \tag{3.56}$$

By selecting a periodic boundary condition along the z-axis (and thus turning the waveguide into an equivalent resonator) we can obtain a discrete series of values for k_z

$$k_z = \frac{2\pi}{L} n_z. \tag{3.57}$$

Using (3.56) and (3.57) we find for the number of modes in the range from ω_ν to $\omega_\nu + d\omega_\nu$ with given $k_\perp^2$

$$dn_z = \frac{L}{2\pi c^2} \frac{\omega_\nu \, d\omega_\nu}{\sqrt{(\omega_\nu^2/c^2) - k_\perp^2}}. \tag{3.58}$$

To conclude this section we shall show briefly how we can describe a radiation field not by an expansion in the fields $\boldsymbol{E}_\nu$ and $\boldsymbol{H}_\nu$ but by an expansion of the vector potential $\boldsymbol{A}$. An electromagnetic field can be described by the vector potential as follows:

$$\boldsymbol{H} = \operatorname{curl} \boldsymbol{A}, \quad \boldsymbol{E} = -\frac{1}{c}\frac{\partial \boldsymbol{A}}{\partial t}, \quad \operatorname{div} \boldsymbol{A} = 0.$$

We expand

$$\boldsymbol{A} = \sum q_\nu \boldsymbol{A}_\nu(\boldsymbol{r}) \tag{3.59}$$

in the eigenfunctions $\boldsymbol{A}_\nu$ that satisfy the equation

$$\nabla^2 \boldsymbol{A}_\nu + k_\nu^2 \boldsymbol{A}_\nu = 0, \tag{3.60}$$

the boundary conditions and the normalization conditions

$$\int (\boldsymbol{A}_\nu \cdot \boldsymbol{A}_{\nu'}) \, dV = 4\pi c^2 \delta_{\nu\nu'}. \tag{3.61}$$

There is the following connection between the eigenfunctions of $\boldsymbol{A}_\nu$ and $\boldsymbol{E}_\nu$, $\boldsymbol{H}_\nu$:

$$\boldsymbol{E}_\nu = -\frac{1}{c}\boldsymbol{A}_\nu, \quad \boldsymbol{H}_\nu = -\frac{\operatorname{curl} \boldsymbol{A}_\nu}{\omega_\nu}. \tag{3.62}$$

In future we shall use both the expansion in $\boldsymbol{A}_\nu$ and in the fields $\boldsymbol{E}_\nu$ and $\boldsymbol{H}_\nu$.

We have now given the quantum theory of fields in ideal undamped resonators. The quantum theory of the fields in a real resonator allowing for the different losses will be given in the third chapter. However, we point out here that the generalization that appears at first sight to be made by introducing the damping operator $\gamma\dot{\hat{q}}_\nu$ into the operator equation (3.21)

$$\ddot{\hat{q}}_\nu + \gamma\dot{\hat{q}}_\nu + \omega_\nu^2 \hat{q}_\nu = 0, \tag{3.63}$$

is not valid. In fact it follows from the last equation and the connection between $\hat{p}_\nu$ and $\hat{q}_\nu$ that as $t \to \infty$ both $\hat{p}_\nu$ and $\hat{q}_\nu \to 0$ as $e^{-\gamma t}$. From this we find also that the uncertainty product $\Delta p_\nu \Delta q_\nu \sim e^{-2\gamma t}$. But this contradicts the uncertainty relation (3.43). Although equation (3.63) gives a correct description of the behaviour of the mean $\bar{q}_\nu$, allowing for the losses, it describes the fluctuations incorrectly.

4. The Interaction of Matter with a Field

4.1. The Classical Hamiltonian Function

Let us move on to the formulation of the quantum theory of the interaction of matter with a field. We shall proceed from the classical theory of the interaction of radiation with matter. In the classical theory part of the interaction of the matter with the radiation is generally allowed for in the properties of the medium such as the dielectric constant and the magnetic permeability. For example, when discussing the Vavilov–Cherenkov effect the interaction of the radiation field with the medium in which electrons are moving is discussed in terms of the macroscopic refractive index of this medium, and the interaction of radiation with electrons moving in the medium at velocities greater than light by use of the Maxwell equations and the equations of motion of the electrons (see, e.g., Ginzburg, 1959). In the present chapter, in order not to complicate the exposition, we shall consider that interaction with radiation takes place in a vacuum. An isotropic medium with field quantization is discussed by Ginzburg (1940), and the generalization to the case of an arbitrary medium can be made by using the expansion in plane waves (Ginzburg, 1940; Kolomenskii, 1953; Ryzhov, 1959). In the present book we shall allow for the properties of the medium in Chapter XI when discussing non-linear effects in optics.

In the development of the quantum theory we shall proceed from the Hamiltonian of the system which consists of the field and the particles of matter interacting with the field. This Hamiltonian can easily be obtained from the classical Hamiltonian function by replacing the classical quantities by the appropriate operators. Let us find the Hamiltonian function of a system of particles interacting with radiation field. To do this we change from the Maxwell equations to the equations for the canonical variables p_ν and q_ν. The Maxwell equations with current and charge distributions are of the form

$$\operatorname{curl} \boldsymbol{E} + \frac{1}{c}\frac{\partial \boldsymbol{H}}{\partial t} = 0, \tag{4.1}$$

$$\operatorname{div} \boldsymbol{H} = 0, \tag{4.2}$$

$$\operatorname{curl} \boldsymbol{H} - \frac{1}{c}\frac{\partial \boldsymbol{E}}{\partial t} = \frac{4\pi}{c}\boldsymbol{j}, \tag{4.3}$$

$$\operatorname{div} \boldsymbol{E} = 4\pi\varrho. \tag{4.4}$$

We expand $\boldsymbol{E}$ and $\boldsymbol{H}$ in terms of the fields $\boldsymbol{E}_\nu$ and $\boldsymbol{H}_\nu$ (or in $\boldsymbol{A}_\nu$)

$$\boldsymbol{H} = -\sum_\nu \omega_\nu q_\nu \boldsymbol{H}_\nu = \sum_\nu q_\nu \operatorname{curl} \boldsymbol{A}_\nu, \tag{4.5}$$

$$\boldsymbol{E} = \sum_\nu p_\nu \boldsymbol{E}_\nu - \operatorname{grad} \varphi = -\frac{1}{c}\sum_\nu p_\nu \boldsymbol{A}_\nu - \operatorname{grad} \varphi$$

(the term grad φ is necessary since div $\boldsymbol{E} \neq 0$ but div $\boldsymbol{E}_\nu = 0$). Substituting $\boldsymbol{E}$ in (4.4) we find that

$$\nabla^2 \varphi = -4\pi\varrho = -4\pi \sum_k e_k \delta(\boldsymbol{r} - \boldsymbol{r}_k) \tag{4.6}$$

(where e_k is the charge on the kth particle and it is assumed that the particles are points). It follows from (4.6) that φ is the electrostatic potential of the system of particles

$$\varphi(\boldsymbol{r}) = \sum_k \frac{e_k}{|\boldsymbol{r}_k - \boldsymbol{r}|}. \tag{4.7}$$

Eliminating $\boldsymbol{H}$ from (4.1) and (4.3) we find

$$\operatorname{curl} \operatorname{curl} \boldsymbol{E} + \frac{1}{c^2}\frac{\partial^2 \boldsymbol{E}}{\partial t^2} = -\frac{4\pi}{c^2}\frac{\partial \boldsymbol{j}}{\partial t}.$$

Substituting (4.5) and using (3.8) we find

$$\sum_\nu \left(\frac{d^2 p_\nu}{dt^2} + \omega_\nu^2 p_\nu\right) \boldsymbol{E}_\nu = \nabla \frac{\partial^2 \varphi}{\partial t^2} - 4\pi \frac{\partial \boldsymbol{j}}{\partial t}.$$

Multiplying this equation by one of the eigenfunctions of $\boldsymbol{E}_\nu$ we integrate over the whole volume of the resonator (in which the field is contained). Remembering here the conditions for orthogonality and normalization and the boundary conditions (3.6) we obtain

$$\ddot{p}_\nu + \omega_\nu^2 p_\nu = -\sum_k e_k(\dot{\boldsymbol{v}}_k \cdot \boldsymbol{E}_\nu(k)) = \frac{1}{c}\sum_k e_k(\boldsymbol{v}_k \cdot \boldsymbol{A}_\nu(k)),$$

where $\boldsymbol{v}_k$ is the velocity of the kth particle.

This equation, the equations for q_ν and the equations of motion of the particles are contained in the following Hamiltonian function:

$$H = \frac{1}{2}\sum_\nu (p_\nu^2 + \omega_\nu^2 q_\nu^2) + \sum_k \frac{1}{2m_k}\left(\boldsymbol{p}_k - \frac{e_k}{c}\boldsymbol{A}(k)\right)^2$$

$$\times \frac{1}{2}\sum_{i \neq k} \frac{e_i e_k}{r_{ik}}, \tag{4.8}$$

where m_k, $\boldsymbol{p}_k$ are the mass and canonical momentum of the kth particle;
r_{ik} is the distance between the ith and kth particles;
$\boldsymbol{A}(k)$ is the vector potential at the kth particle

$$\boldsymbol{A}(k) = \sum_{\nu} q_{\nu} \boldsymbol{A}_{\nu}(\boldsymbol{r}_k), \quad \text{div}\, \boldsymbol{A} = 0. \tag{4.9}$$

The form of the potential is chosen so that the vector potential describes only a transverse field. The canonical equations that follow from the Hamiltonian function (4.8) are of the form

$$\dot{q}_{\nu} = p_{\nu}, \quad \dot{p}_{\nu} = -\omega_{\nu}^2 q_{\nu} + \frac{1}{c} \sum_k e_k (\boldsymbol{v}_k \cdot \boldsymbol{A}_{\nu}(k)), \tag{4.10}$$

$$m_k \frac{d\boldsymbol{v}_k}{dt} = e_k \boldsymbol{E}(k) + \frac{e_k}{c} [\boldsymbol{v}_k \wedge \boldsymbol{H}(k)], \tag{4.11}$$

where $\boldsymbol{v}_k$ is connected with the canonical momentum by the relation

$$\boldsymbol{v}_k = \frac{1}{m_k} \left(\boldsymbol{p}_k - \frac{e_k}{c} \boldsymbol{A}(k) \right).$$

Equations (4.10) are equivalent to the Maxwell equations. In particular after eliminating q_{ν} we obtain the equation obtained above for p_{ν}. Equations (4.11) are the equations of motion of the particles (in the non-relativistic approximation).

4.2. The Hamiltonian of the Field and Particle System

In quantum theory the Hamiltonian of a system of particles interacting with an electromagnetic field can be obtained from (4.8) by replacing the classical quantities by the appropriate operators

$$\hat{H} = \frac{1}{2} \sum_{\nu} (\hat{p}_{\nu}^2 + \omega_{\nu}^2 \hat{q}_{\nu}^2) + \sum_k \frac{1}{2m_k} \left(\hat{\boldsymbol{p}}_k - \frac{e_k}{c} \hat{\boldsymbol{A}}(k) \right)^2 + \frac{1}{2} \sum_{i \neq k} \frac{e_i e_k}{r_{ik}}. \tag{4.12}$$

The quantum equations of motion that follow from this Hamiltonian are of the same form as (4.10) and (4.11).

In order to solve actual problems it is convenient to extract from the Hamiltonian the energy of the interaction with the radiation field

$$\hat{H} = \hat{H}_0 + \hat{V},$$

where $\hat{H}_0$ is the unperturbed Hamiltonian of the system consisting of particles interacting with each other in accordance with Coulomb's law plus the free radiation field

$$\hat{H}_0 = \sum_k \frac{\hat{p}_k^2}{2m_k} + \frac{1}{2}\sum_{i \neq k} \frac{e_i e_k}{r_{ik}} + \frac{1}{2}\sum_\nu (\hat{p}_\nu^2 + \omega_\nu^2 \hat{q}_\nu^2), \tag{4.13}$$

whilst the energy of the interaction with the field is

$$\hat{V} = -\frac{1}{2}\sum_k \frac{e_k}{m_k c}[(\hat{\boldsymbol{p}}_k \cdot \hat{\boldsymbol{A}}(k)) + (\hat{\boldsymbol{A}}(k) \cdot \hat{\boldsymbol{p}}_k)] + \sum_k \frac{e_k^2}{2m_k c^2}\hat{\boldsymbol{A}}^2(k).$$

The commutator of $\hat{\boldsymbol{p}}_k$ with $\hat{\boldsymbol{A}}$ is of the form (Landau and Lifshitz, 1963).

$$(\hat{\boldsymbol{p}}_k \cdot \hat{\boldsymbol{A}}) - (\hat{\boldsymbol{A}} \cdot \hat{\boldsymbol{p}}_k) = -i\hbar \operatorname{div} \hat{\boldsymbol{A}}.$$

With the particular potential distribution chosen, div $\hat{\boldsymbol{A}} = 0$, and $\hat{\boldsymbol{p}}_k$ commutes with $\hat{\boldsymbol{A}}$ so the interaction energy can be rewritten in the form

$$\hat{V} = -\sum_k \frac{e_k}{m_k c}(\hat{\boldsymbol{p}}_k \cdot \hat{\boldsymbol{A}}(k)) + \sum_k \frac{e_k^2}{2m_k c^2}\hat{\boldsymbol{A}}^2(k). \tag{4.14}$$

In the solution of a number of problems (in particular those connected with spontaneous and stimulated emission of radiation) we can neglect the second term in the right-hand side of (4.14). We rewrite the first term in the form

$$\hat{V} = -\sum_\nu \hat{B}_\nu \hat{q}_\nu, \tag{4.15}$$

where $\hat{B}_\nu$ is an operator acting only on the particle variables

$$\hat{B}_\nu = \sum_k \frac{e_k}{m_k c}(\hat{\boldsymbol{p}}_k \cdot \boldsymbol{A}_\nu(k)). \tag{4.16}$$

4.3. The Interaction Energy in the Dipole Approximation

Let us now examine the case when the dimensions of the system of particles are small compared with the wavelength of the radiation (or compared with the characteristic dimensions of the inhomogeneity of the vector potential $\hat{\boldsymbol{A}}(\boldsymbol{r}_k)$). This is generally the case for atoms and molecules over a very wide range of frequencies, including the optical range. In this case we expand the vector potential with respect to a certain point $\boldsymbol{r}_0$ in the system (e.g. its centre of gravity), placing this, for the sake of simplicity, at the origin ($\boldsymbol{r}_0 = 0$)

$$\hat{\boldsymbol{A}}(\boldsymbol{r}_k) = \hat{\boldsymbol{A}}(0) + \tfrac{1}{2}[\operatorname{curl} \hat{\boldsymbol{A}}|_{r=0} \wedge \boldsymbol{r}_k] + \hat{\Phi}|_{r=0} : \boldsymbol{r}_k, \tag{4.17}$$

where $\hat{\Phi}$ is a symmetric tensor and the colon denotes the scalar product of the tensor and the vector.

The expansion (4.17) corresponds to the expansion in multipoles used in radiation theory. The first term gives the electric dipole radiation, the second the magnetic dipole and the third the electric quadrupole. The last two terms, generally speaking, are of the same order. Let us examine the electric and magnetic dipole terms. (We shall not consider the quadrupole term further in order to simplify the discussion.) Substituting (4.17) (without the quadrupole term) in (4.14) and remembering the properties of a vector product we can write the interaction energy in the form

$$\hat{V} = -\frac{1}{c}\left(\hat{A}(0)\cdot\sum_k \frac{e_k\hat{\boldsymbol{p}}_k}{m_k}\right) - \sum_k \frac{e_k}{2m_kc}\left(\left[\hat{\boldsymbol{r}}_k \wedge \left(\hat{\boldsymbol{p}}_k - \frac{e_k}{c}\hat{A}(0)\right)\right]\cdot\hat{\boldsymbol{H}}(0)\right) + \hat{A}^2\sum_k \frac{e_k^2}{2m_kc^2} + \sum_k \frac{e_k^2}{8m_kc^2}[\hat{\boldsymbol{H}} \wedge \hat{\boldsymbol{r}}_k]^2, \tag{4.18}$$

where we have allowed for the fact that $\hat{\boldsymbol{H}} = \operatorname{curl}\hat{\boldsymbol{A}}$ is the magnetic field. In future we shall neglect the last term, which is quadratic in the magnetic field. The weak diamagnetic susceptibility of the atom (if $\boldsymbol{H}$ is constant) is connected with this term (Landau and Lifshitz, 1963). As usual the connection between the kinetic and canonical momenta follows from the original Hamiltonian (4.12)

$$m_k\dot{\hat{\boldsymbol{r}}} = \hat{\boldsymbol{p}}_k - \frac{e_k}{c}\hat{A}(k). \tag{4.19}$$

Therefore the expression forming the scalar product with $\hat{\boldsymbol{H}}(0)$ in equation (4.18) is the angular momentum in the presence of a field. If we ignore the difference between the kinetic and canonical momenta (and neglect other terms of the order of e_k^2), then expression (4.18) can be written in the form (we are omitting the argument zero)

$$\hat{V} = -\frac{1}{c}(\hat{\boldsymbol{A}}\cdot\dot{\hat{\boldsymbol{d}}}) - (\hat{\boldsymbol{\mu}}\cdot\hat{\boldsymbol{H}}), \tag{4.20}$$

where

$$\hat{\boldsymbol{d}} = \sum_k e_k\boldsymbol{r}_k \quad \text{and} \quad \hat{\boldsymbol{\mu}} = \sum_k \frac{e_k}{2m_kc}[\boldsymbol{r}_k \wedge \hat{\boldsymbol{p}}_k] \tag{4.21}$$

are the operators of the electric and magnetic moments of a molecule (or atom) that it has when there is no field. If e_k and m_k are the charge and mass of an electron, then by introducing the Bohr magneton

$$\beta_0 = \frac{e\hbar}{2mc},$$

the magnetic moment can be written in the form

$$\hat{\boldsymbol{\mu}} = \frac{\beta_0}{\hbar} \sum_k [\boldsymbol{r}_k \wedge \hat{\boldsymbol{p}}_k] = \beta_0 \hat{\boldsymbol{L}},$$

where $\hbar\hat{\boldsymbol{L}}$ is the angular momentum of the system.

Up to now we have been allowing only for the orbital motion of the electrons; allowing for the spin (the intrinsic angular momentum) leads to the additional interaction energy

$$-\sum_k 2\beta_0 \hat{\boldsymbol{s}}_k \hat{\boldsymbol{H}}(k), \tag{4.22}$$

where $\hat{\boldsymbol{s}}_k$ is the spin operator of the kth electron (if there are nuclei in the system the interaction energy is of the same form but β_0 must be replaced by the nuclear magneton and $\boldsymbol{s}_k$ by the spin of the nucleus). If we take $\hat{\boldsymbol{\mu}}$ to be the total magnetic moment of the system

$$\beta_0 \hat{\boldsymbol{L}} + 2\beta_0 \hat{\boldsymbol{s}}, \tag{4.23}$$

where $\boldsymbol{s}$ is the total spin of the electrons, then (4.20) can be used without limiting the generality.

Let us examine in greater detail the case of the electric dipole interaction (if the electric dipole moment of the system is not zero the magnetic dipole interaction can as a rule be ignored). The interaction energy (4.18) is now

$$\hat{V} = -\frac{1}{c}\left(\hat{\boldsymbol{A}} \cdot \sum_k \frac{e_k \hat{\boldsymbol{p}}_k}{m_k}\right) + \hat{\boldsymbol{A}}^2 \sum_k \frac{e_k}{2m_k c^2}. \tag{4.18'}$$

In the case when the electric component of the radiation field is a given function of time we can use the interaction energy in the form

$$\hat{V} = -(\hat{\boldsymbol{d}} \cdot \boldsymbol{E}) = -\left(\left(\sum_k e_k \hat{\boldsymbol{r}}_k\right) \cdot \boldsymbol{E}(t)\right), \tag{4.24}$$

and instead of the Hamiltonian (4.12) use the Hamiltonian

$$\hat{H} = \sum_k \frac{\hat{\boldsymbol{p}}_k^2}{2m_k} + \frac{1}{2} \sum_{i \neq k} \frac{e_i e_k}{r_{ik}} - \sum_k e_k (\hat{\boldsymbol{r}}_k \cdot \boldsymbol{E}(t)), \tag{4.25}$$

where $\hat{\boldsymbol{p}}_k$ and $\hat{\boldsymbol{r}}_k$ are the canonical momenta and coordinates.

We notice that in (4.25) the canonical momenta are the same as the kinetic momenta. From (4.25), using rule (2.6), we obtain the quantum equations of motion of $\boldsymbol{p}_k$ and $\boldsymbol{r}_k$, which are the same as (4.11) (when $\boldsymbol{H} = 0$). This also proves the correctness of choosing the interaction energy (4.24) and the Hamiltonian (4.25).

On the other hand, if we define the motion of the dipole moment we can take as the interaction energy the quantity

$$\hat{V} = -\frac{1}{c}(\dot{\boldsymbol{d}}(t)\cdot\hat{\boldsymbol{A}}); \tag{4.26}$$

then from the Hamiltonian

$$\hat{H} = \frac{1}{2}\sum_{\nu}(\hat{p}_{\nu}^{2} + \omega_{\nu}^{2}\hat{q}_{\nu}^{2}) - \frac{1}{c}\left(\dot{\boldsymbol{d}}(t)\cdot\sum_{\nu}\boldsymbol{A}_{\nu}\hat{q}_{\nu}\right) \tag{4.27}$$

follow the correct Maxwell equations (4.10) (for the operators $\hat{p}_{\nu}$ and $\hat{q}_{\nu}$), including the currents. (Unlike the case in (4.20), terms of the order of e_k^2 are not neglected in the interaction energy (4.26).)

If we do not limit ourselves to the approximation of given currents or given fields, then instead of the Hamiltonian (4.12) it is sometimes convenient to use another Hamiltonian which can be derived from (4.12) by the following canonical transformations

$$\begin{aligned} \hat{p}'_{\nu} &= \hat{p}_{\nu} - \sum_{k}\frac{e_k}{c}(\hat{\boldsymbol{r}}_k\cdot\boldsymbol{A}_{\nu}), \quad \hat{q}'_{\nu} = \hat{q}_{\nu}, \\ \hat{\boldsymbol{p}}'_k &= m\dot{\hat{\boldsymbol{x}}}_k = \hat{\boldsymbol{p}}_k - \frac{e_k}{c}\hat{\boldsymbol{A}}, \quad \hat{\boldsymbol{r}}'_k = \hat{\boldsymbol{r}}_k. \end{aligned} \tag{4.28}$$

(In the dipole approximation, when we can neglect the derivatives of $\boldsymbol{A}$, the commutation relations between the new variables are the same as those between the old ones; this is a characteristic property of the canonical transformation in quantum theory.) After the canonical transformation the Hamiltonian (4.12) becomes

$$\begin{aligned} \hat{H}' = {} & \frac{1}{2}\sum_{\nu}(\hat{p}_{\nu}'^{2} + \omega_{\nu}^{2}\hat{q}_{\nu}'^{2}) + \sum_{k}\frac{\hat{\boldsymbol{p}}_k'^{2}}{2m_k} + \frac{1}{2}\sum_{i\neq k}\frac{e_i e_k}{r_{ik}} \\ & + \frac{1}{2}\left(\sum_{k}\frac{e_k}{c}(\hat{\boldsymbol{r}}_k\cdot\boldsymbol{A}_{\nu})\right)^{2} + \sum_{k,\nu}\frac{e_k}{c}(\hat{\boldsymbol{r}}_k\cdot\boldsymbol{A}_{\nu}(k)\,\hat{p}'_{\nu}). \end{aligned} \tag{4.29}$$

The last term in the right-hand side can be taken to be the interaction energy and the penultimate term can be included in the unperturbed Hamiltonian of the system of particles. In this case the unperturbed Hamiltonian of the system can be taken only approximately to be the energy of the system; likewise the first term in (4.29) is only approximately the field energy since according to (4.28) $\hat{p}'_{\nu} \neq \dot{\hat{q}}_{\nu}$. We also notice that the electric field is determined (see (3.23)) by $\hat{p}_{\nu}$ and not by $\hat{p}'_{\nu}$, so even in the dipole approximation the last term in (4.29) is not the same as (4.24).

4.4. The Schrödinger Equation

We shall now discuss the Schrödinger equation (2.1) with the Hamiltonian (4.12). The method generally used to find a solution for this equation is to change to the interaction representation and use perturbation theory. The change to the interaction representation is made by expanding the wave function of the system in the eigenfunctions of the Hamiltonian operator $\hat{H}_0$ (4.13)

$$\Psi = \sum_n b_n \Psi_n \, e^{-\frac{i}{\hbar}\mathscr{E}_n t}. \tag{4.30}$$

The wave functions Ψ_n are assumed to be known. They are given in the form of products of the stationary functions $\Psi_{n_1 n_2 \ldots}$ of the free radiation field and the stationary wave functions Ψ_{An} of the system (e. g. the stationary functions of the atoms or molecules)

$$\Psi_n = \Psi_{n_1 n_2 n_3 \ldots n_\nu \ldots} \Psi_{An}. \tag{4.31}$$

The eigenvalues of the energy are

$$\mathscr{E}_n = \sum_\nu (n_\nu + \tfrac{1}{2}) \hbar\omega_\nu + E_{An}, \tag{4.32}$$

where E_{An} are the eigenvalues of the energy of the system of particles. Substituting Ψ in the form of the expansion (4.30) in the Schrödinger equation (2.1) and (4.12), multiplying by one of the eigenfunctions Ψ_n and integrating over all the field and particle coordinates we come to the interaction representation

$$i\hbar \dot{b}_n = \sum_m V_{nm} \, e^{i\omega_{nm}t} b_m. \tag{4.33}$$

The corresponding equation for the density matrix in the interaction representation takes the form

$$i\hbar \frac{\partial \varrho_{nm}}{\partial t} = [\hat{V}, \hat{\varrho}]_{nm} = \sum_k V_{nk}\varrho_{km} \, e^{i\omega_{nk}t} - \varrho_{nk} V_{km} \, e^{i\omega_{km}t}, \tag{4.34}$$

where

$$\omega_{nm} = \frac{\mathscr{E}_n - \mathscr{E}_m}{\hbar}.$$

The time dependence of the matrix $\hat{V}$ in the Heisenberg representation is explicit in the formulae (4.33) and (4.34).

4.5. Matrix Elements

Let us write out the matrix elements of the interaction energy operator (4.15). By using the wave functions (4.31) and relations (3.27) it can be shown that the non-zero matrix elements of (4.15) are of the form†

$$\langle a, n_1, n_2, n_3, \ldots n_\nu, \ldots |\hat{V}| b, n_1, n_2, n_3, \ldots n_\nu + 1 \ldots\rangle$$

$$\equiv V_{a,n_\nu;b,n_\nu+1} = -\sqrt{\frac{\hbar}{2\omega_\nu}}\sqrt{n_\nu+1}\sum_k \frac{e_k}{m_k c}(p_k \cdot A_\nu)_{ab}$$

$$= -\sqrt{\frac{\hbar}{2\omega_\nu}}\sqrt{n_\nu+1}\,(B_\nu)_{ab} = V^*_{b,n_\nu+1;a,n_\nu}, \qquad (4.35)$$

where a and b are indices numbering the states of the system of particles. In the dipole approximation the matrix elements (4.35) become

$$V_{a,n_\nu;b,n_\nu+1} = -\sqrt{\frac{\hbar}{2\omega_\nu}}\sqrt{n_\nu+1}\,\frac{i\omega_{ab}}{c}(d_{ab}\cdot A_\nu(r_0)) = V^*_{b,n_\nu+1;a,n_\nu}. \qquad (4.36)$$

5. Non-stationary Perturbation Theory. Transition Probability

5.1. *Allowing for External Forces*

Up to now we have been dealing with closed systems whose Hamiltonian is not time-dependent. In order to allow for the external forces acting on the system we must introduce an explicit time dependence into the Hamiltonian of the system.

For example, the effect of an external alternating magnetic field $\boldsymbol{H}$ in the dipole approximation is allowed for by the interaction energy

$$-(\hat{\mu}\cdot \boldsymbol{H}(t)),$$

where $\boldsymbol{H}(t)$ is a given time function. The change to the interaction representation from the Schrödinger representation of the equation for the density matrix

$$i\hbar\frac{\partial\hat{\varrho}}{\partial t} = [\hat{H}_0(t) + \hat{V}(t), \hat{\varrho}] \qquad (5.1)$$

($\hat{H}_0$ and $\hat{V}$, generally speaking, may be explicitly time-dependent) is achieved by means of the unitary matrix that satisfies the equations

$$i\hbar\frac{\partial\hat{S}^{-1}}{\partial t} = -\hat{S}^{-1}\hat{H}_0, \quad i\hbar\frac{\partial\hat{S}}{\partial t} = \hat{H}_0\hat{S}. \qquad (5.2)$$

† We are using the Dirac form of matrix element notation.

Since $\hat{H}_0$ may be explicitly time-dependent, $\hat{S}$ is generally not the same as $\hat{U}_0$ from (2.12b). Using the $\hat{S}$ matrix to transform the left- and right-hand sides of the equation we obtain for the density matrix in the interaction representation

$$i\hbar \frac{\partial \hat{\varrho}}{\partial t} = [\hat{V}, \hat{\varrho}]. \tag{5.3}$$

We shall not introduce any new notation here, but it must be borne in mind that $\hat{V}$ and ϱ in (5.3) are connected with the Schrödinger operators by the relations

$$\hat{S}^{-1}\hat{V}\hat{S} \quad \text{and} \quad \hat{S}^{-1}\hat{\varrho}\hat{S}.$$

5.2. *Non-Stationary Perturbation Theory*

We now proceed to examine perturbation theory. We start from equation (5.3) and consider $\hat{V}$ to be a small quantity. We expand the density matrix $\hat{\varrho}$ into a series

$$\hat{\varrho} = \sum_k \hat{\varrho}^{(k)}, \tag{5.4}$$

where $\hat{\varrho}^{(k)}$ is a quantity of the kth order of smallness with respect to $\hat{V}$. Substituting (5.4) in (5.3) we obtain the system of recurrence relations

$$i\hbar \frac{\partial \hat{\varrho}^{(k)}}{\partial t} = [\hat{V}, \hat{\varrho}^{(k-1)}]. \tag{5.5}$$

As the zero approximation we shall take the value of the density matrix at time $t = 0$. This means that the corrections of the first and subsequent approximations of $\hat{\varrho}^{(k)}$ should be little different from $\hat{\varrho}(0)$. We write the density matrix explicitly with an accuracy up to terms of the second order of smallness

$$\hat{\varrho}(t) = \hat{\varrho}^{(0)} + \hat{\varrho}^{(1)} + \hat{\varrho}^{(2)} + \cdots = \hat{\varrho}(0) - \frac{i}{\hbar}\int_0^t [\hat{V}(t_1), \hat{\varrho}(0)]\, dt_1$$
$$- \frac{1}{\hbar^2}\int_0^t dt_1 \int_0^{t_1} dt_2 [\hat{V}(t_1), [\hat{V}(t_2), \hat{\varrho}(0)]] + \ldots \tag{5.6}$$

5.3. *Transition Probability (Constant Perturbation)*

Let us now examine the special case when initially the system was known to be in the stationary state n_0 with an unperturbed Hamiltonian (assuming that the latter is explicitly time-independent)

$$\varrho_{mn}(0) = \delta_{mn_0} \cdot \delta_{nn_0}. \tag{5.7}$$

Substituting the expression (5.7) in the right-hand side of (5.6) and examining the case when $\hat{V}$ is explicitly time-independent we find

$$\varrho_{nn}(t) = \frac{2\,|V_{nn_0}|^2}{(E_{n_0} - E_n)^2}\left[1 - \cos\,(E_{n_0} - E_n)\,\frac{t}{\hbar}\right] \quad (n \neq n_0), \tag{5.8}$$

where E_n, E_{n_0} are the energy levels of the unperturbed system. The expression $\varrho_{nn}(t)$ is the probability of finding (when measuring) the system in a state n if it was in a state n_0 initially. In other words $\varrho_{nn}(t)$ is the probability of a transition from state n_0 to state n in a time t. It follows from (5.8) that the probability of the transition is finite if the energies of the initial and final states are approximately the same. It should be mentioned that the law of the conservation of energy can be checked with an accuracy up to the uncertainty relation for the energy (see Landau and Lifshitz, 1963, section 44). This can also be seen from the expression (5.8). The transition probability is essentially finite if the difference between the initial and final states ΔE satisfies the uncertainty relation

$$\Delta E\, \Delta t \sim \hbar. \tag{5.9}$$

The energy levels of the initial and final states often form a continuous spectrum. Let us examine in greater detail the case (which is of physical interest) when the energy of the final state is part of a continuous spectrum.

We shall assume that the variables n that describe the final state consist of E (the energy) and a certain set of variables (including the continuous set as well) which we shall denote by the suffix u. Since by assumption E varies continuously we can introduce a number of states with fixed values u in the energy range dE

$$dZ = \eta_u(E)\, dE. \tag{5.10}$$

We are generally interested in the probability of a transition into a state with a fixed value u but with any value E (of course E will very probably be near the initial state). The required probability becomes

$$W_{E_0E} = \int \varrho_{uE,uE}\eta_u(E)\, dE = \int \frac{2\,|\langle E_0, u_0|\, V\, |E, u\rangle|^2}{(E_0 - E)^2} \times \left[1 - \cos\,(E_0 - E)\,\frac{t}{\hbar}\right] \eta_u(E)\, dE.$$

It follows from Appendix I that for large enough t this expression becomes

$$W_{E_0E} = \frac{2\pi}{\hbar}\, |\langle E_0, u_0|\, V\, |E_0, u\rangle|^2 \eta_u(E_0)\, t. \tag{5.11}$$

The term "large enough t" must be taken in the sense that

$$t \gg \frac{\hbar}{\Delta E} \equiv \frac{1}{\omega^*}, \tag{5.12}$$

where ΔE is the characteristic range of variation of the energy in the expression

$$|\langle E_0, u_0| V |E, u\rangle|^2 \eta_u(E).$$

Of course, the quantity $\Delta E \equiv \hbar\omega^*$ is defined by the actual properties of the system under investigation and the interaction energy. It should be stressed that the applicability of (5.11) depends not only on condition (5.12); we must also satisfy the condition concerning the smallness of t compared with the characteristic time required for a significant change to occur in the state of the system. This is the condition of applicability of the perturbation theory. The quantity

$$w_{E_0E} = \frac{2\pi}{\hbar} |\langle E_0, u_0| V |E, u\rangle|^2 \eta_u(E_0) \tag{5.13}$$

is called the transition probability. For very small values of the time

$$t \ll \frac{\hbar}{\Delta E} \tag{5.14}$$

there is no time-independent transition probability. In this case the probability of a transition is proportional to t^2.

Sometimes the transition probability can be conventionally expressed in the form

$$w_{E_0E} = \frac{2\pi}{\hbar} |V_{n_0u}|^2 \delta(E - E_0). \tag{5.13'}$$

It should, however, always be remembered that this expression has meaning only after integration over the energy.

Let us examine the example (which is important for the later discussion) of the probability of a spontaneous transition with the emission of a photon. The Hamiltonian of the system in this case takes the form (4.12), the unperturbed Hamiltonian $\hat{H}_0$ takes the form (4.13) and the perturbation energy is (4.15). Initially let the "matter" (for the sake of definition we shall speak of a molecule) be in an excited state with an energy E_a and let there be no photons in the radiation field. We are interested in the probability of a transition into a state with a lower molecule energy $E_b < E_a$ with the emission of a photon of energy $\hbar\omega_\nu$. Therefore the total energy of the initial state is

$$E_{a0} = E_a, \tag{5.15}$$

and of the final state

$$E_{b1_\nu} = E_b + \hbar\omega_\nu. \tag{5.16}$$

Using the expression for the matrix element of the energy of the interaction with radiation (4.35), and (5.13′), we can write the probability of spontaneous emission of a photon characterized by the suffix ν†:

$$w_{a0\to b1_\nu} = \frac{2\pi}{\hbar} |\langle a; 0, 0, 0, \ldots | \hat{V} | b, 0, 0, \ldots, 1_\nu, 0, 0\rangle|^2 \quad \delta(E_a - E_b - \hbar\omega_\nu)$$

$$= \frac{2\pi}{\hbar} |\langle a, 0 | \hat{V} | b, 1_\nu\rangle|^2 \quad \delta(E_a - E_b - \hbar\omega_\nu).$$

The density of the final states in the energy range $dE_{b1} = \hbar d\omega_\nu$ can be determined by using (3.55) (for the case of emission into free space)

$$dZ = \frac{\mathscr{E}_\nu^2 d\Omega L^3}{(2\pi c)^3 \hbar^3} d\mathscr{E}_\nu, \quad \mathscr{E}_\nu = \hbar\omega_\nu.$$

Therefore in accordance with (5.13) the transition probability is‡

$$w_{a0;b1_\nu} = \frac{L^3\omega_0^2\, d\Omega}{(2\pi)^2 c^3 \hbar^2} |\langle a, 0 | \hat{V} | b, 1_\nu\rangle|^2, \tag{5.17}$$

where $\hbar\omega_0 = E_a - E_b$. In order to obtain the total probability of the emission of a photon of frequency ω_0 with any direction of propagation this expression must be integrated over all directions (for the solid angle $d\Omega$). In the general case (not necessarily in free space) the probability of spontaneous emission of a photon of frequency $\hbar\omega_0$ is

$$W_{ab} = \frac{2\pi}{\hbar} \sum_\nu |\langle a, 0 | \hat{V} | b, 1_\nu\rangle|^2\, \delta(E_a - E_b - \hbar\omega_\nu)\, t. \tag{5.18}$$

We notice that in the case of a continuous spectrum the matrix elements $V_{a0;b1_\nu}$ approach zero. In fact, in accordance with (4.35), (3.62) and (3.53)

† The index ν characterizes the direction of propagation, the polarization, etc.

‡ In the case of emission into free space the appropriate quantity ω^* in the condition for the applicability of the expression for the transition probability is $\omega_0 = (E_a - E_b)/\hbar$. This can easily be checked by examining the expression $|\langle a, 0|\hat{V}|b, 1_\nu\rangle|^2\, \eta_u(E)$. It follows from (4.35) and (3.55) that this expression changes when the energy changes by the amount $\hbar\omega_\nu$. At the same time near the energy of the initial state $\hbar\omega_\nu \approx \hbar\omega_0$. It follows from this and the definition of ω^* (see (5.12) and Appendix I) that $\omega^* \approx \omega_0$.

the matrix elements are inversely proportional to $\sqrt{L^3}$ (where L^3 is the volume of the resonator equivalent to the free space). Ultimately this quantity should approach infinity, i.e. the matrix elements approach zero. The finite value of the spontaneous transition probability arises from the summation over the states ν. In the case of free space this summation (when the spectrum is continuous) is reduced to integration over the energies and solid angles. Here the density of states $\eta_u(E)$ is itself proportional to L^3. This dependence just compensates for the inverse proportionality of the quantity L^3 in the quantity $|V_{a0;b1_\nu}|^2$ and gives a finite value for the transition probability when $L^3 \to \infty$.

5.4. *Transition Probability (Harmonic Perturbation)*

Let us now examine the probability of a transition from a state n_0 described by the density matrix (5.7) into another state of the unperturbed Hamiltonian (explicitly not time-dependent) under the action of a harmonic perturbation

$$\hat{V} = \hat{V}^0 \cos(\omega t + \varphi). \tag{5.19}$$

Substituting (5.19) and (5.7) in (5.6) we obtain for the diagonal elements of the density matrix the expression

$$\varrho_{nn}(t) = \frac{2}{\hbar^2}\int_0^t dt_1 \int_0^{t_1} dt_2 \cos(\omega t_1 + \varphi)\cos(\omega t_2 + \varphi)\cos\omega_{nn_0}(t_1 - t_2)\,|V^0_{nn_0}|^2.$$

After integration we obtain

$$\begin{aligned}\varrho_{nn}(t) = \frac{|V^0_{nn_0}|^2}{2\hbar^2}\Bigg\{&\pi t\,\frac{1-\cos(\omega_{nn_0}-\omega)\,t}{\pi(\omega-\omega_{nn_0})^2\,t} + \pi t\,\frac{1-\cos(\omega_{nn_0}+\omega)\,t}{\pi(\omega_{nn_0}+\omega)^2\,t}\\ &+\frac{1}{(\omega+\omega_{nn_0})(\omega-\omega_{nn_0})}\,[\cos 2\varphi - \cos(2\omega t + 2\varphi)]\\ &+\frac{1}{(\omega+\omega_{nn_0})(\omega-\omega_{nn_0})}\\ &\times\{-2\cos 2\varphi + \cos[(\omega+\omega_{nn_0})\,t + 2\varphi] + \cos[(\omega-\omega_{nn_0})\,t+2\varphi]\Bigg\}.\end{aligned}$$

In the approximation given in Appendix I (see also the formula (5.12)) the first two terms in the expression $\varrho_{nn}(t)$ give a contribution proportional to the time. The remaining terms are negligible provided that

$$\omega t \gg 1. \tag{5.20}$$

Using this approximation and bearing in mind the meaning of the quantity ϱ_{nn} we can find the transition probability from the state n_0 to the state n under the action of the harmonic perturbation (5.19)

$$w_{n_0 n} = \frac{\pi}{2\hbar} |V^0_{nn_0}|^2 \{\delta(E_{n_0} - E_n - \hbar\omega) + \delta(E_{n_0} - E_n + \hbar\omega)\}. \quad (5.21)$$

We notice that this expression does not depend on the phase φ of the perturbation. The terms which depend on φ are small when (5.21) is satisfied, and have been neglected. We also notice that (5.21) cannot be used when $\omega = 0$ since in this case condition (5.20) is not satisfied. When $\omega = 0$ (and $\varphi = 0$) $w_{n_0 n}$ does not become (5.13′) but half the value of this expression, the other half being connected with the previously neglected terms which are not small when $\omega \to 0$.

CHAPTER II

The Quantum Theory of Relaxation Processes

IF A QUANTUM system is closed the evolution of its state with time can be described by the equation for the density matrix (2.3b). However, we more often meet the case in which the system of interest to us is not closed but is in contact with its surroundings. In particular the system may be in contact with a constant temperature bath. The problem is how to describe this kind of non-closed ("open") system. It is obvious that the solution is of major significance in quantum electronics. When we are interested in the steady state of a system (with no external forces) the solution is well known. According to the basic principles of statistical physics a system in contact with a constant temperature bath (with which it interacts weakly) and in a steady state (state of thermal equilibrium) can be described by the density matrix

$$\hat{\varrho} = \exp\left(-\frac{\hat{H}}{kT}\right)\Big/\mathrm{Tr}\exp\left(-\frac{\hat{H}}{kT}\right),$$

where T is the temperature of the bath and $\hat{H}$ is the Hamiltonian of the "open" system we are interested in.

Now let the system be in a non-equilibrium state. In the course of time this system will approach an equilibrium state. The processes which occur during this time are called relaxation processes. Generalizing, we shall give the name of relaxation processes to any decay processes, in particular those in which a system approaches its stationary state (in the presence of an external force). Relaxation processes play an important part in quantum electronics. The following examples may be given of relaxation processes that are met in practice: the decay of an electromagnetic field in a lossy resonator, spontaneous emission in free space, relaxation of a spin system due to interaction with the crystal lattice (spin–lattice relaxation), etc. In all these examples relaxation takes place as the result of interaction with systems which have, in the limit, an infinite number of degrees of freedom—the electrons in the resonator walls, the radiation field in free space, the lattice molecules in a solid, etc. It will be convenient to use the following terminology. We shall call the system we are interested in a *dynamic system* and its surroundings a *dissipative system* (it is a source of dissipation). It is understood here that the dynamic system (or sub-system) has a finite number of degrees of freedom and discrete energy levels, whilst the dissipative system has, in the limit, an infinite number of degrees of freedom and a continuous spectrum.

The dynamic and dissipative sub-systems, interacting with each other, together form a closed system. The behaviour of this closed system can be described by equation (2.3b), using the density matrix of the whole system $\varrho_{m\alpha;n\alpha'}$, where the Latin letters are the discrete indices of the dynamic sub-system and the Greek letters are the indices of the dissipative system (at least part of which run through a continuous series of values). Since we are interested in the behaviour of the dynamic sub-system we need know only that part of the density matrix $\varrho_{m\alpha;n\alpha}$ which is diagonal in the indices α, or its trace in the indices α

$$\sigma_{mn} = \sum_{\alpha} \varrho_{m\alpha;n\alpha}.$$

The present chapter basically covers the derivation of the equations giving $\varrho_{m\alpha;n\alpha}$ (section 7) and σ_{mn} (section 8). We refer to these as transport equations. The derivation of these equations, using a number of assumptions (the chief of which relates to the smallness of the interaction energy between the dynamic and dissipative sub-systems), is rather cumbersome. We shall therefore indicate here the basic results of sections 7 and 8 which will be used in subsequent chapters.

Taking as a basis the equation for the density matrix (7.9) in the interaction representation in section 7 we obtain the transport equations (7.27) and (7.28) which describe the change with time of the density matrix $\varrho_{m\alpha;n\alpha}$. These describe both the relaxation of the dynamic system and of the dissipative system. With certain assumptions about the spectrum of the dynamic sub-system, there follows, from equation (7.28), the balance equation (7.36) which connects the diagonal elements of $\varrho_{m\alpha;n\alpha}$ with the diagonal elements of the density matrix, i.e. this equation describes the relaxation of the populations of the energy levels. If the dissipative sub-system is in a state of thermal equilibrium (at constant temperature) the behaviour of the dynamic system can be described by transport equation (8.3) which is written in terms of the density matrix σ_{mn} of the dynamic sub-system. Under certain conditions the balance equations (8.10), for the dynamic sub-system only, follow from these equations

$$\dot{\sigma}_{mm} = \sum_{k} w_{km}\sigma_{kk} - w_{mk}\sigma_{mm}.$$

We notice that, whilst the steady-state solution of the balance equations for $\varrho_{m\alpha;n\alpha}$ gives a microcanonical ensemble, solution (8.10) when $t \to \infty$ gives a canonical ensemble, in agreement with statistical physics. It is often convenient to use in place of the transport equations the equations derived from them which describe the change with time of the mean value of a certain operator $\hat{O}$ relating to the dynamic system. These equations (8.18) and (8.18a) are obtained in sub-section 3 of section 8. At first reading the derivation of all the equations listed may be omitted and they can be used "ready-made". Section 10 of the present chapter may be read without reference to the preceding sections.

6. General Properties of Irreversible Processes

6.1. Systems with an Infinite Number of Degrees of Freedom

Closed mechanical systems have the following feature. Let the system at a certain point in time be in a state A; then after a large enough time interval T_p the system will approach as close as one likes to state A (see, e.g., Chandrasekhar, 1947, Chapter III, section 4). Poincaré proved this for classical systems, but quantum systems have the same property. At first sight this property of physical systems contradicts the concept of irreversibility, that the system moves away from the initial state and does not return to it. In fact there is no contradiction here. The point is that when the number of degrees of freedom of the system increases the time T_p—called the time of the Poincaré cycle—increases. When the number of degrees of freedom approaches infinity the time T_p also approaches infinity. Therefore a system with an infinite number of degrees of freedom can behave irreversibly. Of course, ascribing an infinite number of degrees of freedom to a system is an idealization. However, this kind of idealization is well-founded if we bear in mind that a macroscopic body has a number of molecules or atoms of the order of 10^{22}–10^{23} per cm^3 (of the same order as the number of degrees of freedom). Under these conditions the time of the Poincaré cycle is so great that it can be taken as infinite in all practical problems.

The irreversible behaviour of a system is associated with an increase in its entropy. Let us examine in greater detail the question of the entropy of a non-equilibrium system.

6.2. Entropy and Information

In thermodynamics the entropy of an equilibrium system is usually defined by

$$\Delta S = \frac{\Delta Q}{T},$$

where ΔS is the increment of the entropy of a system which has a temperature T and receives an amount of heat ΔQ.

The thermodynamic definition can be connected with the statistical definition by the Boltzmann principle†

$$S = k \ln \Delta\Gamma, \tag{6.1}$$

where k is the Boltzmann constant,

$\Delta\Gamma$ is the thermodynamic probability or statistical weight of the macroscopic state of the system (Sommerfeld, 1955; Landau and Lifshitz, 1951).

We shall now look into the behaviour of non-equilibrium systems. The above relations for non-equilibrium system scannot be used directly. We must therefore adduce a more general definition of entropy which is valid for any system. This general definition of entropy is used in information theory (Brillouin, 1956).

Let a probability distribution describe a certain statistical set (ensemble) (see section 1). The entropy of the ensemble is a measure of the statistical scatter or chaotic nature of the probability distribution in the ensemble. By definition the entropy satisfies the following conditions. It is a functional of the probability distribution with its maximum value in the most chaotic ensemble in which all the states (members of the ensemble) are found with equal probability. The entropy has its minimum value (zero) when the system is definitely in a given state. And, lastly, the entropy must be additive: the entropy of a system consisting of two statistically independent subsystems is equal to the sum of the entropies of each of the sub-systems. All these conditions (except for the inclusion of a constant factor) are satisfied by the quantity

$$\mathscr{E} = -\sum_i p_i \ln p_i, \tag{6.1a}$$

where p_i is the probability with which the ith term of the statistical ensemble appears $\left(\sum_i p_i = 1\right)$. The suffix i denotes the combination of suffixes characterizing the given state. The suffixes i can vary continuously. In this case we must change from a sum to an integral. In thermodynamics and statistical physics the expression

$$S = k\mathscr{E} = -k \sum p_i \ln p_i, \tag{6.1b}$$

† As is well known (Sommerfeld, 1955), Boltzmann himself did not derive this expression. The term "Boltzmann's principle" was introduced by Einstein who also used an inverted form of Boltzmann's principle $\Delta\Gamma = \exp (S/k)$ in the study of fluctuations.

which has the correct dimensions, is also used as the entropy. The Boltzmann principle (6.1) for equilibrium distributions follows from this expression. In a micro-canonical ensemble all the $\Delta\Gamma$ of the states in the range of energies ΔE are equally probable and the entropy (6.1 b) is

$$S = -k \sum_{1}^{\Delta\Gamma} \frac{1}{\Delta\Gamma} \ln \frac{1}{\Delta\Gamma} = k \ln \Delta\Gamma .$$

For a canonical ensemble the statistical weight can be defined as (Landau and Lifshitz, 1951)

$$\Delta\Gamma p(\bar{E}) = 1,$$

where $p(E)$ is the probability distribution in the canonical ensemble. Using (6.1 b) we find

$$S = -k \sum_{E} p(E) \ln p(E) = -k \overline{\ln p(E)} = k \ln \Delta\Gamma,$$

this last equality following from the fact that in the canonical distribution $\ln p(E)$ is linearly dependent on E.

Besides the entropy we can introduce a quantity which measures the "ordering" of a given probability distribution. This quantity is called *information* and is (Wiener, 1948)†

$$I = \sum_{i} p_i \ln p_i = -\mathscr{E}. \tag{6.2}$$

Let a system be in a quantum state (generally speaking mixed) which can be described by the density matrix $\hat{\varrho}$. We shall attempt to find the connection between the entropy or information and the density matrix of the system. This problem, generally speaking, has no unique solution. The point is that defining the quantum state described by the density matrix $\hat{\varrho}$ does not mean defining a definite statistical ensemble. In order to define the ensemble we must say which measurements must be made on the system in the state $\hat{\varrho}$. Let measurements of a quantity described by the operator $\hat{A}$, whose eigenvalues are denoted by the suffix n, be made on the system. Then the probability distribution in this ensemble is given by the diagonal elements of the density matrix ϱ_{nn}, and the entropy of the ensemble is‡

$$\mathscr{E}_A = -\sum_{n} \varrho_{nn}^{(A)} \ln \varrho_{nn}^{(A)} \tag{6.3}$$

† Here and later we use Wiener's definition of information. This measure of information is different from that introduced by Shannon (for greater detail see Ashby: 1956).

‡ The density matrix is, of course, characterized by other suffixes as well as n. Summation must obviously be carried out for these suffixes as well, i.e.

$$\mathscr{E}_A = - \sum_{n,l} \varrho_{nl;nl}^{(A)} \ln \varrho_{nl;nl}^{(A)}.$$

(here the matrix elements of $\hat{\varrho}$ are taken in a representation in which the operator $\hat{A}$ is diagonal). Measurement of another quantity $\hat{B}$ which does not commute with $\hat{A}$ leads to another ensemble and therefore to another entropy value

$$\mathscr{E}_B = -\sum_k \varrho_{kk}^{(B)} \ln \varrho_{kk}^{(B)} \neq \mathscr{E}_A, \tag{6.3'}$$

where the matrix elements of $\hat{\varrho}$ are taken in a representation in which the operator $\hat{B}$ is diagonal. From all the ensembles corresponding to a given state we select those ensembles which, when measuring the set of quantities $\hat{L}, \hat{M}, \hat{N}, \ldots$, have a density matrix that is diagonal in a representation in which these quantities are diagonal. We have called ensembles of this kind "complete" (section 1) and the corresponding measurement a complete measurement. The entropy of a complete ensemble is

$$\mathscr{E} = -\sum_l \varrho_{ll} \ln \varrho_{ll} = -\mathrm{Tr}\, \hat{\varrho} \ln \hat{\varrho}, \tag{6.4}$$

where the latter equality defines the entropy of a complete ensemble if the density matrix in an arbitrary ensemble is given. Equation (6.4) allows us to find the entropy or information of quantum states described by the density matrix $\hat{\varrho}$

$$\mathscr{E}_m = -I_m = -\mathrm{Tr}\, \hat{\varrho} \ln \hat{\varrho}. \tag{6.5}$$

It must be borne in mind, however, that this entropy (information) is not that of all the ensembles that appear when measuring in a given state but only that of the complete ensemble.

6.3. *Klein's Lemma*

The following statement can be made. The information defined by (6.5) is the maximum information possessed by the state $\hat{\varrho}$; in other words the information $I_m = \mathrm{Tr}\, \hat{\varrho} \ln \hat{\varrho}$ is greater than that of any ensemble realized in an incomplete measurement:

$$I_m > I_A \quad \text{and} \quad I_B. \tag{6.5A}$$

This statement follows from the fact that

$$\mathrm{Tr}\, \hat{\varrho} \ln \hat{\varrho} \geqslant \sum_n \varrho_{nn} \ln \varrho_{nn}, \tag{6.6}$$

where the equals sign holds if the non-diagonal elements $\varrho_{nm} = 0$. Formula (6.6) is a statement of the so-called Klein's lemma (Klein, 1931). The proof of this lemma follows from the properties of the unitary transformation which connects the matrix elements of ϱ_{nn} and the matrix elements of $\hat{\varrho}$ in an arbitrary representation. Elsasser (1937) has called the quantity

$$I_m = \operatorname{Tr} \hat{\varrho} \ln \hat{\varrho}$$

the mixture index. This name is connected with the fact that this quantity makes it possible to determine whether the system is in a pure or mixed state. In fact, in a pure state in a representation in which the density matrix is diagonal the density matrix has only one non-zero matrix element equal to unity. Hence in a pure state

$$I_m = \operatorname{Tr} \hat{\varrho} \ln \hat{\varrho} = 0,$$

and in any mixed state $I_m < 0$. From now on we shall call the quantity I_m (or $\mathscr{E}_m$) the information of the state as opposed to the information I (entropy $\mathscr{E}$) characterizing the ensemble. The suffix m means that I_m is the maximum information (obtained by a complete measurement)†.

We shall now move on to investigate the evolution of closed systems.

6.4. The General Properties of the Evolution of Closed Systems

It follows from the invariance of the trace under the unitary transformation (and the evolution of the density matrix with time can be looked upon as a unitary transformation) that the entropy of the state (information of the state) of a closed dynamic system is not time-dependent

$$I_m = -\mathscr{E}_m = \text{const.}$$

This means that the state of the system changes in such a way that the information (or entropy) obtained from a complete measurement does not change in the ensemble. Thus if the whole system is originally in a pure state it will stay in this state the whole time‡.

Let a complete measurement be made at time t_0 on a system and let the probability distribution be given by the diagonalized density matrix

$$\varrho_{mn}(t_0) = \varrho_{nn}(t_0)\, \delta_{mn}.$$

The entropy of this ensemble is

$$\mathscr{E}(t_0) = -\sum_n \varrho_{nn}(t_0) \ln \varrho_{nn}(t_0) = -\operatorname{Tr} \hat{\varrho}(t_0) \ln \hat{\varrho}(t_0).$$

† It must be stressed that we are speaking here of the maximum information (or the minimum entropy) in relation to the other ensembles that appear when measuring in a given state at a definite time. The question of the change in entropy with time will be discussed later.

‡ Everywhere here we assume, as usual, that we have a combination of identical systems and that if a measurement is made at some time on some representatives of the ensemble we shall not continue to take them into consideration since they move into a different state as a result of the measurement.

Let a measurement be made at a time $t > t_0$ of the same quantities in the system as was made at time t_0, i.e. we are looking at the time-evolution of the ensemble. The ensemble (originally complete) is then generally not complete since a complete measurement produces at each point in time another ensemble corresponding to the measurement of another set of quantities†. The entropy of the ensemble originally chosen (complete at the time t_0) is at the time t

$$\mathscr{E}(t) = -\sum_n \varrho_{nn}(t) \ln \varrho_{nn}(t) \geqslant \mathscr{E}(t_0) = -\mathrm{Tr}\, \hat{\varrho} \ln \hat{\varrho}, \tag{6.7}$$

where the inequality sign follows from Klein's lemma since at time t the density matrix ϱ_{nm}, generally speaking, is not diagonal. Of course, the inequality (6.7) is not the law of the growth of the entropy in its ordinary thermodynamic formulation since it does not follow from (6.7) that at time $t > t_1$ $\mathscr{E}(t_1) = -\sum_n \varrho_{nn}(t_1) \ln \varrho_{nn}(t)$ should be greater than $\mathscr{E}(t)$; all that can be stated is that $\mathscr{E}(t_1) \geqq \mathscr{E}(t_0)$. The time t_0 is distinguished by the fact that the density matrix is diagonal at this time.

We notice that everything that has been said can easily be translated into classical terms. The part of the complete ensemble is played by the ensemble of measurements of all the coordinates and momenta of the system. This ensemble has the probability distribution $W(x_i, p_i)$. The entropy of this ensemble, as can easily be shown by Liouville's theorem, is independent of time. On the other hand, the entropy of any incomplete ensemble, for example an ensemble described by the momentum distribution only with an arbitrary value of the coordinates, may be time-dependent. The question of the increase in this entropy can be answered by solving the classical dynamical equations for a closed system.

Let us now continue with the general investigation into the behaviour of closed systems. We shall show (see Golden and Longuet-Higgins, 1960; Fain, 1963a) that if $\langle A(t) \rangle$ is the mean value of a certain quantity A at time t, then in a finite closed dynamic system $\langle A(t) \rangle$ has a limit when $t \to \infty$ only if $\langle A(t) \rangle$ does not depend on t. On the other hand, in an infinite closed system, whose energy levels form a continuous spectrum, $\langle A(t) \rangle$ approaches a limit when $t \to \infty$ when general enough assumptions are made. The mean value of the quantity A at the time t is

$$\langle A(t) \rangle = \mathrm{Tr}\, (\hat{\varrho}(t)\, \hat{A}), \tag{6.8}$$

† A complete ensemble appears when measuring a set of quantities $L, M, N, \ldots$, such that the density matrix is diagonal in a representation diagonal for these quantities. This property of the quantities $L, M, N, \ldots$, generally speaking, is not preserved in time. At the time $t > t_0$ the density matrix ϱ is no longer diagonal in the $L, M, N, \ldots$ -representation.

where $\hat{\varrho}$ is subject to equation (2.3b)

$$i\hbar \frac{\partial \hat{\varrho}}{\partial t} = \hat{H}\hat{\varrho} - \hat{\varrho}\hat{H}. \tag{6.9}$$

Remembering that the general solution of (6.9) can be written as (see section 2)

$$\hat{\varrho}(t) = e^{-i\hat{H}t/\hbar}\, \hat{\varrho}(0)\, e^{i\hat{H}t/\hbar},$$

we find the mean value of $\langle A(t)\rangle$ in the form

$$\langle A(t)\rangle = \sum_{nmuu'} \varrho_{nu;mu'}(0)\, A_{mu';nu}\, e^{-i\omega_{nm}t} = \int_{-\infty}^{\infty} g(\omega)\, e^{i\omega t}\, d\omega, \tag{6.10}$$

where n, m are suffixes denoting the energy levels; u, u' are other quantum numbers; the quantity $g(\omega)$, as can easily be checked, is defined by

$$g(\omega) = \sum_{uu'nm} \varrho_{nu;mu'}(0)\, A_{mu';nu}\, \delta\left(\frac{E_n - E_m}{\hbar} + \omega\right). \tag{6.11}$$

If the system has a discrete energy spectrum, $g(\omega)$ is a discontinuous function, $\langle A(t)\rangle$ is equal to the discrete sum of the harmonic functions and has no limit when $t \to \infty$. After a finite time the system will approach as close as one likes to the original state. For an infinite system with a continuous spectrum, for $\langle A(t)\rangle$ to have a limit when $t \to \infty$ it is sufficient for the function $g(\omega)$ to have the form

$$g(\omega) = G\delta(\omega) + h(\omega),$$

where $h(\omega)$ has no δ-type singularities and is absolutely integrable in the range $[-\infty, +\infty]$. Then on the basis of the Lebesque–Riemann theorem

$$\lim_{t\to\infty} \langle A(t)\rangle = \lim_{t\to\infty} \left\{G + \int_{-\infty}^{\infty} h(\omega)\, e^{-i\omega t}\, d\omega\right\} = G. \tag{6.12}$$

Using now equations (6.10) and (6.11) we find that the asymptotic value of $\langle A(\infty)\rangle$ is of the form

$$\langle A(\infty)\rangle = \sum_{nuu'} \varrho_{nu;nu'}(0)\, A_{nu';nu}. \tag{6.13}$$

It must be stressed that the proof given here is based essentially upon the assumption that $g(\omega)$ contains no δ-functions when $\omega \neq 0$. As can be seen from (6.11) this assumption relates both to the density matrix $\hat{\varrho}$ and to the operator $\hat{A}$ which is of interest to us. In principle we can imagine idealized situations when $g(\omega)$ contains δ-functions with $\omega \neq 0$. For example if the matrix elements of $\hat{A}$ are non-zero only when $(E_n - E_m)/\hbar = \omega_0$†, then, as can be seen easily from (6.11)

$$g(\omega) = h_1\delta(\omega - \omega_0) + h_2\delta(\omega + \omega_0)$$

and therefore

$$\langle A \rangle = h_1 \mathrm{e}^{i\omega_0 t} + h_2 \mathrm{e}^{-i\omega_0 t}$$

has no limit when $t \to \infty$.

*7. The Quantum Transport Equation in Γ-Space

7.1. Dynamic and Dissipative Systems

When investigating relaxation processes in different physical systems we are generally dealing with the type in which of situation relaxation occurs as the result of the interaction of a dynamic system and a dissipative system. That part of the system which has a finite number of degrees of freedom and discrete energy levels, and can in principle be described by simple dynamic equations will be called the dynamic system (or dynamic sub-system). This dynamic sub-system interacts with a dissipative system which at the limit has an infinite number of degrees of freedom and a continuous spectrum. Spontaneous emission from an atom in free space can be taken as a simple example of a relaxation process. Here the atom plays the part of the dynamic system and the dissipative system is the radiation field in free space.

Let us examine this example in slightly greater detail. Let the atom initially be in an excited state a with an energy E_a and let the field be in a vacuum. We shall denote the total energy of the atom+field system by E_{a0}. The probability of a spontaneous transition into a state b (with an energy $E_b < E_a$) with the emission of a photon of energy $\hbar\omega_\nu$ is (in accordance with (5.18))

$$W_{ab} = \frac{2\pi}{\hbar} \sum_\nu |V_{a0;b1_\nu}|^2 \, \delta(E_{a0} - E_{b1_\nu})\, t$$

$$= \frac{2\pi}{\hbar} \sum_\nu V_{a0;b1_\nu} V_{b1_\nu;a0} \, \delta(E_{a0} - E_{b1_\nu})\, t = w_{ab} t, \qquad (7.1)$$

† It would seem that the matrix elements of a coordinate (or momentum) of a harmonic oscillator satisfy this condition. It must not be forgotten, however, that the Hamiltonian $\hat{H}$ (because of the assumption that the spectrum of the system is continuous) differs from the Hamiltonian of a harmonic oscillator, so the matrix elements of the coordinate of a harmonic oscillator in a representation in which $\hat{H}$ is diagonal do not, generally speaking, satisfy the above condition.

where E_{b1_ν} is the energy of the final state of the field + atom system (with one photon of energy $\hbar\omega_\nu$).

Here the final state has a continuous spectrum and the sum of all the possible final states is reduced to summation for the different radiation oscillators ν (this can easily be obtained by using the properties of the matrix elements [section 4, sub-section 5]).

7.2. Diagonal Singularity

The features of spontaneous emission of interest to us that follow from (7.1) are as follows.

(1) The interaction of an atom with a radiation field leads to a "cumulative" effect—the probability of a transition is proportional to the time.

(2) Expression (7.1) is valid if we can satisfy

$$\tau_c \ll t \ll \tau_0 = \frac{1}{w_{ab}}, \tag{7.2}$$

where $\tau_c = 2\pi/\omega_0$ is the oscillation period of the spontaneous emission (ω_0 is the frequency of the transition $a \to b$) and w_{ab} is the mean life of the excited state of the atom (see Heitler, 1954, and also section 5). Therefore for a time-independent transition probability we must have

$$\tau_0 = \frac{1}{w_{ab}} \gg \tau_c = \frac{2\pi}{\omega_0}. \tag{7.3}$$

(3) It follows from (7.1) that for a non-zero transition probability we must require that the sum

$$\sum_{\nu} V_{a0;b1_\nu} V_{b1_\nu,a0},$$

taken on the constant energy surface $E_{a0} = E_{b1_\nu}$, is not equal to zero. (Here we must remember that each matrix element of $\hat{V}$ approaches zero in the transition to free space $L^3 \to \infty$.) Generalization of the latter property leads us to the requirement of diagonal singularity as a necessary condition for the existence of relaxation processes (Van Hove, 1955). The diagonal singularity condition is as follows. The matrix elements $\langle\alpha|\hat{V}\hat{A}\hat{V}|\alpha'\rangle$ (where $\hat{A}$ is diagonal in the α-representation) that are diagonal in the continuous suffixes are N times greater than the off-diagonal elements of the same matrix. Here N is the number of degrees of freedom of the dissipative subsystem. This number must approach infinity. It is easy to check that the interaction energy of the matter with the radiation field in free space satisfies the condition of diagonal singularity. Let us take, for example, the matrix element of the square of the interaction energy in a vacuum

$$(\hat{V}^2)_{a0;a0} = \sum_{\nu} V_{a0;b1_\nu} V_{b1_\nu;a0}.$$

This quantity is finite, since the sum for the radiation oscillators in free space means integration with a density of states (3.55) proportional to L^3, whilst each matrix element is inversely proportional to $\sqrt{L^3}$. If, however, we look at the off-diagonal matrix element of $\hat{V^2}$, for example,

$$(\hat{V}^2)_{a0;a2_\nu} = V_{a0;b1_\nu} V_{b1_\nu;a2_\nu},$$

we can see that this matrix element is of the order of L^{-3}. Therefore the diagonal element of $\hat{V}^2$ is L^3 times greater than the off-diagonal one, i.e. is N times greater, since L^3 is proportional to the number of degrees of freedom.

The above features of spontaneous emission appear in other relaxation processes. When we take these features into consideration we can proceed with the derivation of the transport equation.

7.3. The Derivation of the Transport Equation in Γ-Space

In order to be able to determine the different averages relating to a dynamic system we must derive the transport equation for the density matrix that is diagonal in the suffixes α of the dissipative sub-system and, generally speaking, is not diagonal in the discrete suffixes, m, n of the dynamic sub-system. In fact the mean value of a certain operator A relating to the dynamic sub-system is

$$\langle A \rangle = \mathrm{Tr}\, \hat{\varrho}\hat{A} = \sum_{n,m,\alpha,\alpha'} \varrho_{n\alpha';m,\alpha} A_{m\alpha;n\alpha'} = \sum_{m,n,\alpha} \varrho_{n\alpha;m\alpha} A_{mn}. \tag{7.4}$$

This latter equation follows from the fact that the matrix A is diagonal in the suffixes α (cf. (1.10)) in the representation in which the Hamilton operator of a system consisting of non-interacting dynamic and dissipative sub-systems is diagonal.

Let us proceed now to the derivation of the equation for the density matrix $\varrho_{m\alpha;n\alpha}$. We call this equation the transport equation in Γ-space. Using the usual terminology we call the phase space of all the degrees of freedom of the system the Γ-space. Also, we give the name μ-space to part of the phase space, in particular the phase space corresponding to the dynamic sub-system.

The Hamiltonian of a system consisting of dynamic and dissipative sub-systems interacting with each other is of the form

$$\hat{H} = \hat{H}_0 + \hat{V}, \quad \hat{H}_0 = \hat{F} + \hat{E}, \tag{7.5}$$

where $\hat{F}$ is the Hamilton operator of the dissipative sub-system;
$\hat{E}$ is the Hamilton operator of the dynamic sub-system;
$\hat{V}$ is the interaction energy.

In order to allow for the external forces acting on the dynamic sub-system we shall consider that $\hat{E}$ is explicitly time-dependent in an arbitrary manner. Equation (2.3b) for the density matrix becomes

$$i\hbar \frac{\partial \hat{\varrho}}{\partial t} = [\hat{F} + \hat{E} + \hat{V}, \hat{\varrho}]. \tag{7.6}$$

We change to the interaction representation. To do this we must carry out a unitary transformation on all the operators

$$\hat{A} \to e^{i\frac{\hat{F}}{\hbar}t} \hat{S}^{-1} A \hat{S} e^{-i\frac{\hat{F}}{\hbar}t}, \tag{7.7}$$

where the matrix of $\hat{S}$ satisfies the equations

$$i\hbar \frac{d\hat{S}}{dt} = \hat{E}\hat{S}, \quad i\hbar \frac{d\hat{S}^{-1}}{dt} = -\hat{S}^{-1}\hat{E}, \quad \hat{S}^{\pm}(0) = 1. \tag{7.8}$$

In this representation (7.6) becomes

$$i\hbar \frac{\partial \hat{\varrho}}{\partial t} = [\hat{V}, \hat{\varrho}]. \tag{7.9}$$

Here $\hat{V}$ and $\hat{\varrho}$ are operators in the interaction representation. Our task will be to derive the transport equation for the density matrix $\varrho_{m\alpha;\, n\alpha}$ by starting with this equation and using a number of assumptions (Fain, 1962, 1963).

Generally speaking, equation (7.9) connects the matrix elements of $\varrho_{m\alpha;\, n\alpha}$ that are diagonal in the continuous suffixes with the off-diagonal elements of $\varrho_{m\alpha;\, m\alpha'}$. However, with certain assumptions about the properties of the interaction energy $\hat{V}$ and the initial conditions $\hat{\varrho}(0)$ we can obtain the transport equation only for the elements of $\hat{\varrho}$ that are diagonal in the continuous suffix. Below we shall examine in detail the assumptions on which the derivation of the transport equation is based. For the moment we shall point out the following. The derivation of the equation for $\varrho_{m\alpha;\, n\alpha}$ is essentially the derivation from equation (7.9)† of the applicability of a *Markov process*. Let us recall the characteristics of a Markov process (see, e.g., Lax, 1960). We call a process Markovian if there is a set of variables $X = x_1, x_2, x_3, \ldots$ (less than the complete set of microscopic variables) possessing the property that the set of values of X at $t = 0$ determines the values of $X(t)$ for

† We are speaking, of course, of the applicability of the idealization in the special case under discussion.

$t > 0$†. It should be pointed out that the majority of systems that are met in physics are Markovian. Only a small number of Markovian variables need be known to predict the behaviour of a system. For example, in an electrical circuit it is sufficient to know the distribution of the currents and charges at the start to predict their future behaviour, without having recourse to the microscopic picture. The Markovian behaviour of the majority of physical systems can be understood as follows. Each system has a small number of approximate integrals of motion. We can therefore introduce a small set of new variables X (instead of the complete set of macroscopic variables), these variables being approximate integrals of motion and changing very slowly, whilst the remaining variables $Y = y_1, y_2, y_3, \ldots$ change quite rapidly (and their rate of change is far greater than the rate of measurement). The equations for the quantities Y can be solved in the adiabatic approximation, i.e. on the assumption that the quantities X are constant. Because of their rapid motion the quantities Y rapidly come into equilibrium when the values of X are fixed. Neglecting the small lag it may be taken that the quantities Y are functions of the instantaneous values of X. Therefore

$$\frac{dX}{dt} = f(X, Y) \approx f[X, Y(X)] = g(X) \tag{7.10}$$

is approximately a function of X only and the quantities X predict their own future. The Markovian nature of a physical system is essentially the property of quickly forgetting the original values of the variables Y. Strictly speaking the equations for the quantities X should be written in the form

$$\frac{\Delta X}{\Delta t} \simeq g(X), \tag{7.10'}$$

where Δt should be macroscopically small but much greater than the characteristic "forgetting" time for the original values of Y. This time is also called the "correlation time" τ_c, i.e. the average time during which a considerable correlation exists between the quantities Y at different times. It is obvious that for equations (7.10) to be applicable the mean relaxation time τ_0 of the quantities X must be much greater than the correlation time τ_c. We recall that when deriving the Boltzmann transport equation for a rarefied gas the part of X is played by the molecular velocity and coordinate distribution function $f_1(x, y, z, v_x, v_y, v_z)$, whilst the complete set of variables is character-

† The name Markovian is more often given to a process in which knowing the probability distribution at the time $t = 0$ enables one to find the probability distribution when $t > 0$. The concept of a Markovian process in the generalized sense, as it is understood by Lax (1960), includes in particular the evolution of the density matrix with time in accordance with equation (7.9). In this case the set of variables X is the same as all the elements (including those that are not diagonal in α) of the density matrix. In future, however, when we speak of a Markovian process or Markovian systems, we shall exclude this case.

ized by the distribution function in the complete phase space of the gas $f_N(x_1, y_1, z_1, x_2, y_2, z_2, \ldots, x_N, y_N, z_N, v_{1x}, v_{1y}, v_{1z}, \ldots, v_{Nz}, v_{Ny}, v_{Nx})$ where N is the number of molecules. Generally speaking, f_1 depends on f_2, f_2 on f_3 and so on; f_{N-1} depends on f_N. Therefore we can write a chain of "coupled" equations which precisely describe the behaviour of the gas (see Bogolyubov, 1947). It can be shown, however, that in a time much greater than the time of collision between the molecules f_1 is effectively dependent only upon f_1 (Bogolyubov, 1947). This relation leads to the transport equation. In this case the part of τ_c is played by the duration of a single collision between molecules. The part of the relaxation time is played by the average time between collisions. For a sufficiently rarefied gas this time is much greater than τ_c.

We also notice that equations of the type (7.10) are widely used in the thermodynamics of irreversible processes (see, e.g., Landau and Lifshitz, 1951, sections 108, 118; de Groot, 1951).

After these remarks let us move on to the derivation of the transport equation for the quantity $\varrho_{m\alpha;n\alpha}$ (obviously the $\varrho_{m\alpha;n\alpha}$ now play the part of the quantities X).

(1) We first assume that the matrix elements of the operator $\hat{V}$ found from the eigenfunctions of the operator H_0 satisfy the condition of diagonal singularity, i.e. the matrix elements of $\hat{V}\hat{A}\hat{V}$ that are diagonal in the continuous suffixes are much larger than the off-diagonal ones.

(2) The second assumption involves the upper limit of the interaction energy. In order to write the appropriate condition we must introduce the correlation time τ_c. This time is defined as follows:

$$\tau_c = \frac{\hbar}{\delta_0 E} \equiv \frac{1}{\omega^*}, \tag{7.11}$$

where $\delta_0 E$ is the energy difference in the dissipative system. This quantity determines the energy scale of the matrix elements, which are functions of V and ϱ. From what follows it will be obvious which functions are involved. If $f(E)$ is such a function, then when $\Delta E \ll \delta_0 E$ the function $f(E)$ differs little from $f(E + \Delta E)$, and when $\Delta E > \delta_0 E$ the function $f(E)$ differs considerably from $f(E + \Delta E)$. The functions $f(E)$ in the integrand are multiplied by $\delta_t(E - E_0)$, i.e. by functions which become δ-functions when $t \gg 1/\omega^*$ (see Appendix I). The quantity ω^* in (7.11) therefore has the same meaning as in Appendix I. It follows that the time τ_c really plays the part of the correlation time in the same sense as was taken above when explaining Markovian processes: when averaging over a time $t \gg \tau_c$ we obtain the transport equation only for the quantities $\varrho_{m\alpha;n\alpha}$ (and the quantities $\varrho_{m\alpha;n\alpha'}$ are eliminated as the result of averaging†). Further, let τ_0 be the characteristic

† The quantities Y were similarly eliminated in the equality (7.10).

relaxation time (in this time the system changes considerably as the result of relaxation processes); then the condition for the upper limit of the interaction energy can be written in the form

$$\tau_0 \gg \tau_c. \tag{7.12a}$$

When this condition (which is a generalization of condition (7.3)) is satisfied τ_0 can be represented as $\tau_0 = v^{-2}\Gamma$, where v is a quantity which defines the order of magnitude of the interaction energy V and Γ is independent of v. Therefore the condition for the upper limit of the interaction energy can be written in the form

$$v^2\Gamma^{-1} \ll \omega^* = \frac{1}{\tau_c}. \tag{7.12b}$$

Apart from these conditions it is necessary to make certain assumptions about the nature of the initial state in order to derive the transport equation. We shall leave the discussion of this question until a little later (sub-section 5) and now proceed immediately to the derivation of the transport equation.

We are interested in the behaviour of the density matrix $\hat{\varrho}$ in a time of the order of the relaxation time $\tau_0 = v^{-2}\Gamma$. This means that when in future we speak of a change in the density matrix in an infinitesimal time we shall have in mind a change in a time much less than τ_0. But, on the other hand, this small time should be very great when compared with a time of the order of τ_c. The increment of the density matrix† in time τ that satisfies the condition

$$\tau_c \ll \tau \ll \tau_0 \tag{7.13}$$

can be reduced, as we shall find below, to the form

$$\hat{\varrho}(t+\tau) - \hat{\varrho}(t) = \hat{A}(t)\, v^2\tau + \hat{B}(t)\, v + \hat{C}(t)\, v^2. \tag{7.14}$$

We shall use $\tilde{t}$ to denote the time measured in the scale of v^{-2}: $\tilde{t} = v^2 t$. The equation becomes

$$\hat{\varrho}(\tilde{t}+\tilde{\tau}) - \hat{\varrho}(\tilde{t}) = \hat{A}(\tilde{t})\, \tilde{\tau} + \hat{B}(\tilde{t})\, v + \hat{C}(\tilde{t})\, v^2. \tag{7.15}$$

Letting $v \to 0$ and $\tilde{\tau} \to 0$ (the latter is measured in the scale of v^{-2}) we find the differential equation‡

$$\frac{\partial \hat{\varrho}}{\partial \tilde{t}} = \hat{A}$$

† Here we always have in mind a matrix with the elements $\varrho_{m\alpha;\, n\alpha}$.

‡ It will be shown below (sub-section 5) that the terms $\hat{B}v + \hat{C}v^2$ can be neglected for a fairly broad range of initial conditions.

or, changing to the previous time scale for t, we obtain

$$\frac{\partial \hat{\varrho}}{\partial t} = v^2 \hat{A}. \tag{7.16}$$

This is the required differential equation. We notice that only in the interaction representation can we consider that the density matrix changes slowly and neglect its change in a time of the order of τ_c. The point is that the change to the interaction representation just means getting rid of the high-frequency dependence. The density matrix $\hat{\varrho}$ in the interaction representation shows, roughly speaking, the amplitude of the density matrix in the Schrödinger representation. The basic change of this "amplitude" with time is connected with the relaxation processes and, by virtue of the small value of v^2, is a relatively slow change. In order to find $v^2\hat{A}$ in (7.16) we use equation (7.9). From this equation, with an accuracy to terms of the order of v^2, we obtain

$$\hat{\varrho}(t+\tau) - \hat{\varrho}(t) = -\frac{i}{\hbar}\int_0^\tau [\hat{V}(t+\tau'), \hat{\varrho}(t)]\, d\tau'$$
$$-\frac{1}{\hbar^2}\int_0^\tau d\tau' \int_0^{\tau'} d\tau''[\hat{V}(t+\tau'), [\hat{V}(t+\tau''), \hat{\varrho}(t)]]. \tag{7.17}$$

It is not difficult to see that without loss of generality we can carry out the expansion of an arbitrary operator $\hat{A}$ in the interaction representation in the form of the sum of harmonic time functions

$$\hat{A} = \sum_r \hat{A}^r(t)\, e^{i\omega_r t}, \quad \hat{A}^r(t) = e^{i\frac{\hat{F}}{\hbar}t} \hat{A}^r\, e^{-i\frac{\hat{F}}{\hbar}t}, \tag{7.18}$$

where the time-independent operators $\hat{A}^r$ and the frequencies ω_r are determined by the transformation

$$\hat{S}^{-1}\hat{A}\hat{S} = \sum_r \hat{A}^r\, e^{i\omega_r t}. \tag{7.19}$$

Using these equalities the first term in (7.17) can be written in the form

$$\Delta_1 \hat{\varrho}_{\alpha\alpha} = \frac{1}{\hbar}\sum_{r\alpha'} [\zeta_\tau(\omega_r + \omega_{\alpha\alpha'})\, \hat{V}^r_{\alpha\alpha'}(t)\, \hat{\varrho}_{\alpha'\alpha} - \zeta_\tau(\omega_r + \omega_{\alpha'\alpha})\, \hat{\varrho}_{\alpha\alpha'} \hat{V}^r_{\alpha'\alpha}(t)]\, e^{i\omega_r t}, \tag{7.20}$$

where $\hat{\varrho}_{\alpha\alpha'}$ is a matrix operator acting on the variables of the dynamic subsystem; in the representation of the operator $\hat{H}_0$ its matrix elements are equal to $\varrho_{m\alpha;n\alpha}$; the function $\zeta_\tau = -i\int_0^\tau e^{ixt}\, dt$ changes when $\tau \gg \tau_c$ into the singular function (see Appendix I) $\zeta(x) = (P/x) - i\pi\delta(x)$. Since α', in accordance with our assumption, changes continuously, we must change from summation to integration with respect to α' in (7.20). When condition (7.13) is satisfied this integral does not depend on the time τ and is of the order of v, i.e. has the structure $\hat{B}(t)\,v$. We discard terms of this kind. We notice that dependence on the time τ could appear if the matrix $\hat{V}_{\alpha\alpha'}$ had a singularity, i.e. if the contribution to the sum made by an individual term of the sum were finite. An individual term of the sum can give a dependence on τ. However, we shall assume further that the matrix $\hat{V}_{\alpha\alpha'}$ is not singular. This means that when the number of degrees of freedom N approaches infinity the matrix element $\hat{V}_{\alpha\alpha'}$ approaches zero.

Let us now examine the second term in the right-hand side of (7.17). It can be written as follows:

$$\begin{aligned}\Delta_2\hat{\varrho}_{\alpha\alpha} &= -\frac{1}{\hbar^2}\int_0^\tau d\tau'\int_0^{\tau'} d\tau''[\hat{V}(t+\tau'),[\hat{V}(t+\tau''),\hat{\varrho}]]_{\alpha\alpha}\\ &= -\frac{1}{\hbar^2}\sum_{r,s}\sum_{\alpha'\alpha''}\int_0^\tau d\tau'\int_0^{\tau'} d\tau''\\ &\quad\times\{(\hat{V}^r_{\alpha\alpha'}(t)\,\hat{V}^s_{\alpha'\alpha''}(t)\hat{\varrho}_{\alpha''\alpha}\,\mathrm{e}^{i(\omega_r+\omega_{\alpha\alpha'})\tau'+i(\omega_s+\omega_{\alpha'\alpha''})\tau''}\\ &\quad-\hat{V}^r_{\alpha\alpha'}(t)\hat{\varrho}_{\alpha'\alpha''}\,\hat{V}^s_{\alpha''\alpha}(t)\,\mathrm{e}^{i(\omega_r+\omega_{\alpha\alpha'})\tau'+i(\omega_s+\omega_{\alpha''\alpha})\tau''}\\ &\quad-\hat{V}^s_{\alpha\alpha'}(t)\,\hat{\varrho}_{\alpha'\alpha''}\,\hat{V}^r_{\alpha''\alpha}(t)\,\mathrm{e}^{i(\omega_r+\omega_{\alpha''\alpha})\tau'+i(\omega_s+\omega_{\alpha\alpha'})\tau''}\\ &\quad+\hat{\varrho}_{\alpha\alpha'}\hat{V}^s_{\alpha'\alpha''}(t)\hat{V}^r_{\alpha''\alpha}(t)\,\mathrm{e}^{i(\omega_r+\omega_{\alpha''\alpha})\tau'+i(\omega_s+\omega_{\alpha'\alpha''})\tau''}\}\,\mathrm{e}^{i(\omega_r+\omega_s)t},\end{aligned}\tag{7.21}$$

where $\omega_{\alpha\alpha'} = (F_\alpha - F_{\alpha'})/\hbar$.

The F_α are the eigenvalues of the energy of the dissipative system. Just as in the case of $\Delta_1\hat{\varrho}_{\alpha\alpha}$, the double sum (double integral) in α' and $\alpha'' \neq \alpha'$ with a τ satisfying the condition (7.13) is time-independent and has the structure $C(t)\,v^2$; we neglect terms of this kind. However, the condition of diagonal singularity separates the single sum in α' from the double sum. Before calculating this single sum we shall make the additional following calculation. The sum similar to that which is obtained from (7.21) after eliminating the diagonally singular terms is of the form

$$I = \sum_{u,v}\int_0^\tau d\tau'\int_0^{\tau'} d\tau''\,\mathrm{e}^{iu\tau'+iv\tau''}f(u,v), \qquad u+v = \text{const.} \tag{7.22a}$$

Changing the integration variables we find

$$I = \int_0^\tau d\tau_1\left\{\frac{\mathrm{e}^{i(u+v)\tau} - \mathrm{e}^{i(u+v)\tau_1}}{i(u+v)\,\tau}\right\}\tau F(\tau_1), \tag{7.22b}$$

where the function

$$F(\tau_1) = \sum_{u,v} e^{-iv\tau_1} f(u, v), \qquad u + v = \text{const.}$$

differs considerably from zero when $\tau_1 \leqslant \tau_c$, whilst the correlation time τ_c is determined by the characteristic range δv of change of the function $f(\text{const.} - v, v)$; $\tau_c = 1/\delta v$. Let us examine various cases. Let

$$(u + v)\,\tau_c \ll 1; \tag{7.23}$$

in this case the term in the braces in (7.22b) becomes

$$\frac{e^{i(u+v)\tau} - 1}{i(u + v)\,\tau},$$

since $\tau_1 \leqslant \tau_c$ and therefore $e^{i(u+c)\tau_1} \approx 1$. We shall also be interested in time intervals that satisfy (7.13), i.e. $\tau \gg \tau_c$. When condition (7.23) is satisfied we can choose a time interval $\tau \gg \tau_c$ such that $(u + v)\,\tau \ll 1$. In this case the brace in (7.22) becomes 1. Let us now examine the case

$$(u + v)\,\tau_c \gtrsim 1. \tag{7.24}$$

Then for $\tau \gg \tau_c$ we obtain

$$(u + v)\,\tau \gg 1$$

and the brace in (7.22) is therefore much less than unity.

Recapitulating the cases examined we can rewrite integral (7.22) for time-intervals that satisfy (7.13) in the form

$$I = \sum_{u,v} \int_0^{\infty} d\tau_1\, e^{-iv\tau_1} f(u, v)\, \Delta(u + v)\,\tau, \tag{7.25}$$

where we have replaced the integration limit in (7.22) by ∞ by virtue of the properties of $F(\tau_1)$, and used $\Delta(x)$ to denote the function

$$\Delta(x) = \begin{cases} 1 & \text{when} \quad x \ll \tau_c^{-1} \\ 0 & \text{in all other cases.} \end{cases} \tag{7.26}$$

Let us return to equation (7.21). Using the condition of diagonal singularity and changing in the sense indicated above to the limit $\tau \to 0$, we obtain the required transport equation in the form

$$\frac{\partial \hat{\varrho}}{\partial t} = -\frac{1}{\hbar^2} \sum_{r,s} e^{i(\omega_r + \omega_s)t}\, [\hat{\ }\,[\hat{V}^s, \hat{\varrho}]]\, \Delta(\omega_r + \omega_s), \tag{7.27a}$$

where $\hat{\varrho}$ is that part of the matrix $\hat{\varrho}$ that is diagonal in the suffixes α

$$\hat{\mathcal{V}}^s = \int_0^\infty e^{-i\frac{\hat{F}}{\hbar}t'}\,\hat{V}^s\,e^{-i\omega_s t'}\,e^{i\frac{\hat{F}}{\hbar}t'}\,dt'$$

and

$$\hat{\mathcal{V}}^s_{\alpha\alpha'} = \int_0^\infty e^{-i(\omega_s+\omega_{\alpha\alpha'})t'}\,dt'\,\hat{V}^s_{\alpha\alpha'} = -i\hat{V}^s_{\alpha\alpha'}\zeta^*(\omega_s+\omega_{\alpha\alpha'})$$

$$= -i\hat{V}^s_{\alpha\alpha'}\left[\frac{P}{\omega_s+\omega_{\alpha\alpha'}} + i\pi\delta(\omega_s+\omega_{\alpha\alpha'})\right].$$

Equation (7.27a) can be transformed into another form that is more convenient to use. To do this we expand the commutators in (7.27a) and use the matrix multiplication rule

$$\frac{\partial\hat{\varrho}_{\alpha\alpha}}{\partial t} = -\frac{\pi}{\hbar^2}\sum_{r,s,\alpha'} e^{i(\omega_r+\omega_s)t}\{\delta(\omega_s+\omega_{\alpha'\alpha})(\hat{V}^r_{\alpha\alpha'}\hat{V}^s_{\alpha'\alpha}\hat{\varrho}_{\alpha\alpha} - \hat{V}^r_{\alpha\alpha'}\hat{\varrho}_{\alpha'\alpha'}\hat{V}^s_{\alpha'\alpha})$$

$$+ \delta(\omega_s+\omega_{\alpha\alpha'})(\hat{\varrho}_{\alpha\alpha}\hat{V}^s_{\alpha\alpha'}\hat{V}^r_{\alpha'\alpha} - \hat{V}^s_{\alpha\alpha'}\hat{\varrho}_{\alpha'\alpha'}\hat{V}^r_{\alpha'\alpha})\}\,\Delta(\omega_r+\omega_s)$$

$$+ i\frac{P}{\hbar^2}\sum_{r,s,\alpha'} e^{i(\omega_r+\omega_s)t}\left\{\frac{1}{\omega_s+\omega_{\alpha'\alpha}}(\hat{V}^r_{\alpha\alpha'}\hat{V}^s_{\alpha'\alpha}\hat{\varrho}_{\alpha\alpha} - \hat{V}^r_{\alpha\alpha'}\hat{\varrho}_{\alpha'\alpha'}\hat{V}^s_{\alpha'\alpha})\right.$$

$$\left. + \frac{1}{\omega_s+\omega_{\alpha\alpha'}}(\hat{\varrho}_{\alpha\alpha}\hat{V}^s_{\alpha\alpha'}\hat{V}^r_{\alpha'\alpha} - \hat{V}^s_{\alpha\alpha'}\hat{\varrho}_{\alpha'\alpha'}\hat{V}^r_{\alpha'\alpha})\right\}\Delta(\omega_r+\omega_s). \qquad (7.27\text{b})$$

Changing the summation suffixes s into r in the second terms and using the fact that $\omega_r + \omega_s \ll \omega^*$† we find the transport equation

$$\frac{\partial\hat{\varrho}}{\partial t} = -\frac{i}{\hbar}[\hat{N}, \hat{\varrho}] + \hat{R}(\varrho), \qquad (7.28)$$

where‡

$$\hat{N}_{\alpha\alpha} = -\frac{1}{\hbar}\sum_{r,s,\alpha'} e^{i(\omega_r+\omega_s)t}\,\frac{\hat{V}^r_{\alpha\alpha'}\hat{V}^s_{\alpha'\alpha}}{\omega_s+\omega_{\alpha'\alpha}}\,\Delta(\omega_r+\omega_s); \qquad (7.29)$$

$$\hat{R}_{\alpha\alpha}(\hat{\varrho}) = \frac{\pi}{\hbar^2}\sum_{r,s,\alpha'} e^{i(\omega_r+\omega_s)t}\,\delta(\omega_s+\omega_{\alpha'\alpha})$$

$$\times\{2\hat{V}^r_{\alpha\alpha'}\hat{\varrho}_{\alpha'\alpha'}\hat{V}^s_{\alpha'\alpha} - \hat{V}^r_{\alpha\alpha'}\hat{V}^s_{\alpha'\alpha}\hat{\varrho}_{\alpha\alpha} - \hat{\varrho}_{\alpha\alpha}\hat{V}^r_{\alpha\alpha'}\hat{V}^s_{\alpha'\alpha}\}\,\Delta(\omega_r+\omega_s). \qquad (7.30)$$

† It is assumed in addition that $\omega^* = \delta_0 E/\hbar$ also characterizes the energy scale of the inhomogeneity of the density matrix $\hat{\varrho}_{\alpha\alpha}$: δE, i.e. it is assumed that $\delta_0 E \leqslant \delta E$. In the opposite case, in particular if the dissipative system is near absolute zero $T \approx 0$, it is impossible to change from (7.27) to (7.28).

‡ Here and later we shall omit the symbol P denoting the principal value.

We now carry out the unitary transformation

$$\hat{\varrho} \to \hat{S}\hat{\varrho}\hat{S}^{-1}, \quad \hat{\hat{V}} = \hat{S}\hat{V}^r\hat{S}^{-1}. \tag{7.31}$$

It is not difficult to see that in this representation equations (7.28)–(7.30) become

$$\frac{\partial \hat{\varrho}}{\partial t} + \frac{i}{\hbar}[\hat{E} + \hat{\hat{N}}, \hat{\varrho}] = \hat{\hat{R}}(\hat{\varrho}), \tag{7.28a}$$

where

$$\hat{\hat{N}}_{\alpha\alpha} = \hat{S}\hat{N}_{\alpha\alpha}\hat{S}^{-1} = -\frac{1}{\hbar}\sum_{r,s,\alpha'} e^{i(\omega_r+\omega_s)t}\frac{\hat{\hat{V}}^r_{\alpha\alpha'}\hat{\hat{V}}^s_{\alpha\alpha'}}{\omega_s + \omega_{\alpha'\alpha}}\Delta(\omega_s + \omega_r); \tag{7.29a}$$

$$\hat{\hat{R}}_{\alpha\alpha} = \frac{\pi}{\hbar^2}\sum_{s,r,\alpha'} e^{i(\omega_r+\omega_s)t}\,\delta(\omega_s + \omega_{\alpha'\alpha})$$

$$\times \{2\hat{\hat{V}}^r_{\alpha\alpha'}\hat{\varrho}_{\alpha'\alpha'}\hat{\hat{V}}^s_{\alpha'\alpha} - \hat{\hat{V}}^r_{\alpha\alpha'}\hat{\hat{V}}^s_{\alpha'\alpha}\hat{\varrho}_{\alpha\alpha} - \hat{\varrho}_{\alpha\alpha}\hat{\hat{V}}^r_{\alpha\alpha'}\hat{\hat{V}}^s_{\alpha'\alpha}\}\,\Delta(\omega_r + \omega_s). \tag{7.30a}$$

Let us examine the special case when the Hamiltonian of the dynamic sub-system consists of a large constant part $\hat{E}_0$ and a small term $\hat{W}(t)$:

$$\hat{E} = \hat{E}_0 + \hat{W}(t). \tag{7.32}$$

The smallness of $\hat{W}$ means that the matrix elements W_{nm} are much less than $\delta_1 E$ (the difference of the terms of the dynamic sub-system) and much less than ω^*†

$$W_{nm} \ll (\delta_1 E; \omega^*). \tag{7.33}$$

In this case we can say that approximately

$$\hat{S}^{\pm 1} \approx e^{\mp i\frac{\hat{E}_0}{\hbar}t}, \quad \hat{\hat{V}}^r_{n\alpha;m\alpha'} = V_{n\alpha;m\alpha'}\,\delta_{\omega_r;\omega_{nm}}\,e^{-i\omega_{nm}t} \tag{7.34}$$

and equations (7.28a)–(7.30a) become

$$\frac{\partial \hat{\varrho}}{\partial t} + \frac{i}{\hbar}[\hat{E}_0 + \hat{W} + \hat{N}, \hat{\varrho}] = \hat{\hat{R}}(\hat{\varrho}), \tag{7.28b}$$

where

$$\tilde{N}_{m\alpha;n\alpha} = -\sum_{k,\alpha'}\frac{V_{m\alpha;k\alpha'}V_{k\alpha';n\alpha}}{E_k - E_n + F_{\alpha'} - F_\alpha}\Delta(\omega_{nm}); \tag{7.29b}$$

$$\tilde{R}_{m\alpha;n\alpha} = \frac{\pi}{\hbar}\sum_{k,l\alpha}\{2V_{m\alpha;k\alpha'}V_{l\alpha';n\alpha}\varrho_{k\alpha';l\alpha'}\,\delta(E_l - E_n + F_{\alpha'} - F_\alpha)$$

$$\times \Delta(\omega_{ln} + \omega_{mk}) - V_{m\alpha;k\alpha'}V_{k\alpha';l\alpha}\varrho_{l\alpha;n\alpha}$$

$$\times \delta(E_k - E_l + F_{\alpha'} - F_\alpha)\,\Delta(\omega_{ml}) - V_{k\alpha;l\alpha'}V_{l\alpha';n\alpha}\varrho_{m\alpha;k\alpha}$$

$$\times \delta(E_l - E_n + F_{\alpha'} - F_\alpha)\,\Delta(\omega_{nk})\}. \tag{7.30b}$$

† A more careful analysis shows that this approximation can be used when the perturbation $\hat{W}(t)$ has the nature of a resonance and is near resonance ($\Delta\omega \ll \omega^*$).

7.4. The Balance Equation in Γ-Space

From the transport equations (7.28)–(7.30), with certain assumptions which we shall clarify later, follow the balance equations in Γ-space

$$\frac{dp_m}{dt} = \sum_n (w_{nm}p_n - w_{mn}p_m), \tag{7.35}$$

where p_m is the probability of finding the system in the state m (population of the m level);
w_{nm} is the transition probability $n \to m$.

This equation describes the relaxation of the populations of the levels in Γ-space when there are no external forces. In order to obtain the balance equations from (7.28)–(7.30) we put the external perturbation $\hat{W} = 0$, and take the diagonal part of the equation (7.28 b)

$$\frac{\partial \varrho_{n\alpha;n\alpha}}{\partial t} + \frac{i}{\hbar}\sum_k (\tilde{N}_{n\alpha;k\alpha}\,\varrho_{k\alpha;n\alpha} - \varrho_{n\alpha;k\alpha}\tilde{N}_{k\alpha;n\alpha})$$

$$= \frac{\pi}{\hbar}\sum_{k,l}\{2V_{n\alpha;k\alpha'}V_{l\alpha';n\alpha}\varrho_{k\alpha';l\alpha'}\,\delta(E_l - E_n + F_{\alpha'} - F_\alpha)\,\Delta(\omega_{lk})$$

$$- V_{n\alpha;k\alpha'}V_{k\alpha';l\alpha}\varrho_{l\alpha,n\alpha}\,\delta(E_k - E_l + F_{\alpha'} - F_\alpha)\,\Delta(\omega_{nl})$$

$$- V_{k\alpha;l\alpha'}V_{l\alpha';n\alpha}\varrho_{n\alpha;k\alpha}\,\delta(E_l - E_n + F_{\alpha'} - F_\alpha)\,\Delta(\omega_{nk})\}.$$

As can be seen from this equation, in the general case the part of the matrix $\hat{\varrho}$ that is diagonal in all the indices (it has the meaning of the probability of the corresponding state) is connected with the off-diagonal elements of $\hat{\varrho}$. This means that to obtain the balance equations we must introduce further additional conditions: first, there should be no term differences ω_{nm} that are non-vanishing and considerably smaller than ω^* and, secondly, that the levels of the dynamic sub-system should be non-degenerate. Then the diagonal part of equation (7.28 b) becomes

$$\frac{\partial \varrho_{n\alpha;n\alpha}}{\partial t} = \sum_{k\alpha'}\frac{2\pi}{\hbar}\,|V_{n\alpha;k\alpha'}|^2\,\delta(E_n - E_k + F_\alpha - F_{\alpha'})\,(\varrho_{k\alpha';k\alpha'} - \varrho_{n\alpha;n\alpha}), \tag{7.36}$$

which is the same as the balance equation since $\varrho_{n\alpha;n\alpha}$ is the population of the level $E_{n\alpha} = E_n + F_\alpha$ and

$$\frac{2\pi}{\hbar}\,|V_{n\alpha;k\alpha'}|^2\,\delta(E_{n\alpha} - E_{k\alpha'}) = w_{n\alpha;k\alpha'}$$

is the transition probability from state $n\alpha$ into state $k\alpha'$. We note that in the case under discussion the balance equations in Γ-space for the probability of the direct and reverse transitions $m \to n$ and $n \to m$ are equal to each other (the property of detailed balancing or "microscopic reversibility")

$$w_{n\alpha;k\alpha'} = w_{k\alpha';n\alpha}.$$

We also note that the term containing $\hat{N}$ has dropped out in the change to the balance equations. This term, as can be seen from its structure, gives a shift of the energy levels of the unperturbed system and does not itself lead to relaxation.

7.5. *The Initial Conditions under which the Transport Equations are Valid*

As has already been pointed out, certain assumptions have to be made about the initial state in order to derive the transport equations. Let the initial states have the density matrix $\hat{\varrho}_{\alpha\alpha'}(0)$. In the Schrödinger representation the density matrix as a function of time behaves as follows:

$$\hat{\varrho}(t) = \hat{U}\hat{\varrho}(0)\,\hat{U}^{+}, \tag{7.37}$$

where $\hat{U}$ satisfies the condition

$$i\hbar\,\frac{d\hat{U}}{dt} = (\hat{F} + \hat{E} + \hat{V})\,\hat{U}, \quad \hat{U}(0) = 1. \tag{7.38}$$

The solution of this equation (in which $\hat{E}$ may be explicitly time-dependent) is of the form

$$\hat{U} = \exp\left\{-\frac{i}{\hbar}(\hat{F} + \hat{V})\,t - \frac{i}{\hbar}\int_0^t \hat{E}\,dt\right\}. \tag{7.39}$$

The relevant part of the density matrix $\hat{\varrho}(t)$ diagonal in the indices α can be written in the form

$$\hat{\varrho}_{\alpha\alpha}(t) = \sum_{\alpha'}\sum_{\alpha''} \hat{U}_{\alpha\alpha'}\hat{\varrho}_{\alpha'\alpha''}(0)\,\hat{U}^{-}_{\alpha''\alpha}. \tag{7.40}$$

The above-mentioned diagonal singularity condition for the interaction energy leads to the diagonal singularity of $\hat{U}\hat{\varrho}(0)\,\hat{U}^{+}$, and this means that $\hat{\varrho}_{\alpha\alpha}(t)$ can be split into two parts†

$$\hat{\varrho}_{\alpha\alpha}(t) = \sum_{\alpha'} \hat{U}_{\alpha\alpha'}\varrho_{\alpha'\alpha'}(0)\,\hat{U}^{+}_{\alpha'\alpha} - \sum_{\alpha' \neq \alpha''}\sum \hat{U}^{+}_{\alpha\alpha'}\hat{\varrho}_{\alpha'\alpha''}(0)\,\hat{U}^{+}_{\alpha''\alpha}, \tag{7.41}$$

† The actual investigation of the behaviour of $\hat{\varrho}_{\alpha\alpha}(t)$ in the form (7.41) was made by Van Hove (1955) for the case when there is no dynamic sub-system, and by Genkin (1962) for the case when the $\hat{E}$-energy of the dynamic sub-system is independent of time.

where the single sum is of the same order in N (the number of degrees of freedom of the dissipative sub-system) as the double sum in α' and α''. By assuming a small interaction energy we found above an equation describing the behaviour of the first sum of (7.41). In this case we discarded the terms connected with the second sum. Therefore the transport equations are valid with initial conditions which made the expression

$$\sum_{\alpha' \neq \alpha''}\sum \hat{U}_{\alpha\alpha'}\hat{\varrho}_{\alpha'\alpha''}(0)\,\hat{U}^{+}_{\alpha''\alpha} \tag{7.42}$$

zero. This expression becomes zero in the first place if the systems are initially in a mixed state with a diagonal density matrix

$$\hat{\varrho}_{\alpha'\alpha''}(0) = \hat{\varrho}_{\alpha'\alpha'}(0)\,\delta_{\alpha'\alpha''}\,. \tag{7.43}$$

In the second place, it becomes zero if the system is originally in a pure state with given α, i.e. $\Psi(0) = \Psi_{\alpha}$, or in a state in which the relaxation time $\tau_0 = v^{-2}\Gamma$ would be approximately constant if the perturbation $\hat{V}$ were not present. To be more precise, if

$$\Psi(0) = \sum c_{\alpha}\Psi_{\alpha}, \tag{7.44}$$

then the energy range δE characterizing this wave packet should satisfy the inequality (Van Hove, 1955)

$$\delta E \ll \frac{\hbar}{\tau_0}\,. \tag{7.45}$$

Thirdly, it can be shown (Van Hove, 1955) that when $v \to 0$ ($v^2 t$ is finite) (7.42) approaches zero if the initial state is the wave packet (7.45) and δE satisfies the inequality

$$\delta E \gg \frac{\hbar}{\tau_0}\,. \tag{7.46}$$

We note that in this case the characteristic correlation frequency ω^* is the same as $\delta E/\hbar$ if $\delta E < \delta_0 E$ (if $\delta_0 E < \delta E$, then $\omega^* = \delta_0 E/\hbar$)†.

Therefore the validity of the transport equation is proved for the two classes of pure initial states represented by the packets (7.45) or (7.46) and for mixed initial states with a diagonal density matrix.

† By $\delta_0 E$ we here understand the same quantity as in (7.11) but without allowing for the dependence of the matrix elements of $\hat{\varrho}$ on E. In the special case when the temperature of the dissipative system is initially zero the two definitions of $\delta_0 E$ are the same.

*8. The Transport Equation in μ-Space

8.1. *The Derivation of the Transport Equation in μ-Space*

We are very often interested in the behaviour of a dynamic system (μ-space) independently of the behaviour of the whole system: dynamic + dissipative. The behaviour of the dynamic system can be described by the density matrix $\hat{\sigma}$

$$\sigma_{mn} = \sum_{\alpha} \varrho_{m\alpha;\, n\alpha}. \tag{8.1}$$

Generally speaking the change of the matrix $\hat{\sigma}$ with time is not determined by its values at some point in time but depends on the state of the dissipative sub-system. In fact it follows from (7.28 a) that

$$\frac{\partial \sigma_{mn}}{\partial t} + \frac{i}{\hbar}[\hat{E}, \hat{\sigma}]_{mn} + \frac{i}{\hbar}\sum_{\alpha}(\tilde{N}_{m\alpha;\, k\alpha}\,\varrho_{k\alpha;\, n\alpha} - \varrho_{m\alpha;\, k\alpha}\,\tilde{N}_{k\alpha;\, n\alpha})$$
$$= \sum_{\alpha} \tilde{R}_{m\alpha;\, n\alpha}. \tag{8.2}$$

It is obvious that the sums on the left-hand and right-hand sides of this equation cannot generally be reduced to functions of $\hat{\sigma}$ only. For this reduction to be possible we make the assumption that the dissipative system is much larger than the dynamic system and is in the given state all the time so the influence of the dynamic sub-system on it can be neglected. In other words, this assumption means that all the mean quantities relating to the dissipative sub-system are independent of time (or change very slowly). For example, let us examine in this approximation the sum on the left-hand side of (8.2)

$$\sum_{\alpha} \tilde{N}_{m\alpha;\, k\alpha}\varrho_{k\alpha;\, n\alpha} = \tilde{N}_{m\bar{\alpha};\, k\bar{\alpha}} \sum_{\alpha} \varrho_{k\alpha;\, n\alpha} = N_{m\bar{\alpha};\, k\bar{\alpha}}\sigma_{\kappa n}(t),$$

where $N_{m\bar{\alpha};\, k\bar{\alpha}}$ is a certain mean value of $N_{m\alpha;\, k\alpha}$. By assumption this mean should not be time-dependent. We can therefore calculate it at the time of initiation $t = t_0$ of the interaction between the dynamic and dissipative sub-systems, assuming that at this time the matrix density of the whole system can be put in the form

$$\varrho_{m\alpha;\, n\alpha}(t_0) = \sigma_{mn}(t_0)\, P_{\alpha}(t_0),$$

where P_{α} is the diagonal part of the matrix density of the dissipative system.

Therefore

$$\tilde{N}_{m\bar{\alpha};k\bar{\alpha}} \equiv \frac{\sum_{\alpha} \tilde{N}_{m\alpha;k\alpha}\varrho_{k\alpha;n\alpha}}{\sum_{\alpha} \varrho_{k\alpha;n\alpha}} = \frac{\sum_{\alpha} \tilde{N}_{m\alpha;k\alpha}P_{\alpha}(t_0)\,\sigma_{kn}(t_0)}{\sigma_{kn}(t_0)} = \sum_{\alpha} \tilde{N}_{m\alpha;k\alpha}P_{\alpha}.$$

In future we shall omit the argument t_0 in the function P_α since according to our assumption the change in the state of the dissipative system can be neglected. In the approximation used the transport equation that defines the behaviour of the dynamic sub-system becomes†

$$\frac{\partial \sigma_{mn}}{\partial t} + \frac{i}{\hbar}[\hat{E} + \hat{\Gamma}, \hat{\sigma}]_{mn} = \sum_{k,l} (2\Gamma_{mkln}\sigma_{kl} - \Gamma_{klmk}\sigma_{ln} - \Gamma_{lnkl}\sigma_{mk}), \tag{8.3}$$

where

$$\hat{\Gamma} = -\frac{1}{\hbar} \sum_{r,s,\alpha,\alpha'} e^{i(\omega_r+\omega_s)t} P_\alpha \frac{\hat{\tilde{V}}^r_{\alpha\alpha'} \hat{\tilde{V}}^s_{\alpha'\alpha}}{\omega_s + \omega_{\alpha'\alpha}} \Delta(\omega_r + \omega_s);$$

$$\begin{aligned}\Gamma_{mk} &= \sum_{\alpha} \hat{N}_{m\alpha;k\alpha}P_\alpha \\ &= -\frac{1}{\hbar} \sum_{l,r,s,\alpha,\alpha'} e^{i(\omega_r+\omega_s)t} P_\alpha \frac{\tilde{V}^r_{m\alpha;l\alpha'} \tilde{V}^s_{l\alpha';k\alpha}}{\omega_s + \omega_{\alpha'\alpha}} \Delta(\omega_r + \omega_s);\end{aligned} \tag{8.4}$$

$$\begin{aligned}\Gamma_{mkln} = \frac{\pi}{\hbar^2} \sum_{\alpha,\alpha';r,s} & e^{i(\omega_r+\omega_s)t} \delta(\omega_s + \omega_{\alpha'\alpha})\, \tilde{V}^r_{m\alpha;k\alpha'} P_{\alpha'} \tilde{V}^s_{l\alpha';n\alpha} \\ & \times \Delta(\omega_r + \omega_s).\end{aligned} \tag{8.5}$$

In the special case when we can use (7.34) (in which the external force was taken to be small) the coefficients Γ are time-independent and take the form

$$\hat{\Gamma} = -\frac{1}{\hbar} \sum_{r,s} \frac{\hat{V}^r_{\alpha\alpha'} \hat{V}^s_{\alpha'\alpha} P_\alpha}{\omega_s + \omega_{\alpha'\alpha}} \Delta(\omega_r + \omega_s),$$

$$\begin{aligned}\Gamma_{mk} &\equiv -\frac{1}{\hbar} \sum_{l,r,s,\alpha,\alpha'} \frac{V^r_{m\alpha;l\alpha'} V^s_{l\alpha';k\alpha}}{\omega_s + \omega_{\alpha'\alpha}} P_\alpha \Delta(\omega_r + \omega_s) \\ &= -\sum_{l,\alpha,\alpha'} P_\alpha \frac{V_{m\alpha;l\alpha'} V_{l\alpha';k\alpha}}{E_l - E_k + F_{\alpha'} - F_\alpha} \Delta(\omega_{mk}),\end{aligned} \tag{8.6}$$

† See also Bloch (1957) and Hubbard (1961) on the subject of this equation.

$$
\begin{aligned}
\Gamma_{mkln} &= \frac{\pi}{\hbar^2} \sum_{\alpha,\alpha',r,s} \Delta(\omega_r + \omega_s)\, V^r_{m\alpha;k\alpha'} P_{\alpha'} V^s_{l\alpha';n\alpha} \delta(\omega_s + \omega_{\alpha'\alpha}) \\
&= \frac{\pi}{\hbar} \sum_{\alpha\alpha'} V_{m\alpha;k\alpha'} P_{\alpha'} V_{l\alpha';n\alpha}\, \delta(E_l - E_n + F_{\alpha'} - F_\alpha) \\
&\quad \times \Delta(\omega_{mk} + \omega_{ln}). \qquad (8.7)
\end{aligned}
$$

If the dissipative sub-system is in a state of thermodynamic equilibrium at a temperature T there is a connection between the coefficients Γ. To derive this connection we change from summation to integration with respect to the energies

$$
\sum_\alpha \to \sum_u \int dF_\alpha \eta_u(F_\alpha),
$$

where $\eta_\alpha(F_\alpha)$ is the density of states with a given quantum number u.

It is also necessary to allow for the fact that if the dissipative system is in a state of thermodynamic equilibrium, then

$$
P_\alpha = P(F_\alpha) = \frac{e^{-F_\alpha/kT}}{\sum_{\alpha''} e^{-F_{\alpha''}/kT}}.
$$

As a result we find

$$
\begin{aligned}
\Gamma_{lnmk} &= \frac{\pi}{\hbar} \sum_{u,u',r,s} e^{i(\omega_r+\omega_s)t} \Delta(\omega_r + \omega_s) \int dF_\alpha \eta_u(F_\alpha)\, \eta(F_\alpha + \hbar\omega_s) \\
&\quad \times \langle m, F_\alpha + \hbar\omega_s, u' | \tilde{V}^r | k, F_\alpha, u \rangle\, P(F_\alpha) \langle l, F_\alpha, u | \tilde{V}^s | n, F_\alpha + \hbar\omega_s, u' \rangle \\
&\quad \times e^{-\hbar\omega_s/kT} = \Gamma_{mkln}\, e^{-\hbar\omega_s/kT} \approx \Gamma_{mkln}\, e^{\hbar\omega_r/kT}. \qquad (8.8)
\end{aligned}
$$

In the case when we can use (7.34), equation (8.8) becomes†

$$
\Gamma_{mkln} = \Gamma_{lnmk}\, e^{(E_l - E_n)/kT} \approx \Gamma_{lnmk}\, e^{-(E_m - E_k)/kT}. \qquad (8.9)
$$

It follows from the last expressions that when there are no external forces (8.3) has its own stationary solution

$$
\sigma^0_{kl} = \delta_{kl} \frac{e^{-E_k/kT}}{\sum_s e^{-E_s/kT}},
$$

i.e. the dynamic sub-system comes into equilibrium with the dissipative sub-system.

† Expressions (8.8) and (8.9) are valid provided that $\omega_r + \omega_s \ll kT/\hbar$. This condition follows from the condition $\omega_r + \omega_s \ll \omega^*$ only when the inequality (7.46) (where, as can easily be checked, the part of δE is played by kT) is satisfied. In this case if $kT < \delta_0 E$, then $kT/\hbar$ plays the part of ω^*. We also note that the change from (7.27a) to (7.28) is valid in a similar approximation. At low temperatures, when $\omega^* > kT/\hbar$, the change to the equations in μ-space must be made by using (7.27) and not the subsequent equations.

8.2. Balance Equation in μ-Space

From equations (8.3), when there are no external forces we can obtain for non-degenerate levels of the dynamic system (i.e. there are no non-zero $|\omega_{nk}| \ll \omega^*$), provided that Δ becomes the Kronecker symbol, the balance equation in μ-space

$$\dot{\sigma}_{mm} = \sum_k (w_{km}\sigma_{kk} - w_{mk}\sigma_{mm}). \tag{8.10}$$

Here w_{km} is the transition probability $k \to m$

$$w_{km} = 2\Gamma_{mkkm}. \tag{8.11}$$

If the dissipative system is in thermal equilibrium it follows from (8.9) that

$$w_{km} = w_{mk}\, e^{-(E_m - E_k)/kT}. \tag{8.12}$$

We notice that the presence of a dissipative system interacting with the dynamic system leads to the situation that the transition probabilities w_{mk} are finite not only from the upper level to the lower but also from the lower level to the upper. In the balance equations in Γ-space the transition probabilities are the probabilities of transitions with conservation of energy. In contradistinction the probabilities w_{mk} are the probabilities with non-conservation of energy of the dynamic sub-system, i.e. $E_m \neq E_k$. (Of course, the total energy of the whole system is still conserved.)

The corrections to the eigenvalues of the energy of the dynamic system can be found from (8.6)

$$\Delta E_m = \Gamma_{mm} = -\sum_{\alpha,\alpha',k} \frac{|V_{m\alpha;k\alpha'}|^2 P_\alpha}{E_k - E_m + F_{\alpha'} - F_\alpha}. \tag{8.13}$$

8.3. The Equations for Averages Derived from the Transport Equation

It is often convenient to use not the transport equations but the equations that follow from them for the averages of the operators relating to the dynamic system. In order not to limit the generality by the condition

$$\omega_r + \omega_s \ll \frac{kT}{\hbar},$$

which is used in the derivation of equations (8.3), we shall obtain the equations for the averages directly from (7.27). First we carry out a unitary transformation of (7.13). As a result we obtain

$$\frac{\partial \hat{\varrho}}{\partial t} + \frac{i}{\hbar}[\hat{E}, \hat{\varrho}] = -\frac{1}{\hbar^2}\sum_{r,s} e^{i(\omega_r+\omega_s)t}[\hat{\tilde{V}}^r, [\hat{\tilde{V}}^s, \hat{\varrho}]]\Delta(\omega_r + \omega_s). \quad (8.14)$$

The mean value of a certain operator $\hat{O}$ relating to the dynamic sub-system can be written in the form

$$\langle O \rangle = \mathrm{Tr}\, \hat{\varrho}\hat{O} = \mathrm{Tr}\, \hat{\sigma}\hat{O}.$$

In the representation we have selected the operator $\hat{O}$ is time-independent. Therefore

$$\frac{d\langle O \rangle}{dt} = \mathrm{Tr}\, \frac{\partial \hat{\varrho}}{\partial t}\hat{O} = \mathrm{Tr}\, \frac{\partial \hat{\sigma}}{\partial t}\hat{O}. \quad (8.15)$$

Using (8.14) and (8.15), after a simple transformation using the properties of the trace, we obtain

$$\frac{d\langle O \rangle}{dt} - \frac{i}{\hbar}\langle [\hat{E}, \hat{O}] \rangle = -\frac{1}{\hbar^2}\sum_{r,s} e^{i(\omega_r+\omega_s)t}\Delta(\omega_r + \omega_s)$$
$$\times\, \mathrm{Tr}\, \{[\hat{O}, \hat{\tilde{V}}^r][\hat{\tilde{V}}^s, \hat{\varrho}]\}. \quad (8.16)$$

For what follows it is convenient to represent the operator $\hat{V}^r$ in the form of the sum of the product of the operators acting respectively on the variables of the dynamic sub-system $\hat{v}_i^r$ and the dissipative sub-system $\hat{W}_i^r$

$$\hat{V}^r = \sum_i \hat{W}_i^r \hat{v}_i^r. \quad (8.17)$$

We substitute (8.17) in the right-hand side of (8.16) and use the same approximation as we used when deriving equation (8.3)†. After some small transformations making use of the properties of a delta-function and on the assumption that the dissipative system is in a state of thermal equilibrium we come to the required equation that gives the change with time of the mean value $\langle O \rangle$

$$\frac{d\langle O \rangle}{dt} = \frac{i}{\hbar}\langle [\hat{E}, \hat{O}] \rangle + \sum_{r,s,i,i'} \Phi_{ii'}^{rs}\Big\{\langle \hat{\tilde{v}}_i^s, \big[\hat{O}, \hat{\tilde{v}}_i^r\big]$$
$$+ e^{-\frac{\hbar\omega_s}{kT}}\big[\hat{\tilde{v}}_i^r, \hat{O}\big]\,\hat{\tilde{v}}_i^s\rangle\Big\} + \frac{i}{\hbar}\Big\{\overline{\varphi}_{ii'}^{rs}\,\langle \hat{\tilde{v}}_{i'}^s\big[\hat{O}, \hat{\tilde{v}}_i^r\big]\rangle + \varphi_{ii'}^{rs}\,\langle \big[\hat{\tilde{v}}_i^r\hat{O}\big]\,\hat{\tilde{v}}_{i'}^s\rangle\Big\},$$
$$(8.18)$$

† This approximation is in effect the replacement of $\varrho_{m\alpha;n\alpha}$ by $P_\alpha \sigma_{mn}$.

where

$$\Phi^{rs}_{ii'} = \frac{\pi}{\hbar^2} \sum_{\alpha,\alpha'} e^{i(\omega_r+\omega_s)t}\, \delta(\omega_s + \omega_{\alpha'\alpha}) \langle \alpha | W^r_i | \alpha' \rangle$$

$$\times \langle \alpha' | W^s_{i'} | \alpha \rangle P_{\alpha'} \Delta(\omega_r + \omega_s), \tag{8.19}$$

$$\varphi^{rs}_{ii'} = -\frac{1}{\hbar} \sum_{\alpha,\alpha'} e^{i(\omega_r+\omega_s)t} \frac{\langle \alpha | W^r_i | \alpha' \rangle \langle \alpha' | W^s_{i'} | \alpha \rangle}{\omega_s + \omega_{\alpha'\alpha}} P_\alpha \Delta(\omega_r + \omega_s); \tag{8.20}$$

$$\overline{\varphi}^{rs}_{ii'} = -\frac{1}{\hbar} \sum_{\alpha\alpha'} e^{i(\omega_r+\omega_s)t} \frac{\langle \alpha | W^r_i | \alpha' \rangle \langle \alpha' | W^s_{i'} | \alpha \rangle}{\omega_s + \omega_{\alpha'\alpha}} P_{\alpha'} \Delta(\omega_r + \omega_s). \tag{8.21}$$

If we also make the additional assumptions

$$\omega_r + \omega_s \ll \frac{kT}{\hbar}, \quad \omega_r \gtrsim \omega^*, \tag{8.22}$$

then, as can easily be obtained†,

$$\Phi^{rs}_{i'i} = \Phi^{sr}_{i'i}\, e^{\hbar\omega_s/kT}, \quad \overline{\varphi}^{sr}_{i'i} = -\varphi^{rs}_{i'i}, \quad \Phi^{rr}_{ii'} = 0. \tag{8.23}$$

In this case (8.18) can be rewritten in the form

$$\frac{d\langle O \rangle}{dt} = \frac{i}{\hbar} \langle [\hat{E} + \hat{\Gamma}, \hat{O}] \rangle + \sum_{\substack{r,s,i,i' \\ (r \neq s)}} \Phi^{rs}_{ii'} \langle \hat{\tilde{v}}^s_{i'} \left[\hat{O}, \hat{\tilde{v}}^r_i\right] + \left[\hat{\tilde{v}}^s_{i'}, \hat{O}\right] \hat{\tilde{v}}^r_i \rangle, \tag{8.18a}$$

where

$$\hat{\Gamma} = \sum_{r,s,i,i'} \varphi^{rs}_{ii'} \hat{\tilde{v}}^r_i \hat{\tilde{v}}^s_{i'}. \tag{8.24}$$

We also note that the equation (8.18a) could be obtained directly from the transport equation (8.3), the coefficients of this equation being expressed in terms of $\Phi^{rs}_{ii'}$ as follows:

$$\Gamma_{mkln} = \sum_{r,s,i,i'} \Phi^{rs}_{ii'} \langle m | \tilde{v}^r_i | k \rangle \langle l | \tilde{v}^s_{i'} | n \rangle. \tag{8.25}$$

In the following chapters various applications of the formula obtained here will be given.

† Of course if $\omega_r = 0$, then $\Phi^{00} \neq 0$. In the second condition in (8.22) we have in mind non-zero ω_r.

9. The Principle of the Increase of Entropy

9.1. The Derivation of the Principle of the Increase of Entropy from the Balance Equations

The principle of an increase of entropy follows from the transport equations obtained above. This principle follows in particular from the balance equations.

According to (6.3) the entropy of an ensemble whose probability distribution is given by the diagonal elements of the density matrix $p_n = \varrho_{nn}$ is

$$\mathscr{E} = -\sum_n \varrho_{nn} \ln \varrho_{nn} = -\sum p_n \ln p_n.$$

Differentiating this equality and using the balance equations (7.35) and (7.36) we find

$$\dot{\mathscr{E}} = -\sum_n (\dot{p}_n \ln p_n + \dot{p}_n) = -\sum_{n,m} (w_{mn} p_m - w_{nm} p_n) \ln p_n$$

$$= -\sum_{n,m} w_{mn}(p_m - p_n) \ln p_n. \tag{9.1}$$

When changing from the first equality to the second we made use of the fact that $\sum_n p_n$ is time-independent and when changing to the latter equality we used the property of detailed balancing ("microscopic reversibility")

$$w_{nm} = w_{mn}, \tag{9.2}$$

which is true for the balance equations in Γ-space. This property follows from the Hermitian nature of the perturbation energy operator $\hat{V}$. In (9.1) we replace the suffixes n by m (the sum, of course, does not change) and we add the expression obtained to (9.1). After dividing by 2 we obtain

$$\dot{\mathscr{E}} = -\frac{1}{2} \sum_{n,m} w_{mn}(p_m - p_n) \ln \frac{p_n}{p_m}. \tag{9.3}$$

The expression on the right-hand side of (9.3) is always positive, i.e. the entropy increases, with the exception of the case

$$p_m = p_n, \tag{9.4a}$$

when the right-hand side of (9.3) is equal to zero. This case corresponds to a state of equilibrium. If we allow for the fact that in the balance equations w_{nm} denotes the probability of a transition (in unit time) between states with the same energy, then the state of equilibrium (9.4) is described by a micro-canonical distribution in which all the states with a given energy exist with an equal probability of

$$p_n = p_m = \frac{1}{\Delta\Gamma(E)}, \tag{9.4b}$$

where $\Delta\Gamma(E)$ is the number of states with a given energy. It is easy to see that the state (9.4b) has maximum entropy.

The principle of an increase in entropy in a closed system thus follows from the balance equations in Γ-space, the maximum entropy value being realized in a micro-canonical ensemble.

Let us now examine the question of the change in the entropy of a dynamic sub-system in contact with a sink (dissipative sub-system) at a constant temperature T. Using the second equality in (9.1) and relation (8.12), after replacing the suffixes n and m and adding the equalities obtained we find

$$\begin{aligned}\dot{\mathscr{E}} &= -\frac{1}{2}\sum_{n,m}(x_{mn} - x_{nm})\ln\frac{x_{nm}}{x_{mn}}\frac{w_{mn}}{w_{nm}} \\ &= -\frac{1}{2}\sum_{m,n}(x_{mn} - x_{nm})\ln\frac{x_{nm}}{x_{mn}} + \frac{1}{2kT}\sum_{m,n}(x_{mn} - x_{nm})(E_n - E_m),\end{aligned} \tag{9.5}$$

where $x_{mn} = w_{mn}p_m$.

We should further point out that the last term in (9.5) can be expressed in terms of the derivative of the mean energy of the system $\bar{E} = \sum_n p_n E_n$:

$$\dot{\bar{E}} = \sum_n \dot{p}_n E_n = \sum_{m,n}(x_{mn} - x_{nm})E_n = \tfrac{1}{2}\sum_{m,n}(x_{mn} - x_{nm})(E_n - E_m).$$

Finally we obtain

$$\frac{d}{dt}\left(k\mathscr{E} - \frac{\bar{E}}{T}\right) = -\frac{k}{2}\sum_{m,n}(x_{mn} - x_{nm})\ln\frac{x_{nm}}{x_{mn}} \geqslant 0, \tag{9.6}$$

which leads to the rate of loss of the free energy $F = E - kT\mathscr{E}$ in irreversible processes occurring at a constant temperature and constant volume (Landau and Lifshitz, 1951).

Let us examine the question of the increase in entropy in greater detail. As has already been pointed out, the entropy in Γ-space

$$\mathscr{E}_m = -\mathrm{Tr}\,\hat{\varrho}\ln\hat{\varrho}$$

is independent of time. To check this statement experimentally a complete measurement must be made at each point in time on the system and the ensemble which corresponds to this complete measurement (generally speaking a different one at each point in time) must be constructed. If, however, we are interested in the behaviour of one and the same ensemble given by one and the same measurements, then the entropy of this ensemble, as has just been shown, rises steadily. This statement is valid upon the same assumptions as those used in deriving the balance equations (7.35). Let us put it in another way. We are generally interested not in the whole density matrix $\hat{\varrho}$ which contains the complete information on the system (the measure of this information is $I_m = -\mathscr{E}_m$)†, but part of the density matrix, its projection $\hat{\varrho}_1 = \hat{P}\hat{\varrho}$ from which the corresponding transport equation is derived, and from which the rise in entropy follows. For example, in the case discussed, the operator $\hat{P}$ isolated the diagonal elements of the density matrix of the whole system in the eigenfunctions of the unperturbed Hamiltonian $\hat{F}$ of the dissipative system. Part of the information contained in the density matrix $\hat{P}\hat{\varrho}$ is transferred to the remaining degrees of freedom $(1 - \hat{P})\,\hat{\varrho}$ so that the total information does not change.

9.2. The General Derivation of the Transport Equations

If the operator $\hat{P}$ is independent of time it is possible to write in the general case an equation for the quantities (Zwanzig, 1960).

$$\hat{\varrho}_1 = \hat{P}\hat{\varrho} \quad \text{and} \quad \hat{\varrho}_2 = (1 - \hat{P})\,\hat{\varrho}.$$

We can write the equation for the density matrix in the following form

$$i\,\frac{\partial \hat{\varrho}}{\partial t} = \hat{L}\hat{\varrho}, \tag{9.7}$$

where

$$\hat{L}\hat{\varrho} = \frac{1}{\hbar}\,[\hat{H}, \hat{\varrho}].$$

The formal solution of this equation takes the form

$$\hat{\varrho}(t) = \exp\left(-\frac{it\hat{H}}{\hbar}\right)\hat{\varrho}(0)\exp\left(\frac{it\hat{H}}{\hbar}\right) \equiv \mathrm{e}^{-it\hat{L}}\,\varrho(0).$$

In future we shall consider $\hat{P}$ to be a linear operator. Then equation (9.7) can be rewritten in the form

$$i\,\frac{\partial \hat{\varrho}_1}{\partial t} = \hat{P}\hat{L}(\hat{\varrho}_1 + \hat{\varrho}_2). \tag{9.8}$$

$$i\,\frac{\partial \hat{\varrho}_2}{\partial t} = (1 - \hat{P})\,\hat{L}(\hat{\varrho}_1 + \hat{\varrho}_2). \tag{9.9}$$

† See section 6.

The solution of (9.9) can be written as follows:

$$\hat{\varrho}_2 = \exp\left[-it(1-\hat{P})\hat{L}\right]\hat{\varrho}_2(0) - i\int_0^t ds \exp\left[-is(1-\hat{P})\hat{L}\right] \times (1-\hat{P})\hat{L}\hat{\varrho}_1(t-s) \tag{9.10}$$

(direct substitution will confirm that this solution satisfies (9.9)). We then substitute (9.10) in (9.8) and find the required equation

$$i\frac{\partial\hat{\varrho}_1}{\partial t} = \hat{P}\hat{L}\exp\left[-it(1-\hat{P})\hat{L}\right]\hat{\varrho}_2(0) + \hat{P}\hat{L}\hat{\varrho}_1 - i\int_0^t ds\hat{P}\hat{L} \times \exp\left[-is(1-\hat{P})\hat{L}\right](1-\hat{P})\hat{L}\hat{\varrho}_1(t-s). \tag{9.11}$$

In the general case no closed equation is obtained for $\hat{\varrho}_1$ since, as can be seen from (9.11), the behaviour of $\hat{\varrho}_1$ at the time t is determined not only by $\hat{\varrho}_1$ at the time $t = 0$ but also by $\hat{\varrho}_2(0)$. The initial conditions discussed above (see section 7, sub-section 5) correspond to

$$\hat{P}\hat{\varrho}(0) = \hat{\varrho}(0), \quad \hat{\varrho}_2(0) = 0. \tag{9.12}$$

In this case equation (9.11) as a whole does not contain $\hat{\varrho}_2$. Equation (9.11) is thus the general form of the transport equations.

9.3. The Possibility of a Decrease of Entropy

As has already been pointed out above, the balance equation from which the principle of a rise in entropy follows can be derived in particular if initially the density matrix $\hat{\varrho}$ is diagonal in the suffixes α which denote the eigenfunctions of the Hamiltonian of the dissipative sub-system. If initially

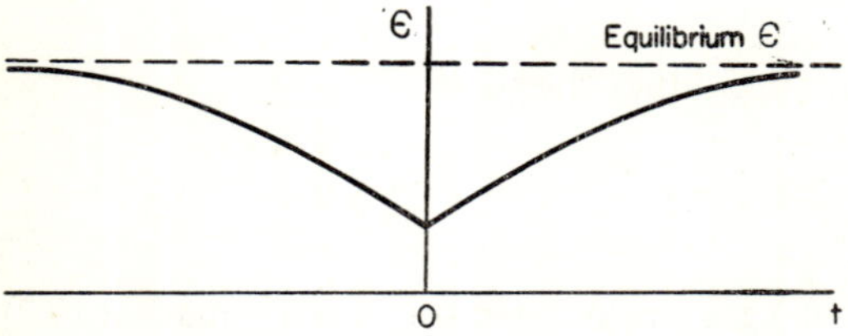

FIG. II.1. Change of entropy $\mathscr{E}$ with time. A non-equilibrium state with a diagonal density matrix is given at time $t = 0$.

the density matrix is not diagonal in the suffixes α, then in principle situations are possible in which the entropy decreases for some space of time. It follows from the work of Van Hove (1955) (see also Adams, 1960; Fain, 1963a) that with negative t a similar balance equation is valid (an anti-Boltzmann equation in the terminology of Adams (1960)), which leads to a micro-canonical distribution when $t \to -\infty$. The behaviour of the entropy is shown diagrammatically in Fig. II.1. When $t = 0$ the entropy has its lowest value. (We should mention that Van Hove (1955) assumes that at time $t = 0$ the density matrix is diagonal.) At a time $t \neq 0$ the density matrix is no longer diagonal and, if we take this time as the initial point in time, then, depending on the value of the non-diagonal elements of $\hat{\varrho}_{\alpha\alpha'}$, the entropy will increase or decrease (the latter will be the case if the matrix elements correspond to $t < 0$). It must, however, be remembered that in a macroscopic experiment we do not know all the details of the initial state and we select *a priori* the most probable, which is, as can be shown (Adams, 1960; Fain, 1963a), the state with a diagonal density matrix. In this case there is a rise in the entropy. A decrease in the entropy is also possible, however. Such cases, since they have a very low probability, relate to fluctuations.

10. The Transport Equation Description of Fluctuations

10.1. The Spectral Density of Fluctuations, and Correlation Functions

Let us examine a certain physical quantity represented by the operator $\hat{A}$. The fluctuations of this quantity can be denoted by the root mean square deviation from the mean (the root mean square dispersion)

$$\Delta A = \sqrt{\langle \hat{A} - \langle A \rangle \rangle^2} = \sqrt{\langle \hat{A}^2 \rangle - \langle A \rangle^2}.$$

However, the spectral distribution of the fluctuations is often of interest. In order to investigate the fluctuation spectrum we introduce the Fourier components of the operator

$$\hat{A}_\omega = \frac{1}{2\pi} \int_{-\infty}^{\infty} \hat{A}(t)\, e^{i\omega t}\, dt; \tag{10.1}$$

an inverse transformation gives

$$\hat{A}(t) = \int_{-\infty}^{\infty} \hat{A}_\omega\, e^{-i\omega t}\, d\omega. \tag{10.2}$$

We shall also introduce (to describe the fluctuations) the correlation functions of the two quantities $\hat{A}$ and $\hat{B}$ at the times t and t'

$$\Psi_{AB}(t, t') = \tfrac{1}{2} \langle \hat{A}(t)\, \hat{B}(t') + \hat{B}(t')\, \hat{A}(t) \rangle. \tag{10.3}$$

We write the correlation in the form of a product that has been made symmetrical since the quantity $\hat{A}\hat{B}$, generally speaking, is not Hermitian and its average is therefore not real.

We shall now examine fluctuations in stationary states. In this case the fluctuation spectrum and the correlation functions are interdependent.

We shall first prove that in a stationary state

$$\Psi_{AB}(t, t') = \Psi_{AB}(t' - t). \tag{10.4}$$

To do this we write in the Heisenberg representation the mean value of the product of the quantities A and B at the times t and t'

$$\langle \hat{A}(t)\, \hat{B}(t') \rangle = \mathrm{Tr}\left\{ \hat{\varrho} \exp\left(\frac{i}{\hbar}\hat{H}t\right) \hat{A} \exp\left(-\frac{i}{\hbar}\hat{H}t\right) \right.$$
$$\left. \times \exp\left(\frac{i}{\hbar}\hat{H}t'\right) \hat{B} \exp\left(-\frac{i}{\hbar}\hat{H}t'\right) \right\}.$$

Then using the characteristic property (2.10) of a stationary state and commuting the operators within the trace we obtain

$$\langle \hat{A}(t)\, \hat{B}(t') \rangle = \mathrm{Tr}\left\{ \hat{\varrho} \exp\left(\frac{i}{\hbar}\hat{H}(t - t')\right) \hat{A} \exp\left(-\frac{i}{\hbar}\hat{H}(t - t')\right) \hat{B} \right\}$$
$$= \langle \hat{A}(t - t')\, \hat{B} \rangle = \langle \hat{A}\hat{B}(t' - t) \rangle. \tag{10.5}$$

Therefore in a stationary state the correlation functions depend only on the time difference $(t - t')$.

In order to relate the correlations to the fluctuation spectrum we substitute the Fourier transforms of $\hat{A}(t)$ and $\hat{B}(t')$ in (10.3)

$$\Psi_{AB}(t, t') = \tfrac{1}{2}\iint_{-\infty}^{\infty} \langle \hat{A}_\omega \hat{B}_{\omega'} + \hat{B}_{\omega'} \hat{A}_\omega \rangle\, \mathrm{e}^{-i(\omega t + \omega' t')}\, d\omega\, d\omega'. \tag{10.6}$$

In a stationary state the integral on the right-hand side will be a function of $t' - t$; the integrand must contain a δ-function of $\omega + \omega'$ for this to be so. This means that we must have

$$\tfrac{1}{2}\langle \hat{A}_\omega \hat{B}_{\omega'} + \hat{B}_{\omega'} \hat{A}_\omega \rangle = (AB)_{-\omega}\, \delta(\omega + \omega'). \tag{10.7}$$

This relation is a definition of the quantity $(AB)_{-\omega}$. We substitute (10.7) in (10.6) and obtain

$$\Psi_{AB}(\tau) = \int_{-\infty}^{\infty} (AB)_\omega\, \mathrm{e}^{-i\omega\tau}\, d\omega, \tag{10.8}$$

where $\tau = t' - t$.

In particular $\Psi_{AB}(0)$ is just the mean value of the symmetrized product $\tfrac{1}{2}(\hat{A}\hat{B} + \hat{B}\hat{A}) = \{\hat{A}\hat{B}\}$

$$\langle \{AB\} \rangle = \int_{-\infty}^{\infty} (AB)_\omega\, d\omega \tag{10.9}$$

and $(AB)_\omega$ is the spectral density of the product AB. Manipulating (10.8) we find

$$(AB)_\omega = \frac{1}{2\pi}\int_{-\infty}^{\infty} \Psi_{AB}(\tau)\, e^{i\omega\tau}\, d\tau. \tag{10.10a}$$

We are often interested in quantities A and B whose mean values are zero. In this case the mean square dispersion of the quantity A is

$$\langle \Delta A\rangle^2 = \langle A^2\rangle$$

and the quantity $(A^2)_\omega$ is the spectral density of the fluctuations $\langle \Delta A^2\rangle$. This spectral density can be expressed in terms of the autocorrelation function Ψ_{AA}

$$(A^2)_\omega = \frac{1}{2\pi}\int_{-\infty}^{\infty} \Psi_{AA}(\tau)\, e^{i\omega\tau}\, d\tau. \tag{10.10b}$$

10.2. Using the Transport Equation to find Correlations

We shall now establish a number of general relations that allow us to use the transport equation to find time correlations (Fain, 1963b). From the solutions of the transport equation

$$\hat{\sigma} = \hat{P}\hat{\varrho} \tag{10.11}$$

(see (7.4)) we can find directly the mean values of the quantities A and B of interest to us

$$\langle A(t)\rangle = \frac{\mathrm{Tr}\, \hat{\sigma}(t)\, \hat{A}}{\mathrm{Tr}\, \hat{\sigma}(t)}, \quad \langle B(t')\rangle = \frac{\mathrm{Tr}\, \hat{\sigma}(t')\, \hat{B}}{\mathrm{Tr}\, \hat{\sigma}(t')}. \tag{10.12}$$

We must use the matrix (10.11) to find quantities of the type (10.3). Let us first examine the non-symmetrized mean of the two quantities $A(t)$ and $B(t)$. Using the density matrix $\hat{\varrho}$ and the operators $\hat{A}(t)$ and $\hat{B}(t')$ in the Heisenberg representation we find†

$$\langle \hat{A}(t)\, \hat{B}(t')\rangle = \frac{\mathrm{Tr}\,(\hat{\varrho}\hat{A}(t)\, \hat{B}(t'))}{\mathrm{Tr}\, \hat{\varrho}}. \tag{10.13}$$

Using (10.13) we can prove the following statements:

(a) Let $\langle A(t)\rangle \neq 0$; then dividing and multiplying the right-hand side of (10.13) by $\mathrm{Tr}\, \hat{\varrho}\hat{A}(t)$ we obtain

$$\langle \hat{A}(t)\, \hat{B}(t')\rangle = \langle A(t)\rangle\, \langle B'(t')\rangle, \tag{10.14}$$

† We recall that the matrix $\hat{\varrho}$ is time-independent.

where

$$\langle B'(t')\rangle = \frac{\mathrm{Tr}\,\hat{\varrho}'_t \hat{B}(t')}{\mathrm{Tr}\,\hat{\varrho}'_t}; \quad \hat{\varrho}'_t = \hat{\varrho}\hat{A}(t). \tag{10.15}$$

Likewise it can be shown that

$$\langle \hat{A}(t)\,\hat{B}(t')\rangle = \langle A''(t)\rangle\,\langle B(t')\rangle, \tag{10.16}$$

where

$$\langle A''(t)\rangle = \frac{\mathrm{Tr}\,\hat{\varrho}''_{t'}\hat{A}(t)}{\mathrm{Tr}\,\hat{\varrho}''_{t'}}; \quad \hat{\varrho}''_{t'} = \hat{B}(t')\,\hat{\varrho} \tag{10.17}$$

(we assume that $\mathrm{Tr}\,\varrho''_{t'} \neq 0$).

It follows from (10.15) that $\langle B'(t')\rangle$ is the mean value of the quantity B as a function of t' (for fixed t).

(b) Let $\langle A\rangle = \langle B\rangle = 0$; then using (10.13) it is easy to arrive at the relations

$$\langle \hat{A}(t)\,\hat{B}(t')\rangle = \frac{\mathrm{Tr}\,\hat{\varrho}_1\hat{B}(t')}{\mathrm{Tr}\,\hat{\varrho}_1} \equiv \langle B_1(t')\rangle = \frac{\mathrm{Tr}\,\hat{\varrho}_2\hat{A}(t)}{\mathrm{Tr}\,\hat{\varrho}_2} \equiv \langle \hat{A}_2(t)\rangle, \tag{10.18}$$

where

$$\hat{\varrho}_1 = \hat{\varrho} + \hat{\varrho}\hat{A}(t); \quad \hat{\varrho}_2 = \hat{\varrho} + \hat{B}(t')\,\hat{\varrho}. \tag{10.19}$$

The meaning of relations (10.14), (10.16) and (10.18) is that with a fixed $t(t')$ the quantity $\langle A(t)\,B(t')\rangle$ behaves as a certain mean value of $B(t')\,A(t)$ (multiplied by $\langle A(t)\rangle$ ($\langle B(t')\rangle$) in case (a)). In order to find these mean values of $B(t')$ (or $A(t)$) we can use the transport equation for the quantity $\hat{\sigma}$. When solving the transport equation we must take as the initial condition

$$\hat{\sigma}'(0) = \hat{P}\hat{\varrho}'_t = \hat{\sigma}(0)\,\hat{A}(t) \tag{10.20}$$

in case (a) with fixed t or

$$\hat{\sigma}''(0) = \hat{P}\hat{\varrho}''_{t'} = \hat{B}(t')\,\hat{\sigma}(0) \tag{10.21}$$

with fixed t' (the same as case (a)). In case (b) we must take the initial conditions in the form

$$\hat{\sigma}_1(0) = \hat{P}\hat{\varrho}_1 = \hat{\sigma}(0) + \hat{\sigma}(0)\,\hat{A}(t) \tag{10.22}$$

with fixed t and

$$\hat{\sigma}_2(0) = \hat{P}\hat{\varrho}_2 = \hat{\sigma}(0) + \hat{B}(t')\,\hat{\sigma}(0) \tag{10.23a}$$

with fixed t'.

It has already been pointed out above that a closed equation for $\hat{P}\hat{\varrho} = \hat{\sigma}$ occurs only for a certain class of initial conditions. It is therefore necessary to check each time whether the transition from $\hat{\varrho}$ to ϱ' has taken the density matrix out of this class.

Let us move on to the special case of a steady-state process. When $\langle A \rangle = \langle B \rangle = 0$ (we shall limit ourselves to this case) we have the relations

$$\Psi_{AB}(t, t') = \Psi_{AB}(t - t') = \tfrac{1}{2}\langle B_1(t - t')\rangle + \tfrac{1}{2}\langle B_2(t - t')\rangle$$

$$= \tfrac{1}{2}\langle A_1(t - t')\rangle + \tfrac{1}{2}\langle A_2(t - t')\rangle, \qquad (10.23\text{b})$$

where $\langle B_1(\tau)\rangle$ and $\langle B_2(\tau)\rangle$ are found from the transport equation as mean values of the quantity B by means of the density matrices $\hat{\sigma}_1$ and $\hat{\sigma}_2$ respectively, which satisfy the initial conditions

$$\hat{\sigma}_1(0) = \hat{\sigma}(0) + \hat{\sigma}(0)\,\hat{A}, \quad \hat{\sigma}_2(0) = \hat{\sigma}(0) + \hat{A}\sigma(0). \qquad (10.24\text{a})$$

$\langle A_1(\tau)\rangle$ and $\langle A_2(\tau)\rangle$ are found similarly. The autocorrelation function of the quantity $\hat{A}$, for which $\langle \hat{A}(t)\rangle = 0$, can be written in the form

$$\Psi_{AA} = \tfrac{1}{2}\langle A'(\tau)\rangle + \tfrac{1}{2}\langle A'(-\tau)\rangle, \qquad (10.25)$$

where $\langle A_1(\tau)\rangle$ is found from the transport equation with the initial condition

$$\hat{\sigma}'(0) = \hat{\sigma}(0) + \hat{\sigma}(0)\,\hat{A}. \qquad (10.24\text{b})$$

In the special case when the quantities $\langle A\rangle$ and $\langle B\rangle$ obey linear equations the correlation function satisfies the same equations. In this case we arrive at the results obtained by Leontovich (1941), who solved the problem of calculating the correlation function in the steady state.

CHAPTER III

Quantum Effects Appearing in the Interaction of Free Electrons with High-frequency Fields in Resonators

---

IN QUANTUM electronics one is generally concerned with bound electrons: these are electrons in atoms, molecules or in a solid. It is obvious that the behaviour of these bound electrons and their interaction with an electromagnetic field must be discussed within the framework of quantum theory.

On the other hand, the interaction of free electrons with electromagnetic fields is generally treated in classical terms. However, recent developments in quantum electronics have been accompanied by the progress of ordinary electronics into regions of higher and higher frequencies when the condition for the validity of a classical approach $\hbar\omega \ll kT$ may not hold. Then the question arises of whether there are quantum effects in the interaction of free electrons with high-frequency fields. Before we discuss these effects we shall devote a section in the present chapter to the quantum theory of the relaxation of a field in a lossy resonator, and give a kinetic explanation of the concept of the Q-factor of a resonator.

11. The Quantum Theory of Fields in Lossy Resonators

11.1. The Relaxation of a Field in a Resonator

As the result of interaction of a field in a resonator with electrons in the walls of the resonator the field relaxes to an equilibrium state. The atoms and electrons of the resonator walls play the part of a dissipative sub-system and the field in the resonator plays the part of a dynamic sub-system (see Chapter II, section 7).

We shall now apply the results of the quantum theory of relaxation given in the preceding chapter to investigate the question of the attenuation of a field in a resonator (see Weber, 1953b, 1954; Fain, 1959a; Senitzky, 1959, 1960a; Fain, 1963a).

The Hamiltonian of the system made up of the field in the resonator and the atoms and electrons of the walls of the resonator can be written in the form (see (4.13), (4.15))

$$\hat{H} = \frac{1}{2} \sum_{\nu} (\hat{p}_\nu^2 + \omega_\nu^2 \hat{q}_\nu^2) + \sum_k \frac{\hat{p}_k^2}{2m_k}$$

$$+ \frac{1}{2} \sum_{i \neq k} \frac{e_i e_u}{r_{ik}} - \sum_{\nu} \hat{q}_\nu \sum_k \frac{e_k}{m_k c} (\hat{p}_k \cdot A_\nu(k)). \tag{11.1}$$

Here we have omitted terms of the order of e_k^2 in the energy of the interaction with the radiation. These terms, generally speaking, are much smaller than terms of the order of e_k. The first sum in (11.1) is the energy of the radiation field, the second and third sums are the non-relativistic energy of the particles making up the resonator walls, and the last sum is the energy of the interaction of the field with the electrons in the walls. Introducing the annihilation and creation operators $\hat{a}_\nu$ and $\hat{a}_\nu^+$ for the field (see (3.28) *et seq.*) the interaction Hamiltonian can be rewritten in the form

$$\hat{V} = - \sum_{\nu} (\hat{a}_\nu + \hat{a}_\nu^+) \hat{F}_\nu;$$

$$\hat{F}_\nu = \sqrt{\frac{\hbar}{2\omega_\nu}} \hat{B}_\nu = \sqrt{\frac{\hbar}{2\omega_\nu}} \sum_k \frac{e_k}{m_k c} (\hat{p}_k A_\nu(k)). \tag{11.2}$$

We can now make use of (8.18a)† where the part of $\hat{v}_r$ is played by $\hat{a}_\nu$ ($\omega_r = -\omega_\nu$) and $\hat{a}_\nu^+$ ($\omega_r = \omega_\nu$) since $\hat{a}_\nu \sim e^{-i\omega_\nu t}$ and $\hat{a}_\nu^+ \sim e^{i\omega_\nu t}$ in the interaction representation. From (8.18a) we obtain for the derivative of the mean value of the operator $\hat{O}$

$$\frac{d\langle O \rangle}{dt} = \frac{i}{\hbar} \langle [\hat{E} + \hat{\Gamma}, \hat{O}] \rangle + \sum_{\nu\nu'} \Phi_{\nu\nu'}^+ \langle \hat{a}_{\nu'} [\hat{O}, \hat{a}_\nu^+] + [\hat{a}_{\nu'} \hat{O}] \hat{a}_\nu^+ \rangle$$

$$+ \sum_{\nu\nu'} \Phi_{\nu\nu'}^- \langle \hat{a}_{\nu'}^+ [\hat{O}, \hat{a}_\nu] + [\hat{a}_{\nu'}^+, \hat{O}] \hat{a}_\nu \rangle, \tag{11.3}$$

where‡

$$\hat{\Gamma} = \sum_{\nu\nu'} \overset{+}{\varphi}_{\nu\nu'} \hat{a}_\nu^+ \hat{a}_{\nu'} + \overset{-}{\varphi}_{\nu\nu'} \hat{a}_\nu \hat{a}_{\nu'}^+;$$

$$\varphi_{\nu\nu'}^{\pm} = \frac{1}{\hbar} \sum_{\alpha\alpha'} \frac{\langle \alpha | F_\nu | \alpha' \rangle \langle \alpha' | F_{\nu'} | \alpha \rangle}{\pm\omega_\nu + \omega_{\alpha\alpha'}} P_\alpha \Delta(\omega_\nu - \omega_{\nu'});$$

$$\Phi_{\nu\nu'}^{\pm} = \frac{\pi}{\hbar^2} \sum_{\alpha\alpha'} \delta(\mp\omega_\nu + \omega_{\alpha'\alpha}) \langle \alpha | F_\nu | \alpha' \rangle \langle \alpha' | F_{\nu'} | \alpha \rangle P_{\alpha'} \Delta(\omega_\nu - \omega_{\nu'}), \tag{11.4}$$

† Here we assume that $\omega_r + \omega_s \ll kT/\hbar$ and $\omega_\nu \gg \omega^*$. The other case, when $\omega_\nu \ll \omega^*$, will be discussed at the end of the section.

‡ Here we are taking into consideration the fact that $(\omega_\nu - \omega_{\nu'})$ can be much less than ω^*, but are assuming that this is impossible for the sum $\omega_\nu + \omega_{\nu'}$.

and if the resonator walls are in a state of thermal equilibrium, then, in accordance with (8.23)

$$\Phi^{+}_{\nu\nu'} = \Phi^{-}_{\nu'\nu}\, e^{-\frac{\hbar\omega_\nu}{kT}}. \tag{11.5}$$

From expression (11.3) we can in particular find the equations for the quantities $a_\nu \equiv \langle a_\nu \rangle$ and $a^+_\nu \equiv \langle a^+_\nu \rangle$ when there are no external forces:

$$\dot{a}_\nu = -i\omega_\nu a_\nu - \sum_{\nu'} \gamma_{\nu'\nu} a_{\nu'}, \quad \dot{a}^+_\nu = i\omega_\nu a^+_\nu - \sum_{\nu'} \gamma_{\nu\nu'} a^+_{\nu'}, \tag{11.6}$$

where

$$\gamma_{\nu\nu'} = \Phi^{-}_{\nu\nu'} - \Phi^{+}_{\nu'\nu} = \Phi^{-}_{\nu\nu'} \left(1 - e^{-\frac{\hbar\omega_\nu}{kT}}\right). \tag{11.7}$$

It is assumed in (11.6) that the frequencies ω_ν are sufficiently great that no allowance need be made for the frequency shift connected with the operator $\hat{\Gamma}$. It must be borne in mind, however, that this shift does generally occur. Therefore, allowing for interaction with the dissipative system leads not only to relaxation but also to a shift in the eigenfrequencies of the system. It can be seen from equations (11.6) that the independence of the radiation oscillators disappears when there is attenuation. They become dependent if their frequencies are close enough. This can be seen from (11.4). Here the $\Phi^{\pm}_{\nu\nu'}$ are large if

$$\omega_\nu - \omega_{\nu'} \ll \frac{1}{\tau_c} = \omega^*, \tag{11.8}$$

where τ_c is the correlation time of the system of particles in the walls of the resonator.

Inequality (11.8) holds in particular for degenerate resonator frequencies. In resonator theory the coefficient $\gamma_{\nu\nu}$ is generally denoted by $\frac{1}{2}(\omega_\nu/Q_\nu)$, where Q_ν is called the Q-factor of the resonator for a given mode. As can be seen from the above analysis the attenuation of the oscillations in a resonator is generally determined not only by the Q-factor but also by the coefficients $\gamma_{\nu\nu'}$ $(\nu \neq \nu')$†.

11.2. The Relaxation of the Energy of a Field in a Resonator

We can use equations (11.3) to determine the attenuation of the energy of the νth oscillator of a field

$$\langle \hat{H}_\nu \rangle = \hbar\omega_\nu \left(\langle \hat{n}_\nu \rangle + \tfrac{1}{2}\right),$$

where

$$\hat{n}_\nu = \hat{a}^+_\nu \hat{a}_\nu .$$

† The latter are connected with the mutual resistance of the resonator (see, e.g., Shteinshleiger, 1955).

From equations (11.3) we find

$$\frac{d}{dt}\langle a_\nu^+ a_\nu\rangle = -\sum_{\nu'}\gamma_{\nu'\nu}\left(\langle \hat{a}_\nu^+ \hat{a}_{\nu'}\rangle - \frac{\delta_{\nu'\nu}}{e^{\hbar\omega_\nu/kT}-1}\right)$$
$$-\sum_{\nu'}\gamma_{\nu\nu'}\left(\langle \hat{a}_{\nu'}^+ \hat{a}_{\nu}\rangle - \frac{\delta_{\nu\nu'}}{e^{\hbar\omega_\nu/kT}-1}\right).$$

It can be seen from this expression that a closed system of equations is not usually obtained for the average occupation numbers $\langle n_\nu\rangle \equiv n_\nu$ (this corresponds to the fact that the balance equation is not valid when there is loss). Also, when we have

$$\gamma_{\nu\nu'} = \gamma_{\nu\nu}\delta_{\nu\nu'}$$

we obtain

$$\frac{dn_\nu}{dt} = -2\gamma_{\nu\nu}(n_\nu - n_\nu^0), \tag{11.9}$$

where

$$n_\nu^0 = \frac{1}{e^{\hbar\omega_\nu/kT}-1}$$

is the equilibrium value of $\langle \hat{n}_\nu\rangle$ corresponding to the temperature of the dissipative system (the mean energy of a Planck oscillator). Therefore the energy of an eigen-oscillation is attenuated as follows:

$$\langle \hat{H}_\nu\rangle = \hbar\omega_\nu(n_\nu + \tfrac{1}{2}) = \hbar\omega_\nu(n_\nu^0 + \tfrac{1}{2}) + \hbar\omega_\nu(n_\nu(0) - n_\nu^0)\, e^{-2\gamma_{\nu\nu}t},$$

where $n_\nu(0)$ is the mean value of $\langle n_\nu\rangle$ at $t = 0$.

11.3. Allowing for Relaxation in the Case $\omega_\nu \ll \omega^*$

The following paradox can be found if we examine equations (11.6) closely. These equations do not satisfy the exact equality

$$\dot{q}_\nu = p_\nu, \tag{11.10}$$

whilst this equality does follow from the original Hamiltonian (11.1). Let us take for the sake of simplicity the case of a single field oscillator. From (11.6) we obtain

$$\dot{a} = -i\omega a - \gamma a; \tag{11.6a}$$

$$\dot{a}^+ = i\omega a^+ - \gamma a^+.$$

By adding and subtracting these two equations and using definitions (3.28) we can find without difficulty the equations

$$\dot{q} = p - \gamma q; \quad \dot{p} = -\omega^2 q - \gamma p. \tag{11.6b}$$

The paradox can be resolved as follows. The difference between the first of the equalities (11.6b) and (11.10) is a term of the order of γ/ω; since $\omega \gtrsim \omega^*$ (see (8.22)), this term $\lesssim \gamma/\omega^*$. At the same time we are neglecting terms of the same order (cf. (7.12)) in the original transport equations.

It will be interesting to examine separately the case

$$\omega \ll \omega^*. \tag{11.11}$$

Here we must use relations (8.18). For one field oscillator we obtain from these relations

$$\frac{d\langle O\rangle}{dt} = \frac{i}{\hbar}\langle[\hat{E}, \hat{O}]\rangle + \Phi^{++}\langle\hat{a}^+[\hat{O}, \hat{a}^+] + e^{-\frac{\hbar\omega}{kT}}[\hat{a}^+, \hat{O}]\, a^+\rangle$$

$$+ \Phi^{+-}\langle\hat{a}[\hat{O}, a^+] + e^{\frac{\hbar\omega}{kT}}[\hat{a}^+, \hat{O}]\,\hat{a}\rangle + \Phi^{-+}\langle\hat{a}^+[\hat{O}, a]$$

$$+ e^{-\frac{\hbar\omega}{kT}}[\hat{a}, \hat{O}]a^+\rangle + \Phi^{--}\langle\hat{a}[\hat{O}, \hat{a}] + e^{-\frac{\hbar\omega}{kT}}[\hat{a}, O]\,\hat{a}\rangle$$

$$+ \frac{i}{\hbar}\{\overline{\varphi}^{++}\langle\hat{a}^+[\hat{O}, \hat{a}^+]\rangle + \varphi^{++}\langle[\hat{a}^+, \hat{O}]\hat{a}^+\rangle + \overline{\varphi}^{+-}\langle\hat{a}[\hat{O}, \hat{a}^+]\rangle$$

$$+ \varphi^{+-}\langle[\hat{a}^+, \hat{O}]\,\hat{a}\rangle + \overline{\varphi}^{-+}\langle\hat{a}^+[\hat{O}, a]\rangle + \varphi^{-+}\langle[\hat{a}, \hat{O}]\, a^+\rangle$$

$$+ \overline{\varphi}^{--}\langle\hat{a}[\hat{O}, \hat{a}]\rangle + \varphi^{--}\langle[\hat{a}, \hat{O}]\, a\rangle\}, \tag{11.12}$$

where (cf. (11.14))

$$\Phi^{+-} = \Phi^+; \quad \Phi^{-+} = \Phi^-; \quad \varphi^{+-} = \varphi^+; \quad \varphi^{-+} = \varphi^-;$$

$$\Phi^{++} = \frac{\pi}{\hbar^2}\sum_{\alpha,\alpha'}\delta(\omega_{\alpha'\alpha} + \omega)\,\langle\alpha\,|F_\nu|\,\alpha'\rangle\,\langle\alpha'\,|F_\nu|\,\alpha\rangle\, P_{\alpha'} = \Phi^{-+};$$

$$\Phi^{--} = \frac{\pi}{\hbar^2}\sum_{\alpha,\alpha'}\delta(\omega_{\alpha'\alpha} - \omega)\,\langle\alpha\,|F_\nu|\,\alpha'\rangle\,\langle\alpha'\,|F_\nu|\,\alpha\rangle\, P_{\alpha'} = \Phi^{+-} \tag{11.13}$$

(the latter equalities follow from condition (11.11) and the definition of the Δ-function). The coefficients $\overline{\varphi}$ and φ can be determined similarly (see (8.20), (8.21)).

Instead of $\hat{O}$ we substitute the operators $\hat{a}$ and $\hat{a}^+$ in (11.12). As a result we obtain

$$\dot{a} + i\omega a = \Phi^{++}a^+\left(1 - e^{-\frac{\hbar\omega}{kT}}\right) + \Phi^{+-}a\left(1 - e^{\frac{\hbar\omega}{kT}}\right)$$

$$+ \frac{i}{\hbar}(\overline{\varphi}^{++} - \varphi^{++})\, a^+ + \frac{i}{\hbar}(\overline{\varphi}^{+-} - \varphi^{+-})\, a,$$

$$\dot{a}^+ - i\omega a^+ = \Phi^{-+}a^+\left(e^{-\frac{\hbar\omega}{kT}} - 1\right) + \Phi^{--}a\left(e^{\frac{\hbar\omega}{kT}} - 1\right)$$

$$+ \frac{i}{\hbar}(\varphi^{-+} - \overline{\varphi}^{-+})\, a^+ + \frac{i}{\hbar}(\varphi^{--} - \overline{\varphi}^{--})\, a. \quad (11.14)$$

Adding these equations, using relations (11.13), the definitions (3.28) and also the fact that when (11.11) holds

$$\overline{\varphi}^{+-} = \overline{\varphi}^{--}, \quad \varphi^{+-} = \varphi^{--}, \quad \varphi^{++} = \varphi^{-+}, \quad \overline{\varphi}^{++} = \varphi^{-+},$$

we obtain (11.10).

If instead of the operator $\hat{O}$ we substitute $\hat{n} = \hat{a}^+\hat{a}$ (the number of photons) and take into consideration the relations $\overline{\varphi}^{-+} = -\varphi^{+-}$ and $\overline{\varphi}^{+-} = -\varphi^{-+}$ (which, as can be checked, are true in our case), then we obtain the equation for the attenuation of the field energy:

$$\frac{dn}{dt} = \left(\Phi^{+-} + e^{-\frac{\hbar\omega}{kT}}\Phi^{-+}\right)\left(n - \frac{1}{e^{\hbar\omega/kT} - 1}\right). \quad (11.15)$$

To conclude this section we should point out that this discussion of the attenuation of a system of oscillators is quite general, and applies not only to fields in resonators but also to any system of oscillators whose energy of interaction with a dissipative system is of the form (11.2).

*12. Quantum Effects in the Interaction of Electrons with the Field in a Resonator (Ginzburg and Fain, 1957; Fain, 1958)

The problem of the interaction of electrons with high-frequency fields in resonators (and in waveguides) is one of the most important problems in electronics. It is therefore relevant to examine in detail possible quantum effects in this kind of problem. Roughly speaking quantum effects connected with the interaction of free electrons with a field in a resonator can be divided into two groups. The first group includes effects connected with the quantum properties of the electromagnetic field, and the second group includes effects connected with the quantum nature of the motion of the actual electrons. The effects in the second group will be discussed in detail in the next section. We shall first use a semi-classical discussion to explain the main features of the effects caused by the quantum properties of a field in a resonator.

12.1. Semi-classical Discussion

An electromagnetic field in a resonator, generally speaking, varies in a random manner, which is described by defining the quantum state. This field causes fluctuations in the quantities governing the interaction of an electron with the field. In particular there is a finite root mean square fluctuation (dispersion) of the electron energy. We proceed as follows to calculate this quantity. We express the kinetic energy dispersion from the classical equations in terms of the mean energy of the field fluctuations, and we use the fluctuation–dissipation theorem† to find the latter in the equilibrium case, thus allowing for the quantum nature of the field.

Let a non-relativistic electron with an energy $K_0 = mv_0^2/2$ enter a resonator at time $t = 0$ and leave at time $t = \tau$ with an energy $K_\tau = mv_\tau^2/2$. Here we shall assume for the sake of simplicity that the electric field E acting on the electron in the resonator is uniform and parallel to the direction of motion of the electron along the x-axis. We take the electric field to be of the form

$$E = E_1 \cos \omega t + (E_2 + E_0) \sin \omega t, \tag{12.1}$$

where E_1 and E_2 are random quantities such that

$$\bar{E}_1 = \bar{E}_2 = 0 \quad \text{and} \quad \overline{E_1^2} = \overline{E_2^2} = \frac{4\pi \bar{H}_f}{V}, \tag{12.2}$$

where V is the volume of the resonator;

$\bar{H}_f$ is the mean energy of the fluctuation field of E_1, E_2.

From the equation of motion of the electron (in the non-relativistic approximation)

$$m\ddot{x} = eE$$

and equation (12.1) we find

$$v_\tau = \dot{x}(\tau) = v_0 + \frac{e}{m\omega} [E_1 \sin \omega\tau + (E_2 + E_0)(1 - \cos \omega\tau)]. \tag{12.3}$$

With an accuracy up to terms in e^2 we find

$$\bar{K}_\tau - \bar{K}_0 = \frac{e^2}{2m\omega^2} \left\{ 4\overline{E_1^2} \sin^2 \frac{\omega\tau}{2} + E_0^2(1 - \cos \omega\tau)^2 \right\} + \frac{2ev_0}{\omega} E_0 \sin \frac{2\omega\tau}{2} \tag{12.4}$$

† There is a detailed discussion of the fluctuation–dissipation theorem in section 17 (see also Ginzburg, 1952).

and

$$\overline{(\Delta K_\tau)^2} = \overline{K_\tau^2} - (\overline{K_\tau})^2 = \overline{(K_\tau - K_0)^2} - [\overline{(K_\tau - K_0)}]^2$$
$$= \frac{4e^2v_0^2}{\omega^2}\overline{E_1^2}\sin^2\frac{\omega\tau}{2}. \qquad (12.5)$$

Here $\overline{K}_0 = K_0 = mv_0^2/2$. The term omitted in (12.5) is of order e^3 and is insignificant if the change in the electron velocity as it passes through the resonator is sufficiently small. In this case the time of flight of the electron through the resonator can be taken as

$$\tau = \frac{d}{v_0},$$

where d is the path travelled by the electron (the width of the resonator).

Remembering (12.2) we can rewrite (12.5) (the expression for the spread in the kinetic energy of the electron) in the following form:

$$\overline{(\Delta K_\tau)^2} = \frac{e^2 4\pi d}{S}\bar{H}_f\left(\frac{\sin\frac{1}{2}\omega\tau}{\frac{1}{2}\omega\tau}\right)^2 = \frac{e^2}{C}\bar{H}_f\left(\frac{\sin\frac{1}{2}\omega\tau}{\frac{1}{2}\omega\tau}\right)^2. \qquad (12.6)$$

It is assumed here that the volume of the resonator can be written in the form $V = Sd$, and the quantity $S/4\pi d$ is denoted by C. This quantity is equal to the capacitance in the LCR-circuit which is equivalent to the resonator. In the case under discussion, when

$$(\Delta v)^2 \equiv \overline{v^2} - (\bar{v})^2 \ll v_0^2,$$

we have the relation

$$\overline{(\Delta v_\tau)^2} = \overline{(\Delta K_\tau)^2}\, m^{-2}v_0^{-2}.$$

The factor $[\sin\frac{1}{2}\omega\tau/(\frac{1}{2}\omega\tau)]^2$ in (12.6) is caused by the change in the field during the time of flight of the electron through the resonator.

We have therefore established classically a connection between the dispersion of the kinetic energy of the electron and the mean energy of the fluctuations. The quantum element of our derivation will be the use of the fluctuation–dissipation theorem to find the mean energy, a theorem based on quantum theory. In the case of weak attenuation the mean energy of the fluctuations is (see section 11)

$$\bar{H}_f = \frac{\hbar\omega}{2} + \frac{\hbar\omega}{e^{\hbar\omega/kT} - 1}.$$

By using the fluctuation–dissipation theorem we can find $\bar{H}_f$ for any attenuation. When only one mode of oscillation is considered we can say for the sake of simplicity that the resonator is equivalent to a circuit consisting of a resistance R, an inductance L and a capacitance C connected in series so that the impedance Z of the circuit when an e.m.f. $\mathscr{E} = ZI$ is applied to it is

$$Z = R + i\left(\omega L - \frac{1}{\omega C}\right).$$

In accordance with Nyquist's theorem the spectral density of the fluctuations of the e.m.f. in this circuit in a state of thermal equilibrium is

$$\overline{|\mathscr{E}_\omega|^2} = \frac{2}{\pi} R(\omega) \left\{\frac{\hbar\omega}{2} + \frac{\hbar\omega}{e^{\hbar\omega/kT} - 1}\right\}.$$

The spectral density of the square of the current fluctuations is

$$\overline{|I_\omega|^2} = \frac{\overline{|\mathscr{E}_\omega|^2}}{|Z(\omega)|^2},$$

and the spectral density of the square of the voltage fluctuations across the capacitor is

$$\overline{|V_\omega|^2} = \frac{\overline{|\mathscr{E}_\omega|^2}}{C^2\omega^2\,|Z(\omega)|^2} = \frac{\overline{|\mathscr{E}_\omega|^2}}{R^2C^2\omega^2 + (LC\omega^2 - 1)^2}.$$

When there is such a frequency spectrum equation (12.6) can be generalized in the form

$$\overline{(\Delta K_\tau)^2} = \frac{e^2}{C}\int_0^\infty |H_{f\omega}| \left(\frac{\sin\frac{1}{2}\omega\tau}{\frac{1}{2}\omega\tau}\right)^2 d\omega,$$

where

$$\overline{|H_{f\omega}|} = C\,\overline{|V_\omega|^2},$$

and we obtain†

$$\overline{(\Delta K_\tau)^2} = e^2 \int_0^\infty \frac{\frac{2}{\pi} R(\omega)\left(\frac{\hbar\omega}{2} + \frac{\hbar\omega}{e^{\hbar\omega/kT} - 1}\right)}{R^2C^2\omega^2 + (LC\omega^2 - 1)} \left(\frac{\sin\frac{1}{2}\omega\tau}{\frac{1}{2}\omega\tau}\right)^2 d\omega. \qquad (12.7)$$

For a circuit with small attenuation and a resonant frequency $\omega_0 = 1/\sqrt{LC}$, it can easily be shown (see, e.g., Ginzburg, 1952) that

$$\overline{(\Delta K_\tau)^2} = \frac{e^2}{C(\omega_0)}\left(\frac{\hbar\omega_0}{2} + \frac{\hbar\omega_0}{e^{\hbar\omega_0/kT} - 1}\right)\left(\frac{\sin\frac{1}{2}\omega_0\tau}{\frac{1}{2}\omega_0\tau}\right)^2, \qquad (12.8)$$

as would be expected.

† The integration to ∞ in (12.7) is arbitrary, but this is not important, as the integral converges strongly (Ginzburg and Fain, 1957a, 1957b).

It can be seen from (12.7) that a beam of electrons passing through a resonator is essentially an instrument that measures the noise of the field. In fact, by measuring the root mean square scatter of the energy of the electrons as they leave the resonator† we can also find the value of

$$\frac{\hbar\omega_0}{2} + \frac{\hbar\omega_0}{e^{\hbar\omega_0/kT} - 1}$$

which is the mean energy of the fluctuation in a weakly attenuated resonator.

It must be stressed that the classical calculation that has just been made with the use of the fluctuation–dissipation theorem is sufficient to calculate only $\overline{(\Delta K_\tau)^2}$. When we wish, for example, to obtain the fluctuation of the electron energy distribution at the exit of the resonator we cannot usually limit ourselves to the classical approach. It is therefore of interest to carry out the formal quantum calculations. The results of this calculation and the comparison with the classical approach will be given in the next section.

12.2. Effects Connected with Field Quantization

We shall now continue by studying the effects connected with the quantization of the electromagnetic field, describing the electron classically. The latter means that we are not allowing for effects connected with the original uncertainty in the velocities and coordinates of the electrons.

The Hamilton operator of the electron + radiation field system is of the form (see section 4)

$$\hat{H} = \frac{\left(\hat{p} - \frac{e}{c}\hat{A}\right)^2}{2m} + \frac{1}{2}\sum_\nu (\hat{p}_\nu^2 + \omega_\nu^2 \hat{q}_\nu^2) = \hat{K} + \hat{H}_{tr}. \tag{12.9}$$

If we neglect the magnetic field, then the quantum equations of motion

$$\frac{d^2\hat{x}}{dt^2} = \frac{e}{m}\hat{E}_x = -\frac{e}{mc}\sum_\nu \hat{p}_\nu A_{\nu x}(x(t))$$

follow from (12.9) with an accuracy up to terms in e. Let the mean initial velocity of the electron be $v_0 \gg \Delta v$, where Δv is the increment in the electron velocity due to the field. Then taking account of terms up to the first order in e we have

$$\frac{d\hat{K}}{dt} = -\frac{ev_0}{c} A_{\nu x}\hat{p}_\nu, \tag{12.10}$$

† In particular when $\omega_i\tau \ll 1$ the factor $[\sin \frac{1}{2}\omega_i\tau/(\frac{1}{2}\omega_i\tau)]^2$ is equal to unity.

where $\hat{K}$ is the electron kinetic energy operator and only a single νth eigen-oscillation is included for the sake of simplicity.

We shall neglect the dependence of $A_{\nu x}$ on x (similar assumptions are made in the example discussed above). The operator equation can be integrated

$$\hat{K} = K_0 - \frac{ev_0}{c}\int_0^t A_{\nu x}\hat{p}_\nu \, dt. \tag{12.11}$$

In the approximation used (including terms up to the first order in e) we must substitute in (12.11) a form of $\hat{p}_\nu$ which corresponds to a field in which no electrons are present (3.45):

$$\hat{p}_\nu = \hat{p}_\nu(0)\cos\omega_\nu t - \omega_\nu \hat{q}_\nu(0)\sin\omega_\nu t. \tag{12.12}$$

Using (12.11) we find that the mean value of the energy of the electron is

$$\langle K_\tau \rangle = K_0 - \frac{ev_0}{c}\int_0^\delta A_{\nu x}\langle p_\nu(t)\rangle \, dt, \tag{12.13}$$

where

$$\langle p_\nu(t)\rangle = \int \Psi^*(0)\,\hat{p}_\nu(t)\,\Psi(0)\, dq_\nu,$$

and $\Psi(0)$ is the wave function of the initial state.

The spread in the kinetic energy of the electron is

$$\begin{aligned}\langle(\Delta K_\tau)^2\rangle &\equiv \langle K_\tau^2\rangle - \langle K_\tau\rangle^2 \\ &= \frac{e^2 v_0^2}{c^2} A_{\nu x}^2 \int_0^\tau dt_1 \int_0^\tau dt_2 \, [\langle \hat{p}_\nu(t_1)\,\hat{p}_\nu(t_2)\rangle \\ &\quad - \langle p_\nu(t_1)\rangle\langle p_\nu(t_2)\rangle].\end{aligned} \tag{12.14}$$

Let us look at two cases:

1. Initially the energy of the field in a resonator has a definite value $(N+\frac{1}{2})\hbar\omega_\nu$. Then (see (3.25b))

$$\Psi(0) = \Psi_N = \left(\frac{\alpha}{\pi^{1/2}2^N N!}\right)^{1/2} e^{-\alpha^2 q_\nu^2/2} H_N(\alpha q_\nu) \tag{3.25c}$$

is the eigenfunction of the harmonic oscillator Hamiltonian.

Using the function for averaging in (12.13) and (12.14), and remembering that $\hat{p}_\nu(0) = -i\hbar(\partial/\partial q_\nu(0))$ we obtain

$$\langle K_\tau\rangle = K_0, \tag{12.15}$$

$$\langle(\Delta K_\tau)^2\rangle = \frac{e^2}{C}\left(N+\frac{1}{2}\right)\hbar\omega_\nu\left(\frac{\sin\frac{1}{2}\omega_\nu\tau}{\frac{1}{2}\omega_\nu\tau}\right)^2. \tag{12.16}$$

Here for the sake of simplicity we discussed the case of a resonator which is a parallelepiped and assumed that (see (3.61))

$$A_{\nu x} = \frac{\sqrt{4\pi c^2}}{\sqrt{V}},$$

so $C = S/4\pi d$ is the capacitance of the plane-parallel capacitor in the equivalent circuit.

In the derivation of (12.15) we calculated the kinetic energy with an accuracy up to terms in e. In the quantum averaging in (12.13), however, the term in e disappears. For the sake of simplicity we did not calculate the terms in e^2. If these terms are not considered the following paradox appears. Let N (the number of photons in the cavity) be zero. Then an electron passing through the cavity may either increase the number of photons or leave it equal to zero. The number of photons cannot decrease, so the mean energy in the cavity should increase, and the mean energy $\langle K_\tau \rangle$ of the electron should decrease, since $\langle K \rangle + \langle H_{tr} \rangle = \text{const}$. At the same time, according to (12.15), the electron energy does not change.

We note here that when terms in e^2 are included in (12.11), terms of an order higher than e^2 appear in (12.16). Expression (12.16) agrees with (12.6) if we remember that under the original conditions chosen $(N + \frac{1}{2})\,\hbar\omega_\nu$ is the same as the energy of the fluctuations since the mean values of the electric and magnetic fields are zero†. The purely quantum term in (12.16) allows for the zero point oscillations of the resonator. When $N = 0$

$$\langle (\Delta K_\tau)^2 \rangle = \frac{e^2}{C}\,\frac{\hbar\omega_\nu}{2}\left(\frac{\sin \frac{1}{2}\omega_\nu \tau}{\frac{1}{2}\omega_\nu \tau}\right)^2. \tag{12.17}$$

2. Another possible initial state corresponds to a field whose properties are closest to a classical coherent field with a definite phase and amplitude (see (3.49)):

$$\Phi = \left(\frac{\omega_\nu}{\pi\hbar}\right)^{1/4} \exp\left\{-\frac{\omega_\nu}{2\hbar}\,(q_\nu - \bar{q}_\nu(0))^2 + \frac{i\bar{p}_\nu(0)}{\hbar}\,q_\nu\right\}.$$

To make this correspond to the field in (12.1) we put $\bar{p}_\nu(0) = 0$. Then

$$\langle p_\nu(t) \rangle = -\omega_\nu \langle q_\nu(0) \rangle \sin \omega_\nu t,$$

and the mean electric field is

$$\langle E \rangle = -\frac{1}{c}\,A_{\nu x}\,\langle p_\nu(t) \rangle \equiv E_0 \sin \omega_\nu t.$$

† We recall (see section 3, sub-section 6) that when $N \gg 1$ the state (3.25b) under discussion is equivalent to an ensemble of classical oscillators with the same frequency but with a completely random phase.

Averaging (12.13) again we have

$$\langle K_i \rangle = K_0 + \frac{2ev_0}{\omega_\nu} E_0 \sin^2 \frac{\omega_\nu \tau}{2}.$$

If we include terms up to those in e this agrees with (12.4), i.e. the quantum mean is the same in this case as the classical average. Using (12.14) we obtain

$$\langle (\Delta K_\tau) \rangle^2 = \frac{e^2}{C} \frac{\hbar\omega_\nu}{2} \left(\frac{\sin \frac{1}{2}\omega_\nu \tau}{\frac{1}{2}\omega_\nu \tau} \right)^2,$$

which is exactly the same as (12.17), i.e. in this case the spread in the velocity of the electron is caused only by the zero point fluctuations of the field.

It is interesting to note that if in the first of the cases examined the initial value of the dispersion of the field energy $H_{tr}(0)$ is zero (the energy has the definite value $(N + \frac{1}{2})\hbar\omega_\nu$), then the field described by the wave function (3.49) has the finite and comparatively large dispersion

$$\sqrt{\langle \Delta H_{tr}^2(0) \rangle} = \sqrt{\hbar\omega_\nu} \cdot \sqrt{H_{tr\,cl}}, \tag{3.52}$$

where $H_{tr\,cl} = \bar{N}\hbar\omega_\nu$.

We notice that this dispersion may in general, be greater than the classical thermal energy kT:

$$\frac{\sqrt{\langle \Delta H_{tr}^2(0) \rangle}}{kT} = \sqrt{\frac{\hbar\omega_\nu}{kT}} \sqrt{\frac{H_{tr\,cl}}{kT}} \gg 1 \tag{12.18}$$

even at frequencies for which $\hbar\omega_\nu \ll kT$ (because of the presence of the second factor).

Let us examine briefly the question of the probability of a change in the field energy by an amount $n\hbar\omega_\nu$ $(n = 0; \pm 1; \ldots)$.

Limiting ourselves to a discussion of a single eigen-oscillation of a field in a resonator, the Hamiltonian (12.9) can be rewritten in the form

$$\hat{H} = \frac{\hat{p}^2}{2m} - \frac{e}{mc} (\hat{A} \cdot \hat{p}) + \frac{1}{2} (\hat{p}_\nu^2 + \omega_\nu^2 \hat{q}_\nu^2),$$

where terms up to those in e have been included, and where

$$\hat{A} = A_\nu \hat{q}_\nu$$

and the term

$$\hat{H}' = -\frac{e}{mc} (\hat{p} \cdot \hat{A})$$

is a small perturbation.

We shall assume that until the beginning of the interaction the system is in a state described by an unperturbed Hamiltonian and the field energy is $(N + \frac{1}{2})\hbar\omega_\nu$.

Using ordinary perturbation theory it is not difficult to show that the probabilities of a change in the field energy by $\hbar\omega_\nu$ are of the form

$$w_k(\tau) = \frac{e^2}{\hbar\omega_\nu C}\left(\frac{\sin\frac{1}{2}\omega_\nu\tau}{\frac{1}{2}\omega_\nu\tau}\right)^2\frac{N+1}{2} \quad \text{for} \quad k = N+1,$$

$$w_k(\tau) = \frac{e^2}{\hbar\omega_\nu C}\left(\frac{\sin\frac{1}{2}\omega_\nu\tau}{\frac{1}{2}\omega_\nu\tau}\right)^2\frac{N}{2} \quad \text{for} \quad k = N-1, \tag{12.19}$$

$$w_k(\tau) = 1 - \frac{e^2}{\hbar\omega_\nu C}\left(\frac{\sin\frac{1}{2}\omega_\nu\tau}{\frac{1}{2}\omega_\nu\tau}\right)^2\frac{2N+1}{2} \quad \text{for} \quad k = N,$$

where $\tau = md/\langle p\rangle = d/v_0$ is the time of interaction of the electron with the field in the resonator.

These expressions (12.19) make it possible to calculate the dispersion of the field energy after a time τ. It is easy to check that the dispersion of the field energy in the case under discussion (and in the approximation used) agrees with the dispersion of the electron energy (12.16).

In the second case discussed above the dispersion of the electron energy is not the same as the dispersion of the field energy (initially the dispersion of the electron energy is zero, whilst the dispersion of the field energy is given by (3.52)).

*13. Effects Connected with the Quantum Nature of the Motion of an Electron. Conclusions and Estimates

13.1. Systematic Quantum Calculations (Senitzky, 1954, 1955)

The systematic quantum-mechanical description of the interaction of an electron with a field in a resonator is as follows. The initial state of the field and the electron is given by the corresponding wave packets (representing the superposition of stationary states). We shall choose these wave packets so as to obtain a description that is as nearly classical as possible. In classical theory the amplitude and phase of the radiation can in principle be determined exactly, and there is no lower limit of the variance of the amplitudes and phases. The same may be said of a coordinate and the corresponding momentum of an electron. In quantum theory, therefore, we select fairly small wave packets. A wave packet of this kind, which describes the state of the radiation field, has been discussed in the first chapter (see (3.49))

$$\Phi_{\nu'}(q_{\nu'}) = \left(\frac{\omega_{\nu'}}{\pi\hbar}\right)^{1/4}\exp\left\{-\frac{\omega_{\nu'}}{2\hbar}(q_{\nu'} - \delta_{\nu'\nu}\bar{q}_\nu(0))^2\right\}. \tag{13.1a}$$

Here (just as in section 12) we take $\langle p_\nu(0)\rangle = 0$ for the sake of simplicity. In addition we have assumed that only the νth eigen-oscillation is excited, the remaining eigen-oscillations being in the ground state. The energy dispersion in the state ν' is defined by (3.52). As has already been shown (see Chapter I), the product of the uncertainties $q_\nu(0)$ and $p_\nu(0)$ has its minimum value in this state and is equal to $\frac{1}{2}\hbar$.

We shall describe the electron by a wave packet, in which the coordinates and momentum of the electron are approximately determined and the product of their uncertainties is a minimum

$$\varphi(\boldsymbol{r}, 0) = b^{-3/2}\pi^{-3/4} \exp\left[(-x^2/2b^2) + ik_0 x\right], \tag{13.1b}$$

where b is a constant which defines the length of the packet,
$\hbar k_0/m = v_0$ is the velocity of the packet.

If the electron moves freely, then the probability of finding the electron at a point $\boldsymbol{r}$ at a time t is

$$|\varphi(\boldsymbol{r}, t)|^2 = b^{-3}\pi^{-3/2}\left[1 + \left(\frac{\hbar t}{mb}\right)^2\right]^{-3/2} \exp\left[\frac{-(\boldsymbol{r} - \boldsymbol{v}_0 t)^2}{b^2 + (\hbar t/mb)^2}\right], \tag{13.2}$$

whilst the probability of obtaining the value $\boldsymbol{p}$ in a measurement is time independent and is given by

$$|\varphi(\boldsymbol{p})|^2 = b^3\hbar^{-3}\pi^{-3/2} \exp\left[-(b/\hbar)^2 (\boldsymbol{p} - m\boldsymbol{v}_0)^2\right]. \tag{13.3}$$

Equation (13.2), expresses the well-known spreading-out of the wave packet of a free electron.

Both distributions (13.2) and (13.3) are Gaussian, each of them having a maximum at the corresponding classical value of the coordinate and momentum.

The initial state of the electron + field system can be written in the form

$$\Phi(\boldsymbol{x}, q_{\nu'}, 0) = \varphi(\boldsymbol{x}, 0) \prod_{\nu'} \Phi_{\nu'}(q_{\nu'}). \tag{13.4}$$

The system's Hamilton operator is of the form of (12.9)

$$\hat{H} = \frac{\left(\hat{\boldsymbol{p}} - \frac{e}{c}\boldsymbol{A}\right)^2}{2m} + \frac{1}{2}\sum_{\nu'} (\hat{p}_{\nu'}^2 + \omega_{\nu'}^2 \hat{q}_{\nu'}^2).$$

Assuming that the velocity of the electron is so small that the action of the magnetic field can be neglected, the quantum equations of motion of the electron can be written in the form

$$\frac{d^2\hat{\boldsymbol{r}}}{dt^2} = -\frac{e}{em}\sum_{\nu'} \hat{p}_{\nu'}(t)\hat{\boldsymbol{A}}_{\nu'}(t), \tag{13.5}$$

where $\hat{\boldsymbol{A}}_{\nu'}(t)$ stands for $\hat{\boldsymbol{A}}_\nu(\boldsymbol{r}(t))$.

Integrating the operator equation (13.5) we obtain

$$\hat{\boldsymbol{v}}(t) = \hat{\boldsymbol{v}}(0) - \frac{e}{mc}\int_0^t dt_1 \sum_{\nu'} \hat{p}_{\nu'}(t_1)\, \hat{\boldsymbol{A}}_{\nu'}(t_1). \tag{13.6}$$

By using this equation we can find the mean value of the velocity and the mean value of the square of the deviation from the mean (the mean square dispersion of the velocity). The operators $\hat{p}_{\nu'}(t)$ and $\hat{\boldsymbol{A}}_{\nu'}(t)$ are found, as usual, in the Heisenberg representation

$$\hat{p}_{\nu'}(t) = \exp\,[(i/\hbar)\,\hat{H}t]\,\hat{p}_{\nu'}(0)\exp\,[-\,(i/\hbar)\,\hat{H}t]. \tag{13.7}$$

$$\hat{\boldsymbol{A}}_{\nu'}(t) = \exp\,[(i/\hbar)\,\hat{H}t]\,\hat{\boldsymbol{A}}_{\nu}(\boldsymbol{r}(0))\exp\,[-\,(i/\hbar)\,\hat{H}t]. \tag{13.8}$$

These expressions are substituted in the right-hand side of (13.6). In order to obtain solutions which include terms up to the first order in e we put $e = 0$ in the exponents of (13.7) and (13.8). Then we immediately obtain (see (3.47))

$$\hat{p}_{\nu'} = \hat{p}_{\nu'}(0)\cos\omega_{\nu'}t - \omega_{\nu'}\hat{q}_{\nu'}(0)\sin\omega_{\nu'}t,$$

$$\hat{q}_{\nu'} = \hat{q}_{\nu'}(0)\cos\omega_{\nu'}t + [\hat{p}_{\nu'}(0)/\omega_{\nu'}]\sin\omega_{\nu'}t \tag{3.47}$$

$$\hat{\boldsymbol{A}}_{\nu'}(t) = \hat{\boldsymbol{A}}_{\nu'}(\boldsymbol{r}_0 + \hat{\boldsymbol{v}}(0)\,t). \tag{13.9}$$

We find for the mean value of the velocity at the time t

$$\langle \boldsymbol{v}(t)\rangle = \langle \boldsymbol{v}(0)\rangle - \frac{e}{mc}\int_0^t dt_1 \sum_{\nu'} \langle \hat{p}_{\nu'}(t_1)\, \boldsymbol{A}_{\nu'}(t_1)\rangle. \tag{13.10}$$

Since we are trying to find a solution which includes terms in e only the mean value of the product can be replaced in the right-hand side of (13.10) by the product of the averages

$$\langle p_{\nu'}(t_1)\, \boldsymbol{A}_{\nu'}(t_1)\rangle = \langle p_{\nu'}(t_1)\rangle\,\langle \boldsymbol{A}_{\nu'}(t_1)\rangle. \tag{13.11}$$

Averaging (13.7) for the initial state (13.4) we find

$$\langle p_{\nu'}(t_1)\rangle = -\delta_{\nu'\nu}\omega_{\nu}\langle q_{\nu}(0)\rangle \sin\omega_{\nu'}t_1. \tag{13.12}$$

To find the mean value of (13.9) we express the argument $\boldsymbol{A}_{\nu}$ in the form

$$\boldsymbol{v}_0 t + [\boldsymbol{r}_0 + (\hat{\boldsymbol{v}}(0) - \boldsymbol{v}_0)\,t],$$

where $\boldsymbol{v}_0 = \langle \boldsymbol{v}(0)\rangle$; these cond term in the average is equal to zero ($\langle \boldsymbol{r}_0\rangle = 0$) and if the fluctuations of the coordinates and velocity are small (as we are assuming), then the second term is much smaller than the first. Expanding

A_ν into a series in $[\boldsymbol{r}_0 + (\hat{\boldsymbol{v}}(0) - \boldsymbol{v}_0)\, t]$ and averaging over the wave function of the initial state we find

$$\langle A_{\nu'}(t)\rangle = [1 + \delta^2(t)\, \nabla^2]\, A_{\nu'}(\boldsymbol{r})|_{r=v_0 t} = \left[1 - \delta^2(t)\, \frac{\omega_{\nu'}^2}{c^2}\right] A_{\nu'}(\boldsymbol{v}_0 t), \quad (13.13)$$

where the latter equality is obtained by using (3.8) and $\delta^2(t) = (b/2)^2 + (\hbar t/2mb)^2$.

Finally we find from (13.10), (13.11), (13.12) and (13.13)

$$\langle \boldsymbol{v}(t)\rangle = \boldsymbol{v}_0 + E_0 \frac{e}{m} \int_0^t dt_1 A_\nu(\boldsymbol{v}_0 t_1) \sin \omega_\nu t_1$$

$$-E_0 \frac{e}{m} \frac{\omega_\nu^2}{c^2} \int_0^t dt_1 \delta^2(t_1)\, A_\nu(\boldsymbol{v}_0 t_1) \sin \omega_\nu t_1, \quad (13.14a)$$

where E_0 denotes the quantity $(\omega_\nu/c)\, \langle q_\nu(0)\rangle$†. Similar but more cumbersome calculations lead to the following form of the velocity dispersion at time t (Senitzky, 1954, 1955):

$$\langle v_x^2(t)\rangle - \langle v_x(t)\rangle^2 = \frac{\hbar^2}{2m^2b^2} \left\{1 - \frac{2E_0 e}{m v_0^2} \int_{x_1}^{x_2} dx \left(x \sin \frac{\omega_\nu}{v_0} x\right) \frac{\partial}{\partial x} A_{\nu x}(x, 0, 0)\right.$$

$$\left. + \frac{E_0^2 e^2}{m^2 v_0^4} \left[\int_{x_1}^{x_2} dx \left(x \sin \frac{\omega_\nu}{v_0} x\right) \nabla A_{\nu x}(\boldsymbol{r})\right]^2_{y=z=0}\right\}$$

$$+ \frac{E_0^2 e^2 b^2}{2m^2 v_0^2} \left[\int_{x_1}^{x_2} dx \left(\sin \frac{\omega_\nu}{v_0} x\right) \nabla A_{\nu x}(\boldsymbol{r})\right]^2_{y=z=0}$$

$$+ \frac{e^2}{2m^2 v_0^2 c^2} \sum_\nu \hbar\omega_{\nu'} \left|\int_{x_1}^{x_2} dx A_{\nu' x}(x, 0, 0) \exp\left[i(\omega_{\nu'}/v_0)\, x\right]\right|^2, \quad (13.15)$$

where the points x_1 and x_2 are the limits of the cavity along the x-axis.

The first two terms in (13.15) result from the quantum nature of the motion of the electron, and the last term from the quantum properties of the radiation field. Let us examine more closely those terms in the electron velocity distribution which are connected with the quantum nature of the motion of the electron. These terms, generally speaking, are greater than the last

† The quantity $E_0 A_\nu$ has the dimensions of an electric field. We have introduced the quantity E_0 so that our formulae and Senitzky's have the same form. Senitzky (1954, 1955) has normalized the vector potential to unity and we (see Chapter I) have normalized it to $4\pi c^2$.

term in (13.15) (and they can be made as large as one likes by an appropriate choice of b^2). Senitzky (1954, 1955), who derived expressions (13.14) and (13.15), states that most of the dispersion of the electron velocity is connected with the wave nature of the motion of electrons. It is easy to see, moreover, that it should not be necessary to allow for the wave properties of the electron. In practice the wavelength of an electron with an energy $K \geqq 10$ eV is $\lesssim 10^{-8}$ cm. The wave packets (bunches) of electrons which can be obtained in practice are always very much longer than this wavelength; because of this the classical treatment of the electron motion would appear to be adequate. In addition the electron motion under discussion is free (uniform and rectilinear) in the first approximation and the blurring of the corresponding wave packets corresponds to the blurring of a "swarm" of classical particles. Senitzky (1954, 1955) nevertheless states that the motion of electrons is not classical. We shall now show (Ginzburg and Fain, 1957a, 1957b; Fain, 1958) why we do not agree with this statement and that there are essentially no quantum effects in the approximation used. In order to do this we shall derive (13.14) and (13.15) (without the last term) by a classical method.

13.2. The Classical Treatment of the Motion of Electrons (Ginzburg and Fain, 1957a, 1957b; Fain, 1958)

Let the initial state of an electron be given by the classical probability distribution

$$R(\boldsymbol{r}_0, t = 0)\, d^3\boldsymbol{r}_0 = \frac{1}{\pi^{3/2} b^3}\, e^{-x_0^2/b^2}\, d^3\boldsymbol{r}_0. \tag{13.16}$$

We write the possible initial velocities of the electron in the form $\boldsymbol{v} = \boldsymbol{v}_0 + \boldsymbol{w}$, where $\boldsymbol{w}$ is a random quantity whose probability distribution is of the form

$$Q(\boldsymbol{p})\, d^3\boldsymbol{p} = \frac{b^3}{f^3 \pi^{3/2}}\, e^{-\frac{b^2 p^2}{f^2}}\, d^3\boldsymbol{p}, \quad \boldsymbol{p} = m\boldsymbol{w}, \tag{13.17}$$

where b and f are constants.

The scatter of the initial positions and momenta of the electrons described by (13.16) and (13.17) may have many causes. In particular this scatter may be due to the cathode temperature, apparatus effects, etc.

In order to find the mean velocity and mean square deviation of the velocity we make use of the classical equations of motion

$$m\frac{dv}{dt} = -\frac{e}{mc}\sum_{\nu'} p_{\nu'}(t_1)\, A_{\nu'}(t_1). \tag{13.18a}$$

Integrating these equations we find

$$\boldsymbol{v}(t) = \boldsymbol{v}_0 - \frac{e}{mc}\int_0^t \sum_{\nu'} p_{\nu'}(t_1)\, \boldsymbol{A}_{\nu'}(t_1)\, dt_1, \tag{13.18b}$$

where

$$\boldsymbol{A}_\nu(t_1) = \boldsymbol{A}_\nu(\boldsymbol{r}(t_1)).$$

Since the interaction of the electron with the field is taken to be small we put $\boldsymbol{A}_\nu(\boldsymbol{r}(t)) = \boldsymbol{A}_\nu(\boldsymbol{r}_0 + (\boldsymbol{v}_0 + \boldsymbol{w})\,t)$. Then, allowing for the very small value of $\boldsymbol{r}_0 + \boldsymbol{w}t$ compared with $\boldsymbol{v}_0 t$ (the initial distributions are assumed to have sufficiently narrow limits), we have

$$\overline{\boldsymbol{A}_\nu(t)} \approx \{[1 + \delta_1^2(t)\,\nabla^2]\,\boldsymbol{A}_\nu(\boldsymbol{r})\}_{\boldsymbol{r}=\boldsymbol{v}_0 t} = \left[1 - \frac{\omega_\nu^2}{c^2}\,\delta_1^2(t)\right]\boldsymbol{A}_\nu(\boldsymbol{v}_0 t),$$

$$\delta_1^2(t) = \frac{b^2}{4} + \left(\frac{ft}{2mb}\right)^2. \tag{13.19}$$

The derivation of (13.19) is similar to the derivation of (13.13) and the bar denotes averaging over the distributions (13.16) and (13.17).

If only one oscillation of the type ν is excited in the resonator, we can write

$$p_\nu(t) = -\delta_{\nu'\nu} c E_0 \sin\omega_\nu t.$$

Remembering (13.19), and using the distributions (13.16) and (13.17) to average (13.18b), we obtain

$$\overline{\boldsymbol{v}(t)} = \boldsymbol{v}_0 + \frac{e}{m}E_0\int_0^t \boldsymbol{A}_\nu(\boldsymbol{v}_0 t_1)\sin\omega_\nu t_1\, dt_1$$

$$- \frac{e}{m}E_0\frac{\omega_\nu^2}{c^2}\int_0^t \delta_1^2(t)\,\boldsymbol{A}_\nu(\boldsymbol{v}_0 t_1)\sin\omega_\nu t_1\, dt_1. \tag{13.14b}$$

This expression is exactly the same as the expression (13.14) obtained by Senitzky (1954, 1955) if we make $\delta_1^2 = \delta^2$ or $f = \hbar$. This means that with the same initial distributions of the coordinates and momenta we obtain the same expressions for the mean value of $\boldsymbol{v}$. Let us now find the velocity dispersion at a time t. We make

$$v_x(t) = v_x(0) + \Delta v_x, \tag{13.20}$$

where†

$$\Delta v_x = -\frac{e}{mc}\int_0^t A_\nu(t_1)\, p_\nu(t_1)\, dt_1.$$

† In future we shall always write A_ν instead of $A_{\nu x}$.

Hence

$$\overline{[\Delta v_x(t)]^2} \equiv \overline{[v_x(t) - \overline{v_x(t)}]^2} = [\overline{v_x^2(0)} - (\overline{v_x(0)})^2]$$
$$+ [\overline{(\Delta v_x)^2} - (\overline{\Delta v_x})^2] + 2[\overline{(v_x(0)\,\Delta v_x)} - \overline{v_x(0)}\,(\overline{\Delta v_x})]. \tag{13.21}$$

From (13.17) we find

$$\overline{v_x^2(0)} - (\overline{v_x(0)})^2 = \frac{f^2}{2m^2b^2}. \tag{13.22}$$

We now calculate the second square bracket:

$$\overline{(\Delta v_x^2)} = \frac{e^2}{m^2c^2}\int_0^t dt_1 \int_0^t dt_2 \overline{A_\nu(t_1)\,A_\nu(t_2)}\;\overline{p_\nu(t_1)\,p_\nu(t_2)}.$$

Similarly to (13.19) we find

$$\overline{A_\nu(t_1)\,A_\nu(t_2)} = \left[1 + \left(\frac{b}{2}\right)^2(\nabla_1 + \nabla_2)^2 + \left(\frac{f}{2mb}\right)^2(t_1\nabla_1 + t_2\nabla)^2\right]$$
$$\times\, A_\nu(\boldsymbol{r}_1)\,A_\nu(\boldsymbol{r}_2)|_{r_1 = v_0 t_1;\, r_2 = v_0 t_2}.$$

Hence

$$\overline{(\Delta v_x)^2} - (\overline{\Delta v_x})^2 = \frac{e^2}{m^2}\int_0^t dt_1 \int_0^t dt_2 \Bigg\{2E_0^2 \sin\omega_\nu t_1 \sin\omega_\nu t_2$$
$$\times \left[t_1 t_2\left(\frac{f}{2mb}\right)^2 + \left(\frac{b}{2}\right)^2\right]\nabla_1\nabla_2 A_\nu(\boldsymbol{r}_1)\,A_\nu(\boldsymbol{r}_2)\Bigg\}_{r_1 = v_0 t_1;\, r_2 = v_0 t_2}.$$

After simplifying we obtain

$$\overline{(\Delta v_x)^2} - (\overline{\Delta v_x})^2 = 2E_0^2\,\frac{e^2}{m^2}\left\{\left[\frac{b}{2v_0}\int_{x_1}^{x_2} dx \sin\left(\frac{\omega_\nu}{v_0}x\right)\nabla A_\nu(\boldsymbol{r})\right]^2_{y=z=0}\right.$$
$$\left. + \left[\frac{f}{2mv_0^2 b}\int_{x_1}^{x_2} dx\left(x \sin\frac{\omega_\nu}{v_0}x\right)\nabla A_\nu(\boldsymbol{r})\right]^2_{y=z=0}\right\}. \tag{13.23}$$

Therefore we can obtain

$$\overline{v_x(0)\,\Delta v_x} = v_0\,\overline{\Delta v_x} + \frac{1}{2}\frac{f^2}{m^2b^2}\frac{e}{m}\int_0^t dt_1 p_\nu(t_1)\,t_1\,\frac{\partial}{\partial x}A_\nu(x, 0, 0) \tag{13.24}$$

and

$$\overline{v_x(0)}\;\overline{\Delta v_x} = v_0\,\overline{\Delta v_x}.$$

Using (13.21), (13.22), (13.23) and (13.24) we obtain the required expression for the velocity dispersion, which is the same as the first two terms of (13.15), derived from quantum theory, if we make $f = \hbar$.

Therefore in the approximation used by Senitzky (1954, 1955) quantum methods are used to solve the classical problem of the effect of the initial scatter of the position and velocity of the electrons on the velocity dispersion when electrons pass through a resonator. The only quantum aspect of this calculation is that if we make $f = \hbar$ in (13.17), as Senitzky (1954, 1955) does from the start, then distributions (13.16) and (13.17) correspond to a wave packet with, by virtue of the uncertainty relation, the minimum possible momentum scatter for a given length. This limitation on the initial distributions does not alter the classical nature of the calculation of the velocity dispersion. At the same time it has no particular value since in all practical cases in the field of electronics the momentum distribution will always be broader than the minimum quantum-mechanical distribution (3.51) with $f = \hbar$. In fact, even when $b = 10^{-3}$ cm the velocity scatter is $\hbar/mb \sim 10^3$ cm/sec, whilst the thermal velocity of the electron is $v \sim \sqrt{3kT/2m} \sim 4 \times 10^5\sqrt{T} \sim 4 \times 10^6$ cm/sec when $T = 10$ °K. We are not allowing for the action of a magnetic field on the electron; this has been done by Fogarassy (1959). It has been shown here that the quantum calculation of the velocity dispersion leads to the same result as the classical calculation if we take the initial scatter of the velocities and coordinates into consideration.

13.3. Estimates of the Quantum Dispersion

We now move on to discuss the last term in (13.15) connected with the quantum properties of the radiation field. It is not difficult to see that with a uniform radiation field, when $A_x(x, 0, 0)$ can be replaced by $\sqrt{4\pi c}/\sqrt{V}$ (from the normalization condition), the last term in (13.15) gives a value of the dispersion of the electron kinetic energy $\overline{(\Delta K_\tau)^2} = m^2 v_0^2 \overline{(\Delta v_\tau)^2}$ which agrees with (12.17), obtained by considering the quantum properties of a single eigen-oscillation. As can be seen from (13.15) all the eigen-oscillations make a contribution to the electron velocity dispersion

$$\overline{(\Delta v_x)^2} = \frac{e^2}{2m^2 v_0^2 c^2} \sum_\nu \hbar\omega_\nu \left| \int_{x_1}^{x_2} dx\, A_{\nu x}(x, 0, 0) \exp\left[i(\omega_\nu/v_0)\, x\right] \right|^2. \qquad (13.15\text{a})$$

The sum on the right-hand side of this equation diverges. This can easily be checked if we remember that at very high frequencies the spectrum of the eigen-oscillations of the resonator is the same as the spectrum of the eigen-oscillations of free space and the resonator eigenfunctions are the same as the eigenfunctions of free space. The divergence of the right-hand side of (13.15a) is connected with the fact that this formula allows for the fictitious interaction with the zero-point oscillations of a vacuum in free space as well as the real interaction with the eigen-oscillations of the field in the resonator. Let the diameter of the hole through which the electron passes be a. Then the resonator eigen-oscillations with wavelengths $\lambda \ll a$ behave as the eigen-oscillations of free space, so we should therefore not allow for the contribution made by these oscillations to (13.15a). On the other hand the eigen-oscillations with wavelengths $\lambda \sim a$ make a significant contribution to the electron velocity dispersion. The interaction with these oscillations is "cut in" at time $t_1 = x_1/v$ and out at time $t_2 = x_2/v$. This cannot be said of the interaction with the oscillations whose wavelengths are $\lambda \ll a$. At these wavelengths the electron does not know of the presence of the resonator. It interacts the whole time with the eigen-oscillations of free space—this interaction cannot be switched on or switched off. A steadily moving electron in free space does not radiate; interaction with a field in a resonator (when $\lambda \gtrsim a$) leads to radiation.

Therefore the summation in (13.15a) must be cut off at frequencies of the order of $\omega \approx 2\pi/a$. Senitzky (1956) gives the following rigorous procedure for doing this. If the resonator is placed in another resonator of volume V, then the eigen-oscillations due to the resonator are divided into three groups. The eigen-oscillations of group I have wavelengths of $\lambda \geqq a$ and are finite inside the smaller resonator. The eigen-oscillations in group II have the same wavelengths but are largely located outside the smaller resonator. Lastly, group III includes the eigen-oscillations with wavelengths of $\lambda < a$. These eigen-oscillations are finite through the volume V of the large resonator (and inside the smaller resonator). When $V \to \infty$ the eigen-oscillations in groups II and III become free space eigen-oscillations. Senitzky (1956) continues by making a canonical transformation when calculating the velocity dispersion in which the contribution of the group III eigen-oscillations approaches zero when $V \to \infty$. With this condition a contribution is made to the electron velocity dispersion by the differences of eigenfunctions $A_\nu^0 - A_\nu$ (where A_ν^0 are the eigenfunctions of free space and A_ν those of the resonator). The contribution made by the eigen-oscillations of group I is described by (13.15a).

We take $A_{\nu x}(x, 0, 0)$ to be constant along the path of the electron in order to estimate the velocity dispersion from (13.15a).

Then

$$\overline{(\Delta v_x)^2} = \frac{e^2}{m^2 v_0^2} \sum \frac{\hbar\omega_\nu}{2} \frac{4\pi d^2}{V} \left(\frac{\sin(\omega_\nu \tau/2)}{\omega_\nu \tau/2} \right)^2$$

$$= \frac{1}{m^2 v_0^2} \frac{e^2}{C} \sum_\nu \frac{\hbar\omega_\nu}{2} \left(\frac{\sin(\omega_\nu \tau/2)}{\omega_\nu \tau/2} \right)^2, \qquad (13.25)$$

$$C = \frac{V}{4\pi d^2}; \quad d = x_2 - x_1.$$

When $\omega_\nu \tau \ll 1$ we can write

$$\overline{(\Delta v_x)^2} = \frac{1}{m^2 v^2} \frac{e^2}{C} \sum_\nu \frac{\hbar\omega_\nu}{2}.$$

In order to estimate the quantity $\sum_{\nu} \hbar\omega_\nu/2$ we replace the sum by an integral as is done for the case of free space

$$\sum \frac{\hbar\omega_\nu}{2} = \frac{\hbar}{2}\int_{\omega_1}^{\omega_2} \frac{\omega_\nu^3 d\omega_\nu}{2\pi^2 c^3} V = \frac{\hbar V}{16\pi^2 c^3}(\omega_2^4 - \omega_1^4),$$

where ω_1 is the fundamental (lowest) eigenfrequency and $\omega_2 \approx 2\pi c/a$. For small enough a, $\omega_1 \ll \omega_2$ and

$$\sum \frac{\hbar\omega_\nu}{2} \approx \frac{\pi V}{2a^3}\hbar\omega_2 .$$

Therefore

$$\overline{(\Delta v_x)^2} \approx \frac{1}{m^2 v_0^2}\frac{e^2}{C}\frac{\pi V}{2a^3}\hbar\omega_2 = \frac{4\pi^2 d^2}{m^2 v_0^2}\frac{e^2}{2a^3}\hbar\omega_2 .$$

With a small enough hole diameter a the velocity dispersion due to the quantum properties of the field in the resonator may be significant. It should, however, be pointed out that allowing for the field fluctuations of black-body radiation at a temperature T leads to our having to make the substitution

$$\frac{h\omega_\nu}{2} \rightarrow \frac{h\omega_\nu}{2} + \frac{h\omega_\nu}{e^{\hbar\omega_\nu/kT} - 1}$$

in (13.25). In this case (when there is black-body radiation in the cavity) the quantum effects are significant when $\hbar\omega > kT$.

Published works sometimes use an expression for the dispersion of the electron energy that is different from the one given above. For example, Smith (1946), Schulman (1951), MacDonald and Kompfner (1949) and Farago and Marx (1955) use

$$\langle(\Delta K_\tau)^2\rangle = (\hbar\omega)^2 (N_+ + N_-), \tag{13.26}$$

where $N_+\hbar\omega = W_+$ and $N_-\hbar\omega = W_-$ are the total classical energies acquired and given up by the electron during acceleration and deceleration respectively, so that the total change in the electron energy in time τ is

$$\overline{K}_\tau - \overline{K}_0 = W_+ - W_- .$$

Here N_+ and N_- are interpreted as the average numbers of absorbed and emitted photons. It is easy to see that expression (13.26) does not agree with the expressions given above. For example, at any time after the beginning of the interaction the right-hand side of (13.26) cannot become zero, whilst $[\sin \frac{1}{2}\omega_\nu\tau/(\frac{1}{2}\omega_\nu\tau)]^2$ in (12.7) is equal to zero when $\omega_\nu\tau = k\pi$. The error in the papers quoted above is that the electron energy dispersion is derived not from quantum theory but by using incorrect statistical assumptions. For example, Farago and Marx (1955) assume that the elementary acts of an electron emitting quanta of an energy $\hbar\omega_\nu$ are statistically independent: then a dispersion $(\Delta(N_+ - N_-))^2 \sim N_+ + N_-$ is obtained and (13.26) follows from it.

To conclude the present chapter we should like to point out that in existing microwave devices quantum effects during the interaction of free electrons with the field in the resonators are obviously small and can be neglected. We cannot exclude the possibility, however, that quantum effects may prove to be significant in future, bearing in mind that work is progressing to higher and higher frequencies and lower and lower temperatures.

CHAPTER IV

The Behaviour of Quantum Systems in Weak Fields

THIS chapter describes the behaviour of quantum systems interacting with an external field. Sections 14–17 discuss the behaviour of quantum systems in the presence of a weak alternating field. The important characteristics of this behaviour are the response functions and susceptibility of the system, which are derived in these sections. The susceptibility is used to express simply such characteristics as the dielectric constant and magnetic permeability of matter. The susceptibility of a system possesses a number of general properties which can be derived without a knowledge of the actual form of the Hamiltonian. In particular the susceptibility can be connected with the spontaneous noise in the system (the fluctuation–dissipation theorem). The problem of the behaviour of quantum systems in strong fields is examined in the next chapter.

In sections 18–19 of the present chapter we derive the equations that describe the behaviour of multi-level, and in particular two-level, systems interacting with an external field. These equations are basic to the quantum electronic systems described in subsequent chapters.

In sections 20–21 we introduce the method of moments which is widely used in quantum electronics, and discuss spin–spin relaxation and cross-relaxation.

14. Susceptibility

14.1. The Response and Relaxation Functions

Let us examine an isolated quantum system whose Hamiltonian we shall denote by $\hat{H}$. We shall not make a distinction between the dynamic and dissipative sub-systems at present; the Hamiltonian $\hat{H}$ relates to the whole system.

Let given external forces $f_a(t)$ act on a system of this kind. In a sufficiently general case the effect of an external force can be represented as the perturbation energy†:

$$\hat{V} = -\sum_a f_a(t)\,\hat{x}_a \tag{14.1}$$

where $\hat{x}_a$ are operators describing the physical quantities of the system in question.

We shall assume that the external forces f_a are sufficiently small so that when solving the equation for the density matrix in the interaction representation (2.14) we can limit ourselves to first order perturbation theory (see section 5). In this case, assuming that the external forces act for long enough on the system, we can find from (2.14) the value of the density matrix at a time t:

$$\hat{\varrho}(t) = \hat{\varrho} - \frac{i}{\hbar}\int_{-\infty}^{t} [\hat{V}(t_1), \hat{\varrho}]\,dt_1, \tag{14.2}$$

where $\hat{\varrho}$ denotes the quantity $\hat{\varrho}(-\infty)$, the value of the density matrix at the time the external forces start to operate. The response of the system to the external forces can be expressed by the mean values of the quantities $\hat{x}_b$ that appear because of the action of the forces $f_a(t)$. We shall further assume that the quantities x_b are chosen so that their mean values in the unperturbed state $\hat{\varrho} \equiv \hat{\varrho}(-\infty)$ are zero:

$$\mathrm{Tr}\,\hat{\varrho}\hat{x}_b = 0. \tag{14.3}$$

Using the definition of the mean (1.8), and expressions (14.2) and (14.3), we find

$$\langle x_b \rangle = \sum_a \int_{-\infty}^{t} \varphi_{ba}(t, t') f_a(t')\,dt' = \sum_a \int_0^{\infty} \varphi_{ba}(t, t-\tau) f_a(t-\tau)\,d\tau, \tag{14.4}$$

where

$$\varphi_{ba}(t, t') = \frac{i}{\hbar}\,\mathrm{Tr}\,\{[\hat{x}_a(t'), \hat{\varrho}]\,\hat{x}_b(t)\} = \frac{i}{\hbar}\,\mathrm{Tr}\,\{[x_b(t), x_a(t')]\,\varrho\}$$
$$= \frac{i}{\hbar}\langle[\hat{x}_b(t), \hat{x}_a(t')]\rangle, \tag{14.5}$$

and $\hat{x}_b(t)$ and $\hat{x}_a(t')$ are operators in the Heisenberg representation (with the Hamiltonian $\hat{H}$)‡.

When writing (14.5) we made use of the commutation property of the operators in the trace sign: $\mathrm{Tr}\,\hat{A}\hat{B} = \mathrm{Tr}\,\hat{B}\hat{A}$. The function φ_{ba} expresses the

† In any case when the forces $f_a(t)$ are small (14.1) is general. We can then eliminate the non-linear terms of the type $f_a f_b x_a x_b$ etc. from the interaction energy.

‡ Or, what is the same thing in this case, the operators in the interaction representation.

response of the system to the action of the forces f_a in the form of the mean value $\langle x_b \rangle$. The function $\varphi_{ab}(t, t')$ will be called the response function or simply the *response*, following Kubo (1957). As can be seen from (14.5), the response is given by the quantities $\hat{x}_a(t)$, $\hat{x}_b(t)$ and $\hat{\varrho}$ relating to the free motion of the system (with no external force). The response determines the relaxation properties of the system. Suppose a certain perturbation f_{a0} acts on the system from time $t = -\infty$ up to $t = 0$. This perturbation leads to a non-equilibrium state. The change in the quantity $\langle x_b \rangle$ connected with the relaxation after removing the external perturbation is, in accordance with (14.4), of the form

$$\langle x_b(t) \rangle = f_{a0} \int_{-\infty}^{0} \varphi_{ba}(t, t')\, dt' = f_{a0} \int_{t}^{\infty} \varphi_{ba}(t, t - \tau)\, d\tau. \tag{14.6}$$

Therefore the function

$$\Phi_{ba}(t) = \int_{t}^{\infty} \varphi_{ba}(t, t - \tau)\, d\tau \tag{14.7}$$

describes the relaxation after the removal of the external perturbation. This function is called the relaxation function (Kubo, 1957).

Until now we have made no assumptions about the state of the system when the interaction is started. The system could be in a non-equilibrium state.

In future we shall assume that at $t = -\infty$, when the interaction starts, the system was in a stationary state, i.e. $[\hat{H}, \hat{\varrho}] = 0$. In a stationary state the correlation functions depend only on the difference of the times they contain (see section 10). It follows in particular from this that the function $\varphi_{ba}(t, t')$, which is expressed in terms of the difference of the correlations $\langle \hat{x}_b(t)\, \hat{x}_a(t') \rangle$ and $\langle \hat{x}_a(t')\, \hat{x}_b(t) \rangle$, will depend only on $t - t'$:

$$\varphi_{ba}(t, t - \tau) = \varphi_{ba}(\tau) = \frac{i}{\hbar} \langle [\hat{x}_b(\tau), \hat{x}_a] \rangle, \tag{14.8}$$

where $\hat{x}_a$ is a time-independent operator.

14.2. Susceptibility

Let us examine the special, but very important, case when the external force changes harmonically

$$f_a(t) = f_{a0} \cos(\omega t + \theta_a) = \operatorname{Re} f_{a0}\, e^{-i\omega t - i\theta_a}. \tag{14.9}$$

The mean value of the quantity x_b that results from this kind of force can be written in the form

$$\langle x_b(t) \rangle = \operatorname{Re} \{\chi_{ba}(\omega) f_{a0}\, e^{-i(\omega t + \theta_a)}\}, \tag{14.10}$$

where

$$\chi_{ba}(\omega) = \int_0^\infty \varphi_{ba}(\tau)\, e^{i\omega\tau} d\tau. \tag{14.11}$$

We shall call the quantity $\chi_{ba}(\omega)$ the susceptibility of the system. This quantity expresses the response of the system to a harmonic perturbation. We note that the susceptibility can be introduced as a function only of the frequency when the unperturbed motion is stationary in nature. On the other hand, φ_{ba} depends not only on τ but also on t and it is impossible to introduce a function only of the frequency $\chi_{ba}(\omega)$† which would express the response of the system to a harmonic perturbation.

Let us express the susceptibility in terms of the matrix elements of the operators $\hat{x}_a$ and $\hat{x}_b$. We shall use the system of eigenfunctions of the operator $\hat{H}$. In this representation the matrix $\hat{\varrho}$ can be made diagonal since, by the assumption of the stationary nature of the unperturbed state, $\hat{\varrho}$ commutes with $\hat{H}$. Using (14.8) and (14.11) we obtain

$$\chi_{ba}(\omega) = \int_0^\infty \frac{i}{\hbar} \operatorname{Tr} \{\hat{\varrho}[\hat{x}_b(\tau), \hat{x}_a]\}\, e^{i\omega t}\, d\tau$$

$$= \frac{1}{i\hbar} \int_0^\infty \sum_{n,k} \{\varrho_{nn} x_{ank} x_{bkn}\, e^{i\omega_{kn}\tau} - \varrho_{nn} x_{bnk} x_{akn}\, e^{i\omega_{nk}\tau}\}\, e^{i\omega\tau}\, d\tau$$

or, after replacing the suffixes k by n in the second term in the integrand, we arrive at the expression

$$\chi_{ba}(\omega) = \frac{1}{\hbar} \sum_{n,k} (\varrho_{nn} - \varrho_{kk})\, x_{ank} x_{bkn} \zeta(\omega - \omega_{nk}), \tag{14.12}$$

where the function $\zeta(x)$ is, as usual (see Appendix I)

$$\zeta(x) = \frac{P}{x} - i\pi\delta(x).$$

14.3. The Dielectric Constant and Magnetic Permeability Tensors. Allowing for Spatial Dispersion

Such characteristics of a medium as the complex dielectric constant and magnetic permeability tensors $\varepsilon_{ik}(\omega)$ and $\mu_{ik}(\omega)$ can be expressed in terms of the susceptibility. Let us examine, for example, the case when an electric field $\boldsymbol{E}(\boldsymbol{r}, t)$ acts upon the system. The interaction energy in this case can be

† It will also depend upon the time t.

written in the form (see (4.24))

$$\hat{V} = -\sum_k (\hat{\boldsymbol{d}}_k \cdot \boldsymbol{E}(\boldsymbol{r}_k, t)), \tag{14.13}$$

where $\boldsymbol{d}_k$ is the magnitude of the dipole moment of the kth particle.

Going over to a continuous distribution of dipole moments and introducing the polarization operator $\hat{\boldsymbol{P}}$ (the density of the dipole moments†) we can write the interaction energy (14.13) as follows:

$$\hat{V} = -\int (\hat{\boldsymbol{P}}(\boldsymbol{r}) \cdot \boldsymbol{E}(\boldsymbol{r}, t))\, dV = -\int \sum_{(i=x,y,z)} \hat{P}_i(\boldsymbol{r})\, E_i(\boldsymbol{r}, t)\, dV. \tag{14.13'}$$

Then by using the general expression (14.10), taking the suffixes a and b as $(i, \boldsymbol{r})$ and $(k, \boldsymbol{r}')$ and changing from summation to integration we can write the mean polarization due to a monochromatic electric field in the form

$$P_i(\boldsymbol{r}) = \sum_k \mathrm{Re}\, \{\textstyle\int \chi_{ik}(\omega, \boldsymbol{r}, \boldsymbol{r}')\, E_k(\boldsymbol{r}')\, \mathrm{e}^{-i\omega t} dV'\}. \tag{14.14}$$

Therefore the value of the polarization at a given point $\boldsymbol{x}$ is generally dependent on the value of the field at all points. In other words, there is a non-local spatial connection between the field and the polarization. This means that the spatial dispersion must be taken into consideration as well as the time-dispersion (which is expressed in terms of the frequency-dependence of the susceptibility). For a medium that is uniform and unbounded in space χ_{ik} in (14.14) depends only on the coordinate difference $\boldsymbol{r} - \boldsymbol{r}'$. In this case it is convenient to expand the field as a Fourier integral in the form of plane waves:

$$E_i(\omega, \boldsymbol{k})\, \mathrm{e}^{i(\boldsymbol{k}\cdot\boldsymbol{r}) - i\omega t}$$

(the phase factor is included in the complex amplitude E_i). For waves of this kind (14.14) becomes (omitting the Re sign)

$$P_i(\omega, \boldsymbol{k}) = \sum_k \chi_{ik}(\omega, \boldsymbol{k})\, E_k(\omega, \boldsymbol{k}),$$

where $P_i(\boldsymbol{k}, \omega)$ is the complex amplitude (when we have $\mathrm{e}^{i(\boldsymbol{k}\cdot\boldsymbol{r}) - i\omega t}$) and

$$\chi_{ik}(\omega, \boldsymbol{k}) = \int dV\, \mathrm{e}^{-i(\boldsymbol{k}\cdot\boldsymbol{r})} \chi_{ik}(\omega, \boldsymbol{r}). \tag{14.15}$$

We can now introduce the dielectric constant tensor $\varepsilon_{ik}(\omega, \boldsymbol{k})$. By definition

$$D_i(\omega, \boldsymbol{k}) = \sum_k \varepsilon_{ik} E_k(\omega, \boldsymbol{k}) = E_i(\omega, \boldsymbol{k}) + 4\pi P_i(\omega, \boldsymbol{k}).$$

It follows from this that

$$\varepsilon_{ik}(\omega, \boldsymbol{k}) = \delta_{ik} + 4\pi \chi_{ik}(\omega, \boldsymbol{k}). \tag{14.16a}$$

† $\int \hat{\boldsymbol{P}}\, dV = \sum_k \hat{\boldsymbol{d}}_k$.

The magnetic permeability tensor can be determined similarly:

$$\mu_{ik}(\omega, \boldsymbol{k}) = \delta_{ik} + 4\pi\chi_{ik}(\omega, \boldsymbol{k}). \tag{14.16b}$$

In the last expression χ_{ik} gives the response of the system to the magnetic field in the form of the mean magnetization. In future, unless otherwise stated, we shall not take the spatial dispersion into consideration. It must be borne in mind, however, that all the conclusions from the general expressions (14.10) and (14.12) can easily be generalized to allow for the spatial dispersion since, as we have seen, it is contained implicitly in these formulae (if we include the spatial coordinates in the suffixes *a*, *b*). (For more detail on taking the spatial dispersion into consideration see Silin and Rukhadze (1961).)

15. Symmetry Relations for the Susceptibility †

The symmetry properties of the susceptibility (14.12) are determined by the properties of the matrix elements. Let us first examine the case when a body has no "magnetic structure", i.e. the mean magnetization at each point is zero (unless a field is applied). We shall also assume that the body is not in a constant external magnetic field. Also let the quantities x_a and x_b be such that they remain unchanged when the sign of the time is changed. (The particle coordinates, and the electric dipole moment expressed in terms of them, possess this property.) In this case we can select a representation such that the matrix elements x_a and x_b are real‡. In fact, the Hamilton operator of a system of particles with a Coulomb interaction

$$\hat{H} = \sum_k \frac{\hat{p}_k^2}{2m_k} + \frac{1}{2} \sum_{i \neq k} \frac{e_i e_k}{r_{ik}}$$

is purely real in the coordinate representation (as are also the boundary conditions or the conditions at infinity). Therefore real eigenfunctions can be chosen for it. In this representation the operators $\hat{x}_a$ and $\hat{x}_b$ are real. It also follows from this that the matrix elements are real. Then by virtue of the Hermitian nature of the operators $\hat{x}_a$ and $\hat{x}_b$ we have

$$x_{ank} = x_{akn}, \quad x_{bnk} = x_{bkn}.$$

From these equalities and from (14.12) it follows that

$$\chi_{ba}(\omega) = \chi_{ab}(\omega). \tag{15.1}$$

† See also Landau and Lifshitz (1957), section 88, where these relations are derived in another way.

‡ It is obvious that the symmetry properties of $\chi_{ba}(\omega)$ do not depend upon the choice of representation, so it is sufficient to prove the symmetry in some arbitrarily selected representation.

The symmetry properties change if the body is in an external magnetic field $\boldsymbol{H}$. In this case, in accordance with (4.20), the Hamiltonian of the system can be written in the form

$$\hat{H} = \hat{H}_0 - \sum_k \frac{e_k}{2m_k c} ([\hat{\boldsymbol{r}}_k \wedge \hat{\boldsymbol{p}}_k] \cdot \boldsymbol{H}).$$

The expression for the eigenfunctions of this Hamiltonian takes the form

$$\hat{H}\Psi(\boldsymbol{H}) = E\Psi(\boldsymbol{H}).$$

We now carry out the following transformation in this equation. We replace both $\boldsymbol{H}$ by $-\boldsymbol{H}$ and i by $-i$. Since in the coordinate representation $p_k = -i\hbar(\partial/\partial x_k)$ the energy of the interaction with the magnetic field in this transformation does not change, so the Hamiltonian and its eigenvalues do not change either.

Therefore

$$\hat{H}\Psi^*(-\boldsymbol{H}) = E\Psi^*(-\boldsymbol{H}),$$

and we can select the eigenfunctions so that

$$\Psi^*(-\boldsymbol{H}) = \Psi(\boldsymbol{H}).$$

In this case we have for the matrix elements of the quantities that do not change sign when the sign of t changes

$$x_{ank}(\boldsymbol{H}) = x_{akn}(-\boldsymbol{H}) \quad \text{and} \quad x_{bkn}(\boldsymbol{H}) = x_{bnk}(-\boldsymbol{H}).$$

Making use once again of (14.12) we find in this case that

$$\chi_{ba}(\omega, \boldsymbol{H}) = \chi_{ab}(\omega, -\boldsymbol{H}). \tag{15.2}$$

It is also easy to find that if only one of the quantities x_a or x_b changes sign when the sign of t changes† (for example $\dot{x}$), then

$$\chi_{ba}(\omega) = -\chi_{ab}(\omega) \tag{15.3}$$

in the absence of a magnetic field and

$$\chi_{ba}(\omega, \boldsymbol{H}) = -\chi_{ab}(\omega, -\boldsymbol{H}), \tag{15.4}$$

when there is a magnetic field.

All the cases that have been discussed can be expressed in the following form:

$$\chi_{ba}(\omega, \boldsymbol{H}) = \varepsilon_a \varepsilon_b \chi_{ab}(\omega, -\boldsymbol{H}), \tag{15.5}$$

† If both the quantities x_a and x_b change sign when the sign of t changes relations (15.1) and (15.2) are still valid.

where the factors ε_a and ε_b are 1 or -1 depending upon whether the quantities x_a or x_b are even or odd when the sign of t changes. The symmetry properties of the susceptibility can easily be connected with the Onsager relations (see Landau and Lifshitz, 1957). When the sign of the frequency changes (Kubo, 1957) we have

$$\mathrm{Re}\,\chi_{ba}(\omega) = \mathrm{Re}\,\chi_{ba}(-\omega), \tag{15.6}$$

$$\mathrm{Im}\,\chi_{ba}(\omega) = -\mathrm{Im}\,\chi_{ba}(-\omega). \tag{15.7}$$

16. The Dispersion Relations

There is a very general relation which must be satisfied by the susceptibility of any stationary system. To obtain this relation we note the identity

$$\int_{\infty}^{-\infty} \frac{\zeta(x - \omega_{nk})}{x - \omega}\, dx = i\pi\zeta(\omega - \omega_{nk}). \tag{16.1}$$

This can easily be obtained if, for example, we write the ζ-function in the form

$$\zeta(x) = \lim_{\sigma \to 0} \left\{ \frac{x}{x^2 + \sigma^2} - i\pi \frac{\sigma}{x^2 + \sigma^2} \right\}$$

and substitute this expression in (16.1) and (14.12). We then obtain the so-called dispersion relation

$$\chi_{ba}(\omega) = \frac{1}{i\pi} \fint_{-\infty}^{\infty} \frac{\chi_{ba}(x)\, dx}{x - \omega}, \tag{16.2}$$

where the barred integral indicates the principal value. From (16.2) follow the expressions for the real χ'_{ba} and imaginary χ''_{ba} parts of the susceptibility ($\chi_{ba} = \chi'_{ba} + i\chi''_{ba}$):

$$\chi'_{ba} = \frac{1}{\pi} \fint_{-\infty}^{\infty} \frac{\chi''_{ba}(x)\, dx}{x - \omega}, \tag{16.3}$$

$$\chi''_{ba} = -\frac{1}{\pi} \fint_{-\infty}^{\infty} \frac{\chi'_{ba}(x)\, dx}{x - \omega}. \tag{16.4}$$

These relations are also called the Kramers–Kronig relations (found by them in 1927), or the dispersion relations. They are a consequence of analytical nature of the function $\chi_{ba}(\omega)$.

17. The Fluctuation–Dissipation Theorem

The fluctuation–dissipation theorem connects the fluctuations of the physical quantities in an unperturbed system, in equilibrium and without external forces, with the susceptibility, which in particular expresses the properties of the system when it is acted upon externally. The spectral density of the fluctuations of the product of two quantities is, in accordance with (10.10),

$$(x_a x_b)_\omega = \frac{1}{2\pi}\int_{-\infty}^{\infty} \Psi_{ab}(\tau)\, e^{i\omega\tau} d\tau, \tag{17.1}$$

where

$$\Psi_{ab}(\tau) = \tfrac{1}{2}\langle \hat{x}_a \hat{x}_b(\tau) + \hat{x}_b(\tau)\hat{x}_a\rangle \tag{17.2}$$

and

$$\tfrac{1}{2}\langle \hat{x}_{a\omega}\hat{x}_{b\omega'} + \hat{x}_{b\omega'}\hat{x}_{a\omega}\rangle = (x_a b_b)_{-\omega}\,\delta(\omega + \omega'). \tag{17.3}$$

The operators $\hat{x}_{a\omega}$ and $\hat{x}_{b\omega'}$ are defined by

$$\hat{x}_{a\omega} = \frac{1}{2\pi}\int_{-\infty}^{\infty} \hat{x}_a(t)\, e^{i\omega t}\, dt, \quad \hat{x}_{b\omega'} = \frac{1}{2\pi}\int_{-\infty}^{\infty} \hat{x}_b(t)\, e^{i\omega' t} dt. \tag{17.4}$$

Let us calculate the mean value $\tfrac{1}{2}\langle \hat{x}_{a\omega}\hat{x}_{b\omega'} + \hat{x}_{b\omega'}\hat{x}_{a\omega}\rangle$. For this we note the matrix element

$$(x_{a\omega})_{nm} = \frac{1}{2\pi}\int_{-\infty}^{\infty} x_{anm}\, e^{i(\omega_{nm}+\omega)t}\, dt = x_{anm}\delta(\omega_{nm} + \omega), \tag{17.5}$$

where x_{anm} is not time-dependent.

Substituting (17.5) and the corresponding expression $(x_{b\omega'})_{nm}$ in (17.3) we obtain after some simple manipulations with δ-functions

$$(x_a x_b)_\omega = \tfrac{1}{2}\sum_{n,k} (\varrho_{nn} + \varrho_{kk})\, x_{ank} x_{bkn} \delta(\omega - \omega_{nk}). \tag{17.6}$$

This formula gives the spectrum of the steady-state fluctuations.

Let us move on now to the dissipative properties. According to Landau and Lifshitz (1951) the change in the mean internal energy of the system is found from the partial time derivative of the complete Hamiltonian (including the perturbation energy):

$$\frac{dU}{dt} = \frac{\partial\langle \mathscr{H}\rangle}{\partial t},$$

where

$$\mathscr{H} = \hat{H} + \hat{V}.$$

Since only the perturbation $\hat{V}$ is explicitly time-dependent

$$\frac{dU}{dt} = \frac{\partial\langle V\rangle}{\partial t}.$$

Then since

$$\hat{V} = -\sum_a f_a \hat{x}_a$$

and x_a is explicitly time-independent

$$\frac{dU}{dt} = -\sum_a \dot{f}_a \langle x_a\rangle. \tag{17.7}$$

By definition of the susceptibility

$$\begin{aligned}\langle x_a\rangle &= \sum_b \mathrm{Re}\,\{\chi_{ab} f_{b0}\, \mathrm{e}^{-i(\omega t+\theta_b)}\}\\ &= \tfrac{1}{2}\sum_b \{\chi_{ab} f_{b0}\, \mathrm{e}^{-i(\omega t+\theta_b)} + \chi^*_{ab} f_{b0}\, \mathrm{e}^{i(\omega t+\theta_b)}\}.\end{aligned} \tag{17.8}$$

Substituting (17.8) in (17.7) and averaging over the period of the external force we obtain the following expression for the energy dissipation per unit time:

$$Q = \frac{i\omega}{4}\sum_{a,b} (\chi^*_{ab} - \chi_{ba}) f_{a0} f_{b0}\, \mathrm{e}^{i(\theta_b-\theta_a)}. \tag{17.9}$$

Therefore the expression $\chi^*_{ab} - \chi_{ba}$ determines the energy dissipation. When χ_{ba} is symmetric $-(\chi^*_{ab} - \chi_{ba})$ is simply twice the imaginary part of the susceptibility. From (14.12) we find that

$$\chi^*_{ab} - \chi_{ba} = \frac{2\pi i}{\hbar}\sum_{n,k} (\varrho_{nn} - \varrho_{kk})\, x_{ank} x_{bkn} \delta(\omega - \omega_{nk}). \tag{17.10}$$

Comparing this expression and (17.6) for $(x_a x_b)_\omega$ we notice that in the case when the system is in a state of thermodynamic equilibrium

$$\varrho_{nn}/\varrho_{kk} = \mathrm{e}^{-\hbar\omega_{nk}/kT} = \mathrm{e}^{-\hbar\omega/kT}$$

there is the following connection between $(x_a x_b)_\omega$ and $\chi^*_{ab}(\omega) - \chi_{ba}(\omega)$:

$$(x_a x_b)_\omega = \frac{i\hbar}{2\pi}(\chi^*_{ab} - \chi_{ba})\left\{\frac{1}{2} + \frac{1}{\mathrm{e}^{\hbar\omega/kT} - 1}\right\}. \tag{17.11}$$

This expression embodies the content of the Callen–Welton fluctuation and dissipation theorem (Callen and Welton, 1951; Bernard and Callen, 1959).

In the special case of the fluctuations of the single quantity x_a we have

$$(x_a^2)_\omega = \frac{\hbar\chi''_{aa}}{\pi}\left\{\frac{1}{2} + \frac{1}{e^{\hbar\omega/kT} - 1}\right\}, \tag{17.12}$$

where χ''_{aa} is the imaginary part of the susceptibility.

In particular it is easy to find the Nyquist formula (Nyquist, 1928) for the current fluctuations in an electric circuit from this

$$(I^2)_\omega = \frac{\hbar\omega}{\pi\,|Z|^2} R\left\{\frac{1}{2} + \frac{1}{e^{\hbar\omega/kT} - 1}\right\}, \tag{17.13}$$

where Z is the circuit impedance;

R is the resistance (Re Z). If we introduce the corresponding e.m.f. $\mathscr{E}_\omega = Z(\omega)\, I_\omega$, then we have

$$(\mathscr{E}_\omega^2) = \frac{\hbar\omega}{\pi} R\left(\frac{1}{2} + \frac{1}{e^{\hbar\omega/kt} - 1}\right). \tag{17.14}$$

In the classical case when $\hbar\omega \ll kT$

$$(\mathscr{E}^2)_\omega = \frac{kT}{\pi} R(\omega). \tag{17.15}$$

The following essential point must be made. As can be seen from the derivation, the fluctuation–dissipation theorem is valid for any value of the temperature T, and in particular for negative T. It can be easily checked from (14.12) that the imaginary part of χ''_{aa} is negative in this case. At the same time the brace in (17.12) is negative. This ensures that the quantity $(x_a^2)_\omega$ is positive. It must be remembered, however, that in a non-equilibrium state with a negative temperature the dynamic and dissipative sub-systems may have different temperatures. Therefore if we know the temperature of the dynamic sub-system, as is generally the case (the spin temperature), we can still say nothing about the temperature of the whole system (which is also contained in the fluctuation–dissipation theorem). The theorem cannot be directly applied to the dynamic sub-system since this system is not closed.

In conclusion we would point out that the spectrum of the equilibrium fluctuations fully determines the susceptibility. The value of $\chi^*_{ab} - \chi_{ba}$ is determined by the fluctuation–dissipation theorem and $\chi^*_{ab} + \chi_{ba}$ can be determined from $\chi^*_{ab} - \chi_{ba}$ via the Kramers–Kronig relation (16.2).

18. Multi-level Systems. The Absorption Line Shape

18.1. Initial Equations

When describing quantum systems we shall proceed from the transport equation in μ-space

$$\frac{\partial \sigma_{mn}}{\partial t} + \frac{i}{\hbar}[\hat{E} + \hat{\Gamma}, \hat{\sigma}]_{mn} = \sum_{k,l} (2\Gamma_{mkln}\sigma_{kl} - \Gamma_{klmk}\sigma_{ln} - \Gamma_{lnkl}\sigma_{mk}). \tag{8.3}$$

We shall then make a number of simplifying assumptions. We shall consider that the levels of the dynamic sub-system are non-degenerate and that the condition $|\omega_{mn}| \gtrsim \omega^*$ is satisfied. In this case the balance equations (8.10) follow from the transport equation (8.3), and the coefficients Γ_{mkln} are finite if

$$\omega_{mk} + \omega_{ln} = 0. \tag{18.1}$$

We shall further consider that the levels of the system are so arranged that condition (18.1) is not satisfied if m, k, l and n are all different levels, or even if three of them are different.

We can write the energy of the dynamic sub-system in the form

$$\hat{E} = \hat{E}_0 + \hat{V}(t), \tag{18.2}$$

where $\hat{V}(t)$ is the energy of interaction with an external field. We assume that the external field is small and that conditions (7.33) are satisfied. In this case the coefficients Γ are time-independent. Using all these assumptions and neglecting the term $\hat{\Gamma}$ in (8.3) (this can be done when $\hat{E}_0$ is sufficiently large) we arrive at the following set of equations:

$$\dot{\sigma}_{mn} + i\omega_{mn}\sigma_{mn} + \frac{i}{\hbar}[\hat{V}, \hat{\sigma}]_{mn} = \begin{cases} \sum_k w_{km}\sigma_{kk} - w_{mk}\sigma_{mm} & \text{when} \quad m = n, \\ -\dfrac{1}{\tau_{mn}}\sigma_{mn} & \text{when} \quad m \neq n, \end{cases} \tag{18.3}$$

where $w_{km} = 2\Gamma_{mkkm} = w_{mk}\, e^{-\hbar\omega_{mk}/kT}$† is the transition probability from the state k to the state m;

$$\frac{1}{\tau_{mn}} = \frac{1}{2}\sum_k (w_{mk} + w_{nk}) - 2\Gamma_{mmnn} + \Gamma_{mmmm} + \Gamma_{nnnn} = \frac{1}{\tau_{nm}}. \tag{18.4}$$

† This equality is valid if the dissipative system is in a state of thermal equilibrium.

18.2. The Shape of the Absorption Line

Using equations (18.3) we can find the susceptibility of the dynamic subsystem. The imaginary part of this susceptibility χ''_{aa} determines the absorption of energy from an external field. Therefore the dependence of χ''_{aa} on frequency determines the shape of the absorption line. If we state $\hat{V}$ in the form (14.1) we can find the density matrix σ_{mn} due to the action of the fields $f_a(t)$. Then, having found the mean value of the quantity x_b, (14.10) can be used to find $\chi_{ba}(\omega)$. According to the results of section 14, however, the susceptibility is entirely determined by the undisturbed motion (when there is no external force). To illustrate the statements proved in section 14 we shall proceed from equations (18.3) with $\hat{V} = 0$. The off-diagonal elements of the matrix σ_{mn} must satisfy

$$\left(\frac{\partial}{\partial t} + i\omega_{mn} + \frac{1}{\tau_{mn}}\right)\sigma_{mn} = 0. \tag{18.5}$$

The solution of this has the form

$$\sigma'_{mn} = \sigma'_{mn}(0) \exp\left(-i\omega_{mn}t - \frac{t}{\tau_{mn}}\right).$$

This solution is valid when $t > 0$. It can be shown (see Chapter II and Van Hove (1955) and Adams (1960)) that when $t < 0$ the coefficients τ_{mn} change their signs. Therefore the solution that is valid for all values of the time is of the form

$$\sigma'_{mn} = \sigma'_{mn}(0) \exp\left(-i\omega_{mn}t - \frac{|t|}{\tau_{mn}}\right). \tag{18.6}$$

We can now find $\chi_{aa}(\omega)$. For this it is sufficient to find the response function (14.8):

$$\varphi_{aa}(\tau) = \frac{i}{\hbar}\langle[\hat{x}_a(\tau), \hat{x}_a]\rangle. \tag{18.7}$$

We are considering the steady state, and the density matrix is diagonal. Let $\hat{x}_a$ be a quantity whose mean value in the steady state is zero (the diagonal elements of x_a are equal to zero). We can then make use of the results of section 10, case (b), and put (18.7) in the form

$$\varphi_{aa}(\tau) = \frac{i}{\hbar}\{\langle x'_a(\tau)\rangle - \langle x'_a(-\tau)\rangle\},$$

where the mean value $\langle x'_a(\tau)\rangle$ is found by using (18.6) and $\sigma'_{mn}(0) = x_{amn}\sigma_{nn}$.

$$\varphi_{aa}(\tau) = -\frac{i}{\hbar}\sum_{m,n}(\sigma_{mm} - \sigma_{nn})\,|x_{amn}|^2 \exp\left(-i\omega_{mn}\tau - \frac{|\tau|}{\tau_{mn}}\right). \tag{18.8}$$

From this, by using (14.11), we can find the susceptibility of the dynamic sub-system:

$$\chi_{aa}(\omega) = -\frac{i}{\hbar}\int_0^\infty \sum_{m,n} (\sigma_{mm} - \sigma_{nn})\, |x_{amn}|^2 \exp\left\{i(\omega - \omega_{mn})\tau - \frac{\tau}{\tau_{mn}}\right\} d\tau$$

$$= -\sum_{m,n} (\sigma_{mm} - \sigma_{nn}) \frac{|x_{amn}|^2}{\hbar} \frac{(\omega_{mn} - \omega) + (i/\tau_{mn})}{(\omega - \omega_{mn})^2 + \tau_{mn}^{-2}}. \qquad (18.9)$$

This expression defines the dispersion and absorption of the dynamic sub-system. We notice that this formula for the susceptibility (and the resulting shape of the absorption line) is a consequence of all the assumptions which led to the transport equation (18.3). Below we shall show that cases are possible (for example spin–spin relaxation) when the shape of the absorption line is quite different.

We shall now check the accuracy with which the shape of the absorption line (18.9) satisfies the fluctuation–dissipation theorem. To do this we use the results of section 10 (equation (10.25)) to find the spectral density of the fluctuations of the quantity x_a:

$$(x_a^2)_\omega = \frac{1}{2\pi}\int_{-\infty}^{\infty} \frac{1}{2}\{\langle x_a(\tau)\, x_a\rangle + \langle x_a x_a(\tau)\rangle\}\, e^{i\omega\tau}\, d\tau$$

$$= \frac{1}{2\pi}\sum_{m,n} (\sigma_{mm} + \sigma_{nn})\, |x_{amn}|^2 \frac{\tau_{mn}^{-1}}{(\omega - \omega_{mn})^2 + \tau_{mn}^{-2}}. \qquad (18.10)$$

Then by using the fluctuation–dissipation theorem (17.12) we find

$$\chi''_{aa}(\omega) = \sum_{m,n} (\sigma_{mm} + \sigma_{nn}) \frac{e^{\hbar\omega/kT} - 1}{e^{\hbar\omega/kT} + 1} \frac{|x_{amn}|^2}{\hbar} \frac{\tau_{mn}^{-1}}{(\omega - \omega_{mn})^2 + \tau_{mn}^{-2}}. \qquad (18.11)$$

It is not difficult to check that in a state of thermodynamic equilibrium (when the fluctuation–dissipation theorem holds) the imaginary parts of (18.9) and (18.11) are the same if

$$|\omega - \omega_{mn}| \ll \frac{kT}{\hbar}. \qquad (18.12)$$

This condition is the condition of applicability of the transport equation (18.3).

We first note that in the approximation we are discussing $kT/\hbar \gtrsim \omega^*$ (see the footnote on p. 77. Therefore condition (18.12) can be rewritten in the form

$$|\omega - \omega_{mn}| \ll \omega^*. \qquad (18.12')$$

If this condition is not satisfied and $|\omega - \omega_{mn}| \gtrsim \omega^*$, then expression (18.11) and the imaginary part of (18.9) are of the order of τ_{mn}^{-1}/ω^*. On the other hand, as can be seen from the derivation of the transport equation (section 7), it is valid with an accuracy up to terms of zero order in τ_{mn}^{-1}/ω^*. This means that if as a result of the solution of the transport equation a term of the order of τ_{mn}^{-1}/ω^* is obtained these terms must be made equal to zero. Therefore in those cases when (18.12) is not satisfied the imaginary part of (18.9) and (18.11) once again agree. In the approximation used they are equal to zero.

19. Two-level Systems

19.1. The Ideal Case of a Two-level System

In many problems connected with radiation it is possible to consider only two energy levels of the system. We therefore come to the ideal case of a two-level quantum object or two-level molecule. This idealization can be made if the object in question has two levels whose difference is

$$E_2 - E_1 \approx \hbar\omega,$$

where ω is the emission frequency and the other levels are spaced so that their difference is not close to $\hbar\omega$.

It must be assumed in addition that the transition probabilities from other levels to the given ones, and from the given levels E_1 and E_2 to others, are sufficiently small. This situation occurs, for example, if 1 and 2 are two low levels of a system and the remaining levels are high enough so that

$$\hbar\omega_{m1}, \quad \hbar\omega_{m2} \gg kT \quad (m > 2). \tag{19.1}$$

In this case the coefficients w_{1m} and w_{2m} (the transition probabilities to all levels m) are sufficiently small (see (8.12)). On the other hand, although w_{m1} and w_{m2} are not small, if condition (19.1) holds the equilibrium populations σ_{mm} are very small and so, as can be seen from (18.3), these levels need not be taken into consideration (unless, of course, the populations σ_{mm} are far from their equilibrium values).

It must be remembered that the reason why this ideal case of a two-level molecule can be considered is due to the purely quantum properties of the system. In fact classical and quasi-classical systems have quasi-equidistant spectra (see, e.g., Landau and Lifshitz, 1963, section 48). Because of this the approximate equality $E_2 - E_1 = \hbar\omega$ will be satisfied for many levels. In the special case of a harmonic oscillator, which has an equidistant energy spectrum, the above relation for the energy of a photon applies to all the levels of the system. Only in the case of quantum systems are there isolated pairs of levels whose difference is not even approximately the same as any other level difference of the same system.

19.2. Energy Spin

A molecule with two energy levels, and also a system of molecules, can be described in the same way no matter what the nature of these molecules. In physics we already have experience of describing systems which can be described by functions of a discrete variable which may take on two values. A spin of $\frac{1}{2}$ is one such variable. In the case of particles with a spin of $\frac{1}{2}$ the projection of the spin onto the z-axis may take up two values: $\frac{1}{2}$ and $-\frac{1}{2}$; the energy of particles of this kind in a magnetic field also takes up two values. There are accordingly two eigenfunctions which describe the states with spin projections $\frac{1}{2}$ and $-\frac{1}{2}$. The physical meaning of the spin operators is that they describe the spin (as opposed to the orbital) angular momentum. It turns out, however, that spin operators can be used when describing functions of such variables which, generally speaking, bear no relation to the spin momentum. For example the nucleon, which may be in two states (proton or neutron), can be described by isotopic spin operators (Fermi, 1951).

We introduce energy spin by analogy with isotopic spin (Dicke, 1954; Feynman, Vernon and Hellwarth, 1957; Fain, 1959a). The energy spin vector $\hat{\boldsymbol{r}}$ is defined by its three components: $\hat{r}_1, \hat{r}_2, \hat{r}_3$. This vector is not in ordinary space but in a space which we shall call energy space. Therefore $\hat{\boldsymbol{r}}$ has no directions in ordinary geometrical space. The components of the vector $\hat{\boldsymbol{r}}$ are defined by the commutation relations

$$\hat{\boldsymbol{r}} \times \hat{\boldsymbol{r}} = i\hat{\boldsymbol{r}}, \quad ([\hat{r}_1, \hat{r}_2] = i\hat{r}_3, [\hat{r}_2, \hat{r}_3] = i\hat{r}_1, [\hat{r}_3, \hat{r}_1] = i\hat{r}_2) \tag{19.2}$$

and the fact that in the representation in which $\hat{r}_3$ is diagonal the components $\hat{r}_1, \hat{r}_2$ and $\hat{r}_3$ take the form

$$\hat{r}_1 = \frac{1}{2}\begin{pmatrix} 0 & 1 \\ 1 & 0 \end{pmatrix}, \quad \hat{r}_2 = \frac{1}{2}\begin{pmatrix} 0 & -i \\ i & 0 \end{pmatrix}, \quad \hat{r}_3 = \frac{1}{2}\begin{pmatrix} 1 & 0 \\ 0 & -1 \end{pmatrix}. \tag{19.3}$$

A group of the four linear Hermitian operators $\hat{r}_1, \hat{r}_2, \hat{r}_3$ and the unit operator $\hat{I} = \begin{pmatrix} 1 & 0 \\ 0 & 1 \end{pmatrix}$ possesses the following property. The linear combinations of these operators are all the possible linear Hermitian operators for any function of two-valued variables. It is unimportant here whether these variables describe the spin, the two states of the nucleon or simply two energy levels of a molecule. We can demonstrate this property of the operators $\hat{r}_1, \hat{r}_2, \hat{r}_3$ and $\hat{I}$ by expressing an arbitrary linear Hermitian operator $\hat{O}$ relating to a two-level molecule in terms of the energy spin operators and $\hat{I}$:

$$\hat{\boldsymbol{O}} = \boldsymbol{a}\hat{I} + \boldsymbol{b}\hat{r}_1 + \boldsymbol{c}\hat{r}_2 + \boldsymbol{d}\hat{r}_3 = \begin{pmatrix} \boldsymbol{a} + \frac{1}{2}\boldsymbol{d} & \frac{1}{2}(\boldsymbol{b} - i\boldsymbol{c}) \\ \frac{1}{2}(\boldsymbol{b} + i\boldsymbol{c}) & \boldsymbol{a} - \frac{1}{2}\boldsymbol{d} \end{pmatrix}, \tag{19.4}$$

where ***a***, ***b***, ***c*** and ***d*** are four arbitrary vectors in ordinary space (c-numbers).

The right-hand side of (19.4) is just the representation of an arbitrary Hermitian operator of a quantity that can be described by only two eigenfunctions.

The operators $\hat{r}_1$, $\hat{r}_2$ and $\hat{r}_3$ possess all the properties of angular momentum operators. In particular if we have a system of two-level molecules we can introduce the total energy spin of this system:

$$\hat{\boldsymbol{R}} = \sum_j \hat{\boldsymbol{r}}_j.$$

The components of this total spin are subject to the ordinary commutation relations between the components of the angular momentum:

$$\hat{\boldsymbol{R}} \times \hat{\boldsymbol{R}} = i\hat{\boldsymbol{R}};\ [\hat{R}_1, \hat{R}_2] = i\hat{R}_3;\ [\hat{R}_2, \hat{R}_3] = i\hat{R}_1;\ [\hat{R}_3, \hat{R}_1] = i\hat{R}_2. \tag{19.5}$$

The non-zero matrix elements of the operators $\hat{\boldsymbol{R}}^2$, $\hat{R}_1$, $\hat{R}_2$ and $\hat{R}_3$ in the representation in which $\hat{R}_3$ and $\hat{\boldsymbol{R}}^2$ are diagonal are of the form:

$$\langle M, R\,|\hat{\boldsymbol{R}}^2|\,M, R\rangle = R(R+1), \tag{19.6}$$

$$\langle M, R\,|\hat{R}_3|\,M, R\rangle = M, \tag{19.7}$$

$$\langle M, R\,|\hat{R}_1 + i\hat{R}_2|\,M-1, R\rangle = \langle M-1, R\,|\hat{R}_1 - i\hat{R}_2|\,M, R\rangle$$

$$= \sqrt{(R+M)(R-M+1)}, \tag{19.8}$$

$$\langle M, R\,|\hat{R}_1|\,M-1, R\rangle = \langle M-1, R\,|\hat{R}_1|\,M, R\rangle$$

$$= \tfrac{1}{2}\sqrt{(R+M)(R-M+1)}, \tag{19.9}$$

$$\langle M, R\,|\hat{R}_2|\,M-1, R\rangle = -\langle M-1, R\,|\hat{R}_2|\,M, R\rangle$$

$$= -\frac{i}{2}\sqrt{(R+M)(R-M+1)}, \tag{19.10}$$

where M and R are the quantum numbers of the operators $\hat{R}_3$ and $\hat{\boldsymbol{R}}^2$ and are defined by relations (19.6) and (19.7).

Equalities (19.6)–(19.10) can be derived from the commutation relations (19.5) and the Hermitian property only. This derivation of the relations (19.6)–(19.10) (for the angular momentum operators) is contained in the book by Landau and Lifshitz (1963) (section 27).

We shall express the different operators relating to a two-level molecule in terms of the energy spin operators. The Hamilton operator of a two-level molecule, as can be easily seen, can be written in the form

$$\hat{E} = \hbar\omega_0 \hat{r}_3; \tag{19.11}$$

since $\hbar\omega_0\hat{r}_3$ and $\hat{E}$ have the identical eigenvalues $\hbar\omega_0/2$ and $-\hbar\omega_0/2$ these operators are the same†. (The quantity ω_0 is defined by the relation $E_2 - E_1 = \hbar\omega_0$.) We then express the operators $(1/c)\,\hat{\dot{d}}$ (the operator of the derivative of the dipole moment) and $\hat{\mu}$ (the magnetic moment operator) in terms of the spin operators:

$$\frac{1}{c}\,\hat{\dot{\boldsymbol{d}}} = \boldsymbol{a}\hat{I} + \boldsymbol{e}_1\hat{r}_1 + \boldsymbol{e}_2\hat{r}_2 + \boldsymbol{e}_3\hat{r}_3,$$
$$\hat{\boldsymbol{\mu}} = \boldsymbol{b}\hat{I} + \boldsymbol{m}_1\hat{r}_1 + \boldsymbol{m}_2\hat{r}_2 + \boldsymbol{m}_3\hat{r}_3. \tag{19.12}$$

The constants $\boldsymbol{a}$, $\boldsymbol{b}$, $\boldsymbol{e}_i$ and $\boldsymbol{m}_i$ can be determined by writing (19.12) in the representation in which $\hat{r}_3$ is diagonal (and the Hamilton operator $\hat{E}$ of the molecule is therefore diagonal). In this energy representation, using (19.3), we can express the constants $\boldsymbol{a}$, $\boldsymbol{b}$, $\boldsymbol{e}_i$ and $\boldsymbol{m}_i$ in terms of the matrix elements of the electric and magnetic dipole moments:

$$\frac{1}{2}(\boldsymbol{e}_1 - i\boldsymbol{e}_2) = -\frac{i\omega_0}{c}\,\boldsymbol{d}_{12}, \quad \frac{1}{2}(\boldsymbol{m}_1 - i\boldsymbol{m}_2) = \boldsymbol{\mu}_{12},$$
$$\frac{1}{2}(\boldsymbol{e}_1 + i\boldsymbol{e}_2) = \frac{i\omega_0}{c}\,\boldsymbol{d}_{21}, \quad \frac{1}{2}(\boldsymbol{m}_1 + i\boldsymbol{m}_2) = \boldsymbol{\mu}_{21}, \tag{19.13}$$
$$\boldsymbol{a} + \frac{1}{2}\boldsymbol{e}_3 = \frac{1}{c}(\dot{\boldsymbol{d}})_{11} = 0, \quad \boldsymbol{b} + \frac{\boldsymbol{m}_3}{2} = \boldsymbol{\mu}_{11},$$
$$\boldsymbol{a} - \frac{1}{2}\boldsymbol{e}_3 = \frac{1}{c}(\dot{\boldsymbol{d}})_{22} = 0, \quad \boldsymbol{b} - \frac{\boldsymbol{m}_3}{2} = \boldsymbol{\mu}_{22}.$$

It follows from these equalities that $\boldsymbol{a} = \boldsymbol{e}_3 = 0$. In future we shall assume that $\boldsymbol{b} = 0$. This does not limit the generality of the discussion. In fact it follows from (19.12) that the term which contains $\boldsymbol{b}$ makes a constant contribution to the dipole moment (since the unit operator $\hat{I}$ of which $\boldsymbol{b}$ is a coefficient does not depend on the choice of representation and is therefore not time-dependent in the Heisenberg representation). In the special case when $\hat{\boldsymbol{\mu}}$ is the spin magnetic moment (associated with the eigenvalue of the spin) (see (4.23)):

$$\boldsymbol{\mu} = 2\beta_0\boldsymbol{s},$$

we have

$$\boldsymbol{m}_1 = 2\beta_0\boldsymbol{i}, \quad \boldsymbol{m}_2 = 2\beta_0\boldsymbol{j}, \quad \boldsymbol{m}_3 = 2\beta_0\boldsymbol{k} \tag{19.14}$$

(here the z-axis is chosen along the external magnetic field or, in the general case, along the quantization axis).

† We have arbitrarily taken the energy of the ground state to be $-\hbar\omega_0/2$.

Let us now move on to discuss a system of two-level molecules. An ideal gas made up of this type of molecule can be described by the Hamiltonian

$$\hat{H} = \sum_j \hat{E}_j = \hbar\omega_0 \sum_j \hat{r}_{3j} = \hbar\omega_0 \hat{R}_3, \tag{19.15}$$

where $\hat{r}_{3j}$ is the third energy spin component of the jth molecule.

The eigenfunctions of this Hamiltonian can be selected in the form of the products of the eigenfunctions of the Hamiltonians of the individual molecules

$$\Psi = \prod_j \Psi_{n_j}(j), \tag{19.16a}$$

where the suffixes n_j can take up the two values: $\pm\frac{1}{2}$, and $\Psi_{n_j}(j)$ are the eigenfunctions of the operators $\hat{r}_{3j}$.

The quantum number M introduced in (19.7) is

$$M = \frac{N_+ - N_-}{2}, \tag{19.17}$$

where N_+ and N_- are the populations of the upper and lower levels of the molecule respectively.

When the operator $\hat{R}_3 = \sum \hat{r}_{3j}$ acts on the function (19.16a) it multiplies it by $(N_+ - N_-)/2$ since N_+ is the number of functions of the type $\Psi_{1/2}(j)$ and N_- is the number of functions of the type $\Psi_{-1/2}(j)$. Therefore the operator $2\hat{R}_3$ is the operator of the population difference of the system (or, as it is sometimes called, of the number of active molecules). Eigenfunctions of the type (19.16a), generally speaking, are not eigenfunctions of the operator $\hat{\boldsymbol{R}}^2 = \hat{R}_1^2 + \hat{R}_2^2 + \hat{R}_3^2$. In what follows we shall be interested in eigenfunctions which are simultaneously eigenfunctions of $\hat{R}_3$ and $\hat{\boldsymbol{R}}^2$:

$$\hat{\boldsymbol{R}}^2 \Psi_{RM} = R(R+1)\,\Psi_{RM}, \quad \hat{R}_3 \Psi_{RM} = M\Psi_{RM}. \tag{19.18}$$

The eigenfunctions Ψ_{RM} can be found by the usual spin operator technique. In the general case these functions are a superposition of functions of the type (19.16a). To illustrate this we consider the particular case of a system of two molecules. The complete system of eigenfunctions of such a system can be expressed in the form:

$$\Psi_1 = \Psi_{1/2}(1)\,\Psi_{1/2}(2), \quad \Psi_3 = \Psi_{-1/2}(1)\,\Psi_{1/2}(2), \tag{19.16b}$$

$$\Psi_2 = \Psi_{1/2}(1)\,\Psi_{-1/2}(2), \quad \Psi_4 = \Psi_{-1/2}(1)\,\Psi_{-1/2}(2).$$

It can be shown that the eigenfunctions Ψ_{RM} are of the form

$$\Psi_{1,1} = \Psi_{1/2}(1)\,\Psi_{1/2}(1),$$

$$\Psi_{1,0} = \frac{1}{\sqrt{2}}\,(\Psi_{1/2}(1)\,\Psi_{-1/2}(2) + \Psi_{1/2}(2)\,\Psi_{-1/2}(1)),$$

$$\Psi_{1,-1} = \Psi_{-1/2}(1)\,\Psi_{-1/2}(2), \tag{19.19}$$

$$\Psi_{0,0} = \frac{1}{\sqrt{2}}\,(\Psi_{1/2}(1)\,\Psi_{-1/2}(2) - \Psi_{1/2}(2)\,\Psi_{-1/2}(1)).$$

It follows from the properties of the spin quantum numbers that the quantum numbers M have the upper limit

$$|M| \leqq R. \tag{19.20}$$

It follows from this that R in its turn must satisfy the relation

$$R \leqslant \frac{N}{2} \equiv \frac{N_+ + N_-}{2}. \tag{19.21}$$

The components $\hat{R}_1$ and $\hat{R}_2$ basically determine the oscillatory part of the dipole moment of the system.

In order to understand clearly the significance of the quantities $\hat{R}_1$, $\hat{R}_2$, $\hat{R}_3$ and $\hat{\boldsymbol{R}}^2$ we need only look at the special case when the energy spin is the same as the ordinary spin—this is the precession in a magnetic field of the magnetic moment associated with the ordinary spin. In this case R_3 is proportional to M_z (the component of the magnetic moment), R_1 and R_2 are proportional to M_x and M_y and $\hat{\boldsymbol{R}}^2$ is proportional to $\boldsymbol{M}^2$. It is well known that the equation of motion of a magnetic moment precessing in a magnetic field $\boldsymbol{H}$ (neglecting loss) is of the form

$$\dot{\boldsymbol{M}} = \gamma[\boldsymbol{M} \wedge \boldsymbol{H}], \tag{19.22}$$

where $\gamma = 2\beta_0/\hbar$, the gyromagnetic ratio†.

A similar equation can be obtained for the energy spin.

19.3. *The Energy Spin Equations of Motion*

We shall find the Hamiltonian of a system of identical two-level molecules interacting with a radiation field. It is assumed that the dimensions of each molecule are small compared with the wavelength of the radiation, i.e. we

† The equation (19.22) can be obtained easily if we remember that $\boldsymbol{M} = \gamma \boldsymbol{L}$, where $\boldsymbol{L}$ is the angular momentum of the system. The change in this momentum per unit time is equal to the couple $\dot{\boldsymbol{L}} = [\boldsymbol{M} \wedge \boldsymbol{H}]$. (19.22) follows from this after multiplication by γ.

shall use the dipole approximation for the individual molecules. The dimensions of the whole system of molecules need not of course be small. In this case, using (4.13), (4.20) and (19.15), we have

$$\hat{H} = \hbar\omega_0 \sum_j \hat{r}_{3j} + \frac{1}{2} \sum_\nu (\hat{p}_\nu^2 + \omega_\nu^2 \hat{q}_\nu^2) - \sum_j \times \left(\frac{1}{c} (\hat{A}(j) \cdot \dot{\hat{d}}_j) + (\hat{\mu}_j \cdot \hat{H}(j)) \right).$$

Then by using (19.12) and (19.13) we can write down the Hamiltonian in terms of the spin variables and the field operators:

$$\hat{H} = \hbar\omega_0 \hat{R}_3 + \tfrac{1}{2} \sum_\nu (\hat{p}_\nu^2 + \omega_\nu^2 \hat{q}_\nu^2) - \sum_j \hat{A}(j)\,(e_1 \hat{r}_{1n} + e_2 \hat{r}_{2j}) - \sum_j (\hat{H}(j) \cdot [m_1 \hat{r}_{1j} + m_2 \hat{r}_{2j} + m_3 \hat{r}_{3j}]), \tag{19.23a}$$

where

$$\hat{A}(j) = \sum_\nu \hat{q}_\nu A_\nu(j); \quad \hat{H}(j) = \operatorname{curl} \hat{A}(j) = \sum \hat{q}_\nu \operatorname{curl} A_\nu(j) = -\sum \omega_\nu \hat{q}_\nu H_\nu(j).$$

The Hamiltonian (19.23a) can also be written in the form

$$\hat{H} = \tfrac{1}{2} \sum_\nu (\hat{p}_\nu^2 + \omega_\nu^2 \hat{q}_\nu^2) - \sum_j \hbar (\hat{r}_j \cdot \hat{K}(j)), \tag{19.23b}$$

where

$$K(j) = \left\{ \frac{(\hat{A}(j) \cdot e_1)}{\hbar} + \frac{(\hat{H}(j) \cdot m_1)}{\hbar};\ \frac{(\hat{A}(j) \cdot e_2)}{\hbar} + \frac{(\hat{H}(j) \cdot m_2)}{h};\ -\omega_0 + \frac{(\hat{H}(j) \cdot m_3)}{\hbar} \right\}. \tag{19.24}$$

The energy spin equations of motion and the field equations follow from the Hamiltonian (19.23b). The derivative of the operator $\hat{O}$ can be found by the rule (2.6):

$$\frac{d\hat{O}}{dt} = \frac{i}{\hbar} [\hat{H}, \hat{O}].$$

Using the commutation relations (19.2) and (3.17) we obtain the following set of operator equations:

$$\dot{\hat{r}}_j = [\hat{r}_j \wedge \hat{K}(j)], \tag{19.25}$$

$$\ddot{\hat{q}}_\nu + \omega_\nu^2 \hat{q}_\nu = \sum_j (A_\nu(j) \cdot (e_1 \hat{r}_{1j} + e_2 \hat{r}_{2j})) - \sum_j \omega_\nu (H_\nu(j) \cdot [m_1 \hat{r}_{1j} + m_2 \hat{r}_{2j} + m_3 \hat{r}_{3j}]). \tag{19.26}$$

In the case of a continuous energy spin distribution we can introduce the density $\hat{s}_1(\boldsymbol{r})$, $\hat{s}_2(\boldsymbol{r})$, $\hat{s}_3(\boldsymbol{r})$ of the spins at a given point $\boldsymbol{r}$. The total spin of the system is then

$$\hat{\boldsymbol{R}} = \int \hat{\boldsymbol{s}}(\boldsymbol{x})\, dV.$$

The equations of motion for the quantities $\hat{\boldsymbol{s}}$ and $\hat{q}_\nu$ become

$$\dot{\hat{\boldsymbol{s}}} = [\hat{\boldsymbol{s}} \wedge \hat{\boldsymbol{K}}], \tag{19.27}$$

$$\ddot{\hat{q}}_\nu + \omega_\nu^2 \hat{q}_\nu = \int (\boldsymbol{A}_\nu(\boldsymbol{x}) \cdot [\boldsymbol{e}_1 \hat{s}_1 + \boldsymbol{e}_2 \hat{s}_2])\, dV - \omega_\nu \int (\boldsymbol{H}_\nu(\boldsymbol{x}) \cdot [\boldsymbol{m}_1 \hat{s}_1 + \boldsymbol{m}_2 \hat{s}_2 + \boldsymbol{m}_3 \hat{s}_3])\, dV, \tag{19.28}$$

where

$$\hat{\boldsymbol{s}} = \hat{\boldsymbol{s}}(\boldsymbol{r}); \quad \hat{\boldsymbol{K}} = \hat{\boldsymbol{K}}(\boldsymbol{r}).$$

Equations (19.25)–(19.28) are the quantum equations of motion. By averaging these equations for the appropriate quantum state we can obtain the equations for the mean values. The equations for the mean values $\langle q_\nu \rangle \equiv q_\nu$ and $\langle \boldsymbol{s} \rangle = \boldsymbol{s}$ have the same form as equations (19.26) and (19.28) which are linear in the operators $\hat{\boldsymbol{s}}$ and $\hat{q}_\nu$. It is possible to make the transition from equations (19.25) and (19.27) to equations for the mean values that are of the same form if we assume that quantum correlations such as $\langle \hat{s}_1 \hat{K}_2 \rangle - s_1 K_2$, $\langle \hat{s}_2 \hat{K}_3 \rangle - s_2 K_3, \ldots$ can be neglected.

We must make one further remark on the derivation of equations (19.25)–(19.28). By using the interaction energy (4.20) we neglect the difference between the kinetic and canonical momenta in the interaction energy (see sub-section 4.3). It is interesting to follow the changes that appear in equations (19.25)–(19.28) when we take into consideration the difference between the kinetic and canonical momenta in the interaction energy. For the sake of simplicity we shall examine the interaction in the electric dipole approximation. In accordance with (4.18′) we have

$$\hat{V} = -\frac{1}{c}\left(\hat{\boldsymbol{A}}(0) \cdot \sum_k \frac{e_k \hat{\boldsymbol{p}}_k}{m_k}\right) + \hat{A}^2(0) \sum_k \frac{e_k^2}{2m_k c^2}, \tag{19.29}$$

where summation is carried out for the time being with respect to the coordinates of one molecule (whose dimensions are much less than a wavelength). Then we express the first term in terms of the energy spin

$$\frac{1}{c} \sum_k \frac{e_k \hat{\boldsymbol{p}}_k}{m_k} = \boldsymbol{e}_1 \hat{r}_1 + \boldsymbol{e}_2 \hat{r}_2 \tag{19.30}$$

and, by introducing summation over all the molecules, we obtain

$$\hat{V} = -\sum_j (\hat{\boldsymbol{A}}(j) \cdot [\boldsymbol{e}_1 \hat{r}_{1j} + \boldsymbol{e}_2 \hat{r}_{2j}]) + \sum_j \hat{A}^2(j) \sum_k \frac{e_{kj}^2}{2m_{kj} c^2}. \tag{19.31}$$

We notice that definition (19.30) is more rigorous than (19.12) since the derivative $\dot{\hat{\boldsymbol{d}}}$ is, generally speaking, a function of quantities relating not only to the two-level molecule but also to the field (see (4.19)), and only quantities relating to the two-level molecule can be expanded in terms of the spin operators, as is done in (19.30). Using the interaction energy (19.31) we find that equations (19.25) and (19.27), which give the motion of the spins in a given field, do not change, and the field equations become

$$\ddot{\hat{q}}_\nu + \omega_\nu^2 \hat{q}_\nu = \sum_j \left(\boldsymbol{A}_\nu(j) \cdot \left[(e_1 \hat{\boldsymbol{r}}_{1j} + e_2 \hat{\boldsymbol{r}}_{2j}) - \hat{\boldsymbol{A}}(j) \sum \frac{e_{kj}^2}{m_{kj} c^2} \right] \right). \tag{19.32}$$

The expression in square brackets is just

$$\frac{1}{c} \dot{\boldsymbol{d}}_j,$$

and can be expressed in terms of the kinetic momentum, whilst the equations of motion (19.25) and (19.27) contain the quantities $\hat{\boldsymbol{r}}_1$ and $\hat{\boldsymbol{r}}_2$ which, in accordance with (19.30), define the canonical momentum. In future we shall use unaltered equations such as (19.25)–(19.28). When these equations are applied to a practical case, however, we must remember that they are approximate in the sense described above.

19.4. Allowing for Dissipation

Dissipative processes are not taken into account in equations (19.25)–(19.28). In real systems there is dissipation because of losses in the resonator (in which the radiation field is contained) and various relaxation processes within the system of molecules. The losses in the resonator were allowed for in section 11. We shall now study dissipation within a system of molecules. First we shall treat the case when the energy spin is the same as the ordinary spin. This leads to the Bloch equations (Bloch, 1946; Wangsness and Bloch, 1953) which were originally applied to the relaxation of a system of nuclear spins. The lattice (or molecular environment) is a dissipative sub-system and the spins are a dynamic sub-system. This kind of spin relaxation is called spin–lattice relaxation to distinguish it from spin–spin relaxation, in which there is an exchange of energies between the spins (and not with the lattice)†. We shall now repeat in full the treatment of Wangsness and Bloch (1953) and give the derivation of the Bloch equations for the case of a system of spins of $\frac{1}{2}$ interacting with a lattice. The nature of the assumptions made by Wangsness and Bloch (1953) will be clear from this derivation.

† We shall discuss spin–spin relaxation in the next section.

The Hamiltonian of a system of nuclei interacting with a lattice is of the form

$$\hat{H} = -\gamma\hbar\Big(\boldsymbol{H}\cdot\sum_i \hat{\boldsymbol{I}}_i\Big) + \hat{F} - \gamma\hbar\sum_i \big(\hat{h}_z(i)\,\hat{I}_{zi} + \tfrac{1}{2}\hat{h}^-(i)\,\hat{I}_i^+ + \tfrac{1}{2}\hat{h}^+(i)\,\hat{I}_i^-\big), \quad (19.33)$$

where the external magnetic field is

$$\boldsymbol{H} = \boldsymbol{k}H_0 + \boldsymbol{H}_1(t) \quad (H_1(t) \ll H_0),$$

$\hat{\boldsymbol{I}}_i$ is the spin operator of the ith nucleus; $\hat{I}_i^{\pm} = \hat{I}_{ix} \pm i\hat{I}_{iy}$;
$\hbar\hat{F}$ is the lattice energy;
$\hat{\boldsymbol{h}}(i)$ is the magnetic field produced by the lattice at the position of the ith nucleus;
$(\hat{h}^{\pm} = \hat{h}_x \pm i\hat{h}_y)$
γ is the gyromagnetic ratio; the last term in (19.33) is the interaction energy of the nuclei with the lattice.

We notice that the interaction energy (19.33) is the most general expression for spins of $\frac{1}{2}$. Any linear Hermitian operator can be expanded in operators of spin $\frac{1}{2}$; the interaction energy (19.33) is just such an expansion (where $\hat{\boldsymbol{h}}(i)$ is the effective magnetic field). The operators $\hat{I}_{zi}$, $\hat{I}_i^{\pm}$ can be identified with the operators $\hat{v}_i^r$ introduced in section 8 (see (8.17)); the operator $\hat{I}_{zi}$ corresponds to the frequency $\omega_r = 0$; the operators $\hat{I}_i^{\pm}$ correspond to the frequencies $\pm\omega_0 = \mp\gamma H_0$. We obtain from (8.18a) the equations of motion for the mean value of the spin operator $\langle\hat{O}\rangle$†:

$$\frac{d\langle\hat{O}\rangle}{dt} = -i\gamma\sum_i \langle[(\boldsymbol{H}\cdot\hat{\boldsymbol{I}}_i), \hat{O}]\rangle + \sum_{r,s,i,i'} \Phi_{ii'}^{rs}\langle\hat{I}_{i'}^s[\hat{O}, \hat{I}_i^r] + [\hat{I}_{i'}^s, \hat{O}]\,\hat{I}_i^r\rangle. \quad (19.34)$$

Here r, s take up the values $+$, $-$ and 0 ($\hat{I} = \hat{I}^0$);

$$\Phi_{ii'}^{-+} = \frac{\pi\gamma^2}{4}\sum_{\alpha\alpha'} \delta(\omega_0 + \omega_{\alpha'\alpha})\,\langle\alpha\,|h^-(i)|\,\alpha'\rangle\,\langle\alpha'\,|h^+(i')|\,\alpha\rangle\,P_{\alpha'}$$

$$= \Phi_{i'i}^{+-}\,e^{\frac{\hbar\omega_0}{kT}}; \quad (19.35)$$

$$\Phi_{ii'}^{00} = \frac{\pi\gamma^2}{4}\sum_{\alpha\alpha'} \delta(\omega_{\alpha\alpha'})\,\langle\alpha\,|h_z(i)|\,\alpha'\rangle\,\langle\alpha'\,|h_z(i')|\,\alpha\rangle\,P_{\alpha'}; \quad (19.36)$$

$$\Phi_{ii'}^{++} = \Phi_{ii'}^{--} = \Phi_{ii'}^{0+} = 0. \quad (19.37)$$

If we substitute the operators $\hat{I}_i^r$ for $\hat{O}$ in the equations (19.34), then it is easy to see that the resultant equations connect the spin of the ith nucleus with the spins of the other nuclei (with the suffix i'). Therefore the individual

† For the sake of simplicity we shall not discuss the level shifts connected with Γ in (8.18a).

spins do not relax independently: there is a certain coherence. It can be seen from (19.35) and (19.36) that this coherence is connected with the correlations in the field produced by the lattice at different points i and i'. If we neglect these correlations, as can clearly be done for a sufficiently rarefied gas, a closed system of equations for the spin of the ith nucleus is obtained. In this approximation it is easy to derive the Bloch equation

$$\frac{\partial}{\partial t}\langle \boldsymbol{I}\rangle = \gamma\langle[\boldsymbol{I}\wedge \boldsymbol{H}]\rangle - \boldsymbol{i}\frac{I_x}{T_2} - \boldsymbol{j}\frac{I_y}{T_2} - \boldsymbol{k}\frac{I_z - I_0}{T_1}, \qquad (19.38)$$

where

$$T_1^{-1} = 2(\Phi_{ii}^{-+} + \Phi_{ii}^{+-}) = \Phi_{ii}^{+-}\left(1 + e^{\frac{\hbar\omega_0}{kT}}\right),$$

$$T_2^{-1} = \frac{1}{2}T_1^{-1} + \Phi_{ii}^{00}; \quad I_0 = \frac{n}{2}\tanh\frac{\hbar\omega_0}{2kT}; \qquad (19.39)$$

$\boldsymbol{I}$ is the spin moment per unit volume and n is the number of spins per unit volume.

The Bloch equation (19.38), as can be seen from the derivation, is valid for weak external alternating fields

$$H_1(t) \ll H_0 \quad \text{and} \quad \gamma H_1(t) \ll \omega^*.$$

When the external fields are strong we must allow for the dependence of the relaxation coefficients Γ_{mkln} (and thus T_1 and T_2) on the amplitude and frequency of the field. This dependence follows from the general formulae in section 8.

The energy spin system can be derived in a way similar to that given above for ordinary spins. To do this the interaction energy of a system of energy spins with a dissipative sub-system must be written in the form

$$\hat{V} = \sum_i (\hat{r}_{3i}\hat{F}_i^0 + \hat{r}_i^+\hat{F}_i^- + \hat{r}_i^-\hat{F}_i^+)$$
$$(\hat{r}^{\pm} = \hat{r}_1 \pm i\hat{r}_2). \qquad (19.40)$$

This form of the interaction energy follows from the general properties of spin operators.

Assuming further that the individual spins relax independently we arrive, just as in the case of ordinary spins, at the generalized Bloch equations:

$$\dot{\boldsymbol{s}} = \langle[\boldsymbol{s}\wedge \boldsymbol{K}]\rangle - \boldsymbol{i}\frac{s_1}{T_2} - \boldsymbol{j}\frac{s_2}{T_2} - \boldsymbol{k}\frac{s_3 - s_3^0}{T_1}. \qquad (19.41)$$

Here T_1 and T_2 are the effective relaxation times (which generally differ from (19.39)) and

$$s_3^0 = \frac{n}{2} \tanh \frac{\hbar\omega_0}{2kT},$$

where n is the number of spins per unit volume.

Therefore (19.41) describes the behaviour of a system of two-level molecules in an external field $\boldsymbol{K}$, taking account of dissipative processes of the spin–lattice relaxation type.

To conclude this section we shall write out the complete set of equations for the energy spins and the field, including the effect of relaxation processes (Fain, 1959a):

$$\dot{s}_1 + \omega_0 s_2 + \frac{1}{T_2} s_1 + \frac{1}{\hbar} s_3 [(\boldsymbol{A} \cdot \boldsymbol{e}_2) + (\boldsymbol{H} \cdot \boldsymbol{m}_2)] - s_2 \frac{(\boldsymbol{H} \cdot \boldsymbol{m}_3)}{\hbar} = 0, \quad (19.42)$$

$$\dot{s}_2 - \omega_0 s_1 + \frac{1}{T_2} s_2 - \frac{1}{\hbar} s_3 [(\boldsymbol{A} \cdot \boldsymbol{e}_1) + (\boldsymbol{H} \cdot \boldsymbol{m}_1)] s_3 + s_1 \frac{(\boldsymbol{H} \cdot \boldsymbol{m}_3)}{\hbar} = 0, \quad (19.43)$$

$$\dot{s}_3 = -\frac{1}{T_1}(s_3 - s_3^0) - \frac{1}{\hbar}(\boldsymbol{A} \cdot [\boldsymbol{e}_1 s_2 - \boldsymbol{e}_2 s_1]) - \frac{1}{\hbar}(\boldsymbol{H} \cdot [\boldsymbol{m}_1 s_2 - \boldsymbol{m}_2 s_1]), \quad (19.44)$$

$$\ddot{q}_\nu + \frac{\omega_\nu}{Q_\nu} \dot{q}_\nu + \omega_\nu^2 q_\nu = \int (\boldsymbol{A}_\nu(\boldsymbol{r}) \cdot [\boldsymbol{e}_1 s_1 + \boldsymbol{e}_2 s_2])\, dV$$

$$- \omega_\nu \int (\boldsymbol{H}_\nu(\boldsymbol{r}) \cdot [\boldsymbol{m}_1 s_1 + \boldsymbol{m}_2 s_2 + \boldsymbol{m}_3 s_3])\, dV. \quad (19.45)$$

The relaxation term in the last equation is written in a simplified form for the case where it can be assumed that $\gamma_{\nu\nu'}$ in (11.6) and (11.7) is finite only when $\nu = \nu'$.

20. The Method of Moments. Spin–spin Relaxation

20.1. The Method of Moments

The problem of finding the susceptibility of a system as a function of the frequency (and thus of finding the shape of the absorption line) proves to be too difficult to solve in a number of cases. The problem of finding the different moments of the absorption line is much simpler.

We shall define the moments of the absorption line by the expression

$$\langle \omega^s \rangle = \int \omega^s \chi''_{ab}(\omega)\, d\omega / [\int \chi''_{ab}(\omega)\, d\omega]. \quad (20.1)$$

In a number of cases it is possible to find these moments without knowing the imaginary part of the system's susceptibility $\chi''_{ab}(\omega)$. Using certain simplifying assumptions we can change (20.1) into a convenient form.

From (14.12) we find for the imaginary part $\chi''_{ab}(\omega)$ the expression†

$$\chi''_{ab} = \frac{\pi}{\hbar} \sum (\varrho_{kk} - \varrho_{nn})\, x_{bnk} x_{akn}\, \delta(\omega - \omega_{nk}).$$

Using the fact that $\hat{\varrho}$ is the equilibrium density matrix

$$\hat{\varrho} = \zeta\, \mathrm{e}^{-\hat{H}/kT}; \quad \zeta = [\mathrm{Tr}\, \mathrm{e}^{-\hat{H}/kT}]^{-1}, \tag{20.2}$$

we transform χ''_{ab} to the form

$$\chi''_{ab} = \frac{\zeta}{2\hbar} \sum_{n,k} \int_{-\infty}^{\infty} \left(1 - \mathrm{e}^{-\frac{h\omega}{kT}}\right) x_{bnk}\, \mathrm{e}^{-E_k/kT} x_{akn}\, \mathrm{e}^{i\omega_{kn}t}\, \mathrm{e}^{i\omega t} dt. \tag{20.3}$$

We shall further consider that the width of the absorption line is much less than the frequency ω_0 of the centre of the line. In this case we can replace $\mathrm{e}^{-\omega/kT}$ by $\mathrm{e}^{-\omega_0/kT}$ and (20.3) becomes

$$\chi''_{ab}(\omega) = \int_{-\infty}^{\infty} \eta_{ab}(t)\, \mathrm{e}^{i\omega t} dt, \tag{20.4}$$

where

$$\eta_{ab}(t) = \frac{\zeta}{2\hbar} \left(1 - \mathrm{e}^{\frac{h\omega_0}{kT}}\right) \mathrm{Tr} \left(\hat{x}_a(t)\, \hat{x}_b\, \mathrm{e}^{-\frac{\hat{H}}{kT}}\right).$$

From (20.4) we find

$$2\pi\eta_{ab}(t) = \int_{-\infty}^{\infty} \chi''_{ab}(\omega)\, \mathrm{e}^{-i\omega t} d\omega. \tag{20.5}$$

The sth moment of the absorption line is defined in terms of $\eta_{ab}(t)$ by

$$\langle \omega^s \rangle = (i)^s \eta^{(s)}_{ab}(0)/\eta_{ab}(0), \tag{20.6}$$

where $\eta^{(s)}_{ab}$ denotes the sth time derivative.

Let us examine the special case of high temperatures when $\mathrm{e}^{-\hat{H}/kT}$ in (20.4) is equal to unity. For the off-diagonal elements of η we have

$$\eta_{aa}(t) = \frac{\zeta\omega_0}{2kT} \mathrm{Tr}\, (\hat{x}_a(t)\, \hat{x}_a). \tag{20.7}$$

† For the sake of simplicity we shall assume that $\chi_{ab} = \chi_{ba}$; in this case the imaginary part χ''_{ab} determines the absorption.

From this and from (20.6), using the quantum equations of motion

$$\dot{\hat{x}}_a = \frac{i}{\hbar}[\hat{H}, \hat{x}_a],$$

we obtain for the second and fourth moments:

$$\langle\omega^2\rangle = -\frac{1}{\hbar^2}\frac{\mathrm{Tr}\,[\hat{H}, \hat{x}_a]^2}{\mathrm{Tr}\,(\hat{x}_a^2)}, \tag{20.8a}$$

$$\langle\omega^4\rangle = \frac{\mathrm{Tr}\,[\hat{H}, [\hat{H}, \hat{x}_a]]^2}{\hbar^4\,\mathrm{Tr}\,(\hat{x}_a^2)}. \tag{20.9a}$$

We can likewise obtain expressions for the higher moments. We shall now apply these results to find the moments of the absorption line in the case when the absorption is determined by a spin–spin interaction (Van Vleck, 1948).

20.2. Spin–spin Relaxation

In addition to spin–lattice relaxation, spin–spin relaxation processes are often significant. This type of relaxation occurs because of the interaction of spins with each other. In this case the dissipative system is the spin system itself, so it cannot be considered to be in equilibrium if we are investigating a relaxation process. The spin–spin relaxation process cannot be described by a simple transport equation because the condition for the upper limit of the inverse relaxation time compared with the characteristic frequency ω^* of the spin motion is not satisfied (see (7.12)). It can be shown that these quantities are of the same order.

The Hamiltonian of a system of spins in an external constant magnetic field $\boldsymbol{H}_0$ directed along the z-axis and an alternating field $h_x^0 \cos \omega t$ is of the form

$$\hat{H} = -\gamma\hbar H_0 \sum_j \hat{I}_{zj} + \sum_{k>j} \tilde{A}_{jk}(\hat{\boldsymbol{I}}_j \cdot \hat{\boldsymbol{I}}_k) + \gamma^2\hbar^2 \sum_{k>j} [r_{jk}^{-3}(\hat{\boldsymbol{I}}_j \cdot \hat{\boldsymbol{I}}_k)$$
$$- 3r_{jk}^{-5}(\boldsymbol{r}_{jk} \cdot \hat{\boldsymbol{I}}_j)(\boldsymbol{r}_{jk} \cdot \hat{\boldsymbol{I}}_k)] - \gamma\hbar h_x^0 \sum_j \hat{I}_{xj} \cos \omega t. \tag{20.10}$$

The first term of this expression is the Zeeman energy (the energy in the external field $\boldsymbol{H}_0$), the second and third terms are the exchange and dipole–dipole interaction operators respectively, and the last term is the energy of interaction with the external alternating field.

$\hat{\boldsymbol{I}}_j$ denotes the spin (electron or nuclear) operator of the atom j; r_{jk} is the distance between the jth and kth atoms (these distances are assumed to be fixed in the crystal lattice); γ is the gyromagnetic ratio. In what follows we shall be interested in absorption near the Larmor frequency

$$\omega_0 = \gamma H_0.$$

Absorption near this frequency is largely determined by that part of the Hamiltonian† (20.10) (without h_x^0) which is diagonal in the magnetic quantum number M (z is the component of the total spin). It can be shown that this part of the Hamiltonian (20.10) (without the external alternating field h_x) is of the form

$$\hat{H} = -\gamma\hbar H_0 \sum_j \hat{I}_{zj} + \sum_{k>j} A_{jk}(\hat{\boldsymbol{I}}_j \cdot \hat{\boldsymbol{I}}_k) + \sum_{k>j} B_{jk}\, \hat{I}_{zj}\hat{I}_{zk}. \tag{20.11}$$

Here

$$A_{jk} = \tilde{A}_{jk} + \gamma^2 r_{jk}^{-3}\left[\frac{3}{2}\gamma_{jk}^2 - \frac{1}{2}\right],$$
$$B_{jk} = -3\gamma^2 r_{jk}^{-3}\left[\frac{3}{2}\gamma_{jk}^2 = \frac{1}{2}\right], \tag{20.12}$$

where γ_{jk} is the direction cosine of $\boldsymbol{r}_{jk}$ with respect to the z-axis.

We are now in a position to find the moments of the line. We first note that it follows from the definition of the susceptibility (14.10) that x_a in (20.8) and (20.9) is the x-component of the total spin‡:

$$\langle\omega^2\rangle = -\frac{1}{\hbar^2}\frac{\mathrm{Tr}\,[\hat{H}, \hat{I}_x]^2}{\mathrm{Tr}\,\hat{I}_x^2}, \tag{20.8 b}$$

$$\langle\omega^4\rangle = \frac{\mathrm{Tr}\,[\hat{H}, [\hat{H}, \hat{I}_x]]^2}{\hbar^4 \mathrm{Tr}\,\hat{I}_x^2}. \tag{20.9 b}$$

The scalar product $(\hat{\boldsymbol{I}}_j \cdot \hat{\boldsymbol{I}}_k)$ commutes with the operator $\hat{I}_x = \sum \hat{I}_{xj}$. Therefore the second term in (20.11) has no effect on the second moment $\langle\omega^2\rangle$. To calculate $\langle\omega^2\rangle$ from (20.8b) we take into consideration the commutation relations for the spins

$$\hat{I}_{xj}\hat{I}_{yk} - \hat{I}_{yk}\hat{I}_{xj} = i\delta_{jk}\hat{I}_{zj}.$$

Omitting the intermediate steps we find (Van Vleck, 1948)

$$\mathrm{Tr}\,(\boldsymbol{I}_x)^2 = \tfrac{1}{3} NI(I+1)(2I+1)^N,$$

$$\hat{H}\hat{I}_x - \hat{I}_x\hat{H} = -\gamma H_0 i\hat{I}_y + i\sum_{k>j} B_{jk}(\hat{I}_{yj}\hat{I}_{zk} + \hat{I}_{yk}\hat{I}_{zj}),$$

$$\langle\omega^2\rangle = \omega_0^2 + \tfrac{1}{3} I(I+1)\hbar^{-2}\sum B_{jk}^2,$$

where N is the total number of atoms;

I is the spin quantum number of an individual atom.

† For more detail see Van Vleck (1948).

‡ These formulae were derived intuitively by Van Vleck (1948).

It follows from this that the mean square deviation of the frequency from the Larmor frequency $\omega_0 = \gamma H_0$ is of the form

$$\langle \Delta\omega \rangle^2 = \tfrac{1}{3} I(I+1) \hbar^{-2} \sum_k B_{jk}^2. \tag{20.13}$$

It is assumed that all the atoms are in identical sites, so the sum in (20.13) does not depend upon j. We shall not give here the calculation of the sum $\sum_k B_{jk}^2$ for different types of lattice (Van Vleck, 1948). We shall content ourselves with saying that for powders, but not for single crystals, (20.13) (allowing for spherical symmetry) becomes

$$\langle (\Delta\omega)^2 \rangle = \tfrac{3}{5} \gamma^4 \hbar^{-2} I(I+1) \sum_k r_{jk}^{-6}. \tag{20.14}$$

It is usually assumed when discussing spin–spin interaction that there is a Gaussian frequency distribution:

$$f(\omega) = \frac{1}{\sqrt{2\pi \langle \Delta\omega \rangle^2}} \exp\left[-\frac{(\omega - \omega_0)^2}{2 \langle \Delta\omega^2 \rangle} \right]. \tag{20.15}$$

Then the total width of the resonance curve, i.e. twice the frequency deviation from the mean when the intensity drops to half its maximum, is

$$\Delta\omega = 2 (\ln \sqrt{2})^{1/2} \sqrt{2} \langle (\Delta\omega)^2 \rangle^{1/2} = 1.65 \langle (\Delta\omega)^2 \rangle^{1/2}. \tag{20.16}$$

To justify the use of (20.15) it must be proved that this distribution leads to the correct expressions for all the moments. It is necessary in particular to obtain the correct quartic deviation which can be calculated by means of (20.9b). In this case, when the exchange interaction can be neglected, the deviation from a Gaussian form because of purely dipole interaction is small. (Of course, this statement is strictly valid only if we are not interested in higher moments of the absorption line than the fourth.) On the other hand taking the exchange interaction into consideration leads to an essentially non-Gaussian line; so-called exchange narrowing occurs (Van Vleck, 1948).

21. Cross-relaxation

The contents of this section do not entirely correspond to the title of the chapter, but we considered it best to concentrate into one chapter the description of all the phenomena connected with relaxation processes in spin systems; cross-relaxation is one of these.

When discussing spin–spin relaxation in section 20 we made use of the abbreviated Hamiltonian (20.11) which allows only for the diagonal part of the dipole–dipole interaction which commutes with the Zeeman energy operator, and for this reason is responsible for the relaxation only of the transverse† components of the magnetic (or spin) moments. We shall denote the characteristic relaxation time of the transverse components of the moment by T_2. The time T_2 is a measure of the time taken to establish equilibrium in a system of interacting spins. The off-diagonal part of the spin–spin interaction makes a negligibly small contribution to the relaxation of the transverse components of the magnetic moment, but it may make an extremely large contribution to the relaxation of the longitudinal component.

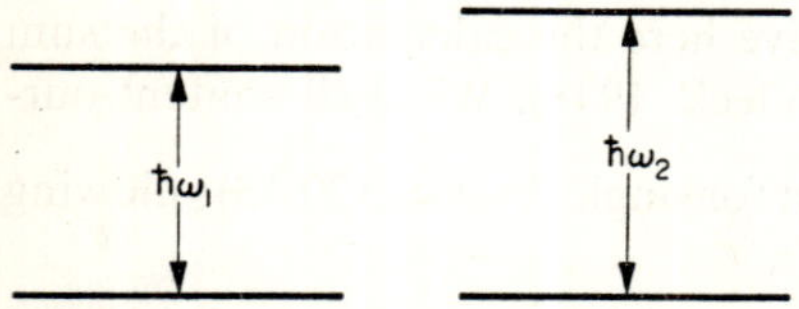

FIG. IV.1. Energy levels of two spins with close eigenfrequencies in an external magnetic field.

Longitudinal relaxation, which leads to a change in the magnetic moment component parallel to the field, is associated, as a rule, with spin–lattice interaction (section 19). We shall denote the characteristic time of this interaction by T_1. Under certain conditions, however, when for example there are two kinds of magnetic atoms with a spin of $\frac{1}{2}$ (Fig. IV.1), and the difference between the Zeeman frequencies is sufficiently small (comparable with T_2^{-1}), the change in the longitudinal component of the magnetic moment that occurs because of the establishment of an equilibrium population in the Zeeman levels of the system takes place in a time T_{12} such that

$$T_2 \ll T_{12} \ll T_1 .$$

The relaxation process consists of simultaneous reversing of the different spins, the change in the Zeeman energy being compensated for by a rearrangement of the dipole lattice (Bloembergen, Shapiro, Pershan and Artman, 1959). The off-diagonal part of the spin–spin interaction is responsible for processes of this kind (they are called cross-relaxation).

Here we shall examine in detail the example of cross-relaxation in a system consisting of two kinds of magnetic particles which have the same spin of $\frac{1}{2}$ but different gyromagnetic ratios γ. The Hamiltonian of such a system in an external field $\boldsymbol{H}_0$, which is parallel to the z-axis, is of the form

$$\begin{aligned} \hat{H} = & -\gamma_1 \hbar H_0 \sum_i \hat{I}_{zi} - \gamma_2 \hbar H_0 \sum_j \hat{I}_{zj} + \sum_{i>i'} A_{ii'}(\hat{\boldsymbol{I}}_i \cdot \hat{\boldsymbol{I}}_{i'}) + \sum_{j>j'} A_{jj'}(\hat{\boldsymbol{I}}_j \cdot \hat{\boldsymbol{I}}_{j'}) \\ & + \tfrac{1}{2} \sum_{ij} A_{ij}(\hat{\boldsymbol{I}}_i \cdot \hat{\boldsymbol{I}}_j) + \gamma_1^2 \hbar^2 \sum_{i>i'} [r_{ii'}^{-3}(\hat{\boldsymbol{I}}_i \cdot \hat{\boldsymbol{I}}_{i'}) - 3r_{ii'}^{-5}(\boldsymbol{r}_{ii'} \cdot \hat{\boldsymbol{I}}_i)(\boldsymbol{r}_{ii'} \cdot \hat{\boldsymbol{I}}_{i'})] \\ & + \gamma_2^2 \hbar^2 \sum_{j>j'} [r_{jj'}^{-3}(\hat{\boldsymbol{I}}_j \cdot \hat{\boldsymbol{I}}_{j'}) - 3r_{jj'}^{-5}(\boldsymbol{r}_{jj'} \cdot \hat{\boldsymbol{I}}_j)(\boldsymbol{r}_{jj'} \cdot \hat{\boldsymbol{I}}_{j'})] \\ & + \tfrac{1}{2}\gamma_1\gamma_2\hbar^2 \sum_{i,j} [r_{ij}^{-3}(\hat{\boldsymbol{I}}_i \cdot \hat{\boldsymbol{I}}_j) - 3r_{ij}^{-5}(\boldsymbol{r}_{ij} \cdot \hat{\boldsymbol{I}}_i)(\boldsymbol{r}_{ij} \cdot \hat{\boldsymbol{I}}_j)]. \end{aligned} \tag{21.1}$$

† We are considering directions relative to the external magnetic field when we speak transverse and longitudinal components of vectors.

The first two terms are the Zeeman energy of the spins in the magnetic field (the suffix i denotes spins of kind 1, and j spins of kind 2). The next three terms describe the exchange interaction of the two kinds of spin with themselves and each other and the last three describe the dipole–dipole interaction energy. The notation in (21.1) is the same as that of the preceding section. We shall use, as usual, a representation in which $\hat{I}_z$ is diagonal. From now on we shall discuss only external fields for which

$$\gamma_{1,2}H_0 \equiv \omega_{1,2} \gg T_2^{-1} \tag{21.2}$$

and

$$|\omega_1 - \omega_2| \geqq 2T_2^{-1}. \tag{21.3}$$

If we change to the variables $\hat{I}^{\pm} = \hat{I}_x \pm i\hat{I}_y$ we can transform the Hamiltonian (21.1) to the form

$$\hat{H} = \hat{H}_3 + \hat{H}_{\text{dip}} + \hat{\ }. \tag{21.4}$$

Here $\hat{H}_3$ is the Zeeman energy (the first two terms in the right-hand side of (21.1));

$$\begin{aligned} \hat{H}_{\text{dip}} = &\sum_{i>i'} [A_{ii'} + \gamma_1^2 r_{ii'}^{-3}(1 - 3\cos^2\theta_{ii'})]\, \hat{I}_{zi}\hat{I}_{zi'} \\ &+ \sum_{j>j'} [A_{jj'} + \gamma_2^2 r_{jj'}^{-3}(1 - 3\cos^2\theta_{jj'})]\, \hat{I}_{zj}\hat{I}_{zj'} \\ &+ \sum_{ij} \tfrac{1}{2}[A_{ij} + \gamma_1\gamma_2 r_{ij}^{-3}(1 - 3\cos^2\theta_{ij})]\, \hat{I}_{zi}\hat{I}_{zj} \qquad (21.5) \\ &+ \sum_{i>i'} [\tfrac{1}{2}A_{ii'} - \tfrac{1}{4}\gamma_1^2 r_{ii'}^{-3}(1 - 3\cos^2\theta_{ii'})]\, (\hat{I}_i^+\hat{I}_{i'}^- + \hat{I}_i^-\hat{I}_{i'}^+) \\ &+ \sum_{j>j'} [\tfrac{1}{2}A_{jj'} - \tfrac{1}{4}\gamma_2^2 r_{jj'}^{-3}(1 - 3\cos^2\theta_{jj'})]\, (\hat{I}_j^+\hat{I}_{j'}^- + \hat{I}_j^-\hat{I}_{j'}^+) \end{aligned}$$

is the diagonal part of the Hamiltonian for the dipole–dipole and exchange interactions; θ_{ij} is the angle between r_{ij} and the z-axis. In (21.4) $\hat{V}$ denotes the non-diagonal part of the spin–spin interaction energy. By virtue of conditions (21.2) and (21.3) we can keep in $\hat{V}$ only that part which is responsible for cross-relaxation:

$$\begin{aligned} \hat{V} &= \tfrac{1}{2}\sum_{i,j} [A_{ij} - \tfrac{1}{4}\gamma_1\gamma_2 r_{ij}^{-3}(1 - \cos^2\theta_{ij})]\, (\hat{I}_j^+\hat{I}_j^- + \hat{I}_i^-\hat{I}_i^+) \\ &= \sum_{ij} B_{ij}(\hat{I}_i^+\hat{I}_j^- + \hat{I}_i^-\hat{I}_j^+). \end{aligned} \tag{21.6}$$

The terms omitted have a structure of the $\sum_{kk'} C_{kk'}\hat{I}_k^{\pm}\hat{I}_{zk}$ and $\sum_{kk'} D_{kk'}\hat{I}_k^{\pm}\hat{I}_{k'}^{\pm}$ type, where k denotes the two kinds of spin. These terms correspond to the

Kronig–Bouwkamp processes (Kronig and Bouwkamp, 1938) whose contribution to the longitudinal spin–spin relaxation can be calculated by the method we use below for the process of cross-relaxation. We notice that the contribution of the Kronig–Bouwkamp processes to the longitudinal relaxation is significant only for values of the external field such that ω is slightly greater than T_2^{-1} in order of magnitude.

We shall use the transport equation in the form in which it was given in section 8 to describe the cross-relaxation. We shall break down the system into dynamic and dissipative parts so that the Hamiltonian of the dynamic sub-system is an operator $\hat{H}_3$ whose eigenvalue spectrum, as is well known, is *discrete*, and the Hamiltonian of the dissipative sub-system is an operator $\hat{H}_{\text{dip}}$ whose eigenvalue spectrum is *quasi-continuous*. This distinction can be made since the operators $\hat{H}_3$ and $\hat{H}_{\text{dip}}$ commute with each other. The characteristic time in which the relaxation processes take place in the dissipative sub-system is T_2 and it can therefore be taken that

$$\omega^* \approx T_2^{-1}. \tag{21.7}$$

We shall now check that the operator $\hat{V}$, which is the interaction of the dynamic and dissipative parts of the system, satisfies the condition of diagonal singularity. As the basic system of functions we shall take the set that consists of all the possible products of the eigenfunctions of the operators $\hat{I}_{zi}$ and $\hat{I}_{zj}$:

$$\Psi_{ln} = \Psi_{l_1}\Psi_{l_2} \cdots \Psi_{l_j} \cdots \Psi_{n_1}\Psi_{n_2} \cdots \Psi_{n_j} \cdots = \prod_{ij} \Psi_{li}\Psi_{nj}. \tag{21.8}$$

Here Ψ_{li} and Ψ_{nj} are the eigenfunctions of the operators $\hat{I}_{zi}$ and $\hat{I}_{zj}$ respectively. Using an arbitrary pair of the functions (21.8) we take the matrix element of the operator $\hat{V}^2$:

$$\hat{V}^2_{ln;n'l'} = \Big\langle \prod_{ij} \Psi_{li}\Psi_{nj} \Big| \sum_{i',j',i'',j''} B_{i'j'}B_{i''j''}(\hat{I}^+_{i'}\hat{I}^-_{j'}\hat{I}^+_{i''}\hat{I}^-_{j''}$$
$$+ \hat{I}^+_{i'}\hat{I}^-_{j'}\hat{I}^-_{i''}\hat{I}^+_{j''} + \hat{I}^-_{i'}\hat{I}^+_{j'}\hat{I}^+_{i''}\hat{I}^-_{j''} + I^-_{i'}I^+_{j'}I^-_{i''}I^+_{j''}) \Big| \prod_{ij} \Psi_{l'_i}\Psi_{n'_j} \Big\rangle. \tag{21.9}$$

Because of the orthogonality of the eigenfunctions of the different operators $\hat{I}^{\pm}_i$, $\hat{I}^{\pm}_j$ in the sum (21.9) not more than one term is finite (provided that for some i and j $l'_i \neq l_i$ and $n_j \neq n'_j$). As can easily be seen from the structure of the expression the non-vanishing terms of the diagonal matrix elements of V^2 (i.e. in the case when for all i and j: $l_i \equiv l'_i$, $n_j \equiv n'_j$) in the sum (21.9), are N times greater than those occurring in any off-diagonal element. The number N is the number of spins of one kind in the system (the smaller number).

It is clear from the above that when $N \to \infty$ the operator $\hat{V}^2$ satisfies the condition of diagonal singularity. The part of the interaction $\hat{V}^2$ responsible for the singularity is of the form

$$\hat{V}_1^2 = \sum_{i,j} B_{ij}^2(\hat{I}_i^+ \hat{I}_i^- \hat{I}_j^- \hat{I}_j^+ + \hat{I}_i^- \hat{I}_i^+ \hat{I}_j^+ \hat{I}_j^-). \tag{21.10}$$

As has already been stated, we shall discuss only systems for which

$$T_{12} \ll T_2 \approx \omega^{*-1}. \tag{21.11}$$

It follows from an analysis of the result obtained below that (21.11) is satisfied if (21.3) holds. We are generally interested in the diagonal elements of the density matrix (the population of the Zeeman levels). These elements, since the interaction satisfies the condition of diagonal singularity, and by virtue of (21.2) and (21.11), are subject to a transport equation of the type (8.10)

$$\dot{\sigma}_{psps} = \sum_{p''s''} 2(\Gamma_{p,s,p''s'';p''s''ps}\, \sigma_{p''s''p''s''} - \Gamma_{p''s''ps;psp''s''}\, \sigma_{psps}). \tag{21.12}$$

The suffixes p and s denote the levels of the Zeeman sub-systems of the spins of the first and second kind respectively. Expression (21.12) differs from (8.10) only in that the single suffix of the dynamic sub-system is replaced by a double suffix:

$$\Gamma_{psp''s'';p''s''ps} = \frac{\pi}{\hbar^2} \sum_{\alpha\alpha'} V_{ps\alpha;p''s''\alpha'}\, P_{\alpha'} V_{p''s''\alpha';ps\alpha}\, \delta(\omega_{psp''s''} + \omega_{\alpha\alpha'}), \tag{21.13}$$

where

$$\omega_{ps;p''s''} = \frac{1}{\hbar}(E_p + E_s - E_{p''} - E_{s''});$$

P_α is the diagonal part of the dissipative system's density matrix, α denotes the eigenvalues of the operator $\hat{H}_{\mathrm{dip}}$; $\hbar\omega_{\alpha\alpha'} = E(\alpha) - E(\alpha')$.

It is clear from the form of the delta-function in (21.13) why, with the condition (21.2), we can limit ourselves to discussing part of the interaction operator $\hat{V}_1$. In fact the whole of the rest of $\hat{V}$ is connected with transitions involving frequencies $\omega_{ps,p''s''} \gtrsim \omega_1, \omega_2$, whilst the density of states α, α', for which $\omega_{\alpha\alpha'} \gg T_2^{-1}$, very rapidly approaches zero as $\omega_{\alpha\alpha'}$ increases.

Equations (21.12) and (21.13) allow us to find the changes in the magnetization of each of the Zeeman sub-systems. For example, for the zth projection of spins of kind 1 we have, allowing for (21.10),

$$\begin{aligned}\frac{\partial}{\partial t}\left\langle \sum_j \hat{I}_{zi} \right\rangle = &- \sum_{i,j,\alpha,\alpha',p,s} \frac{2\pi}{\hbar^2} B_{ij}^2 \sigma_{ps;ps}(t)\, P_\alpha \delta(\omega_{\alpha\alpha'} + \omega_{12}) \\ &\times \{(\hat{I}_j^+ \hat{I}_j^-)_{ps\alpha;p-1,s+1,\alpha'} (\hat{I}_i^- \hat{I}_j^+)_{p-1,s+1,\alpha';ps\alpha} \\ &- (\hat{I}_i^- \hat{I}_j^+)_{ps\alpha',p+1,s-1,\alpha} (\hat{I}_i^+ \hat{I}_j^-)_{p+1,s-1,\alpha;ps\alpha'}\},\end{aligned} \tag{21.14}$$

where $\omega_{12} = \omega_1 - \omega_2$. When deriving (21.14) we used the fact that the coefficients Γ, by virtue of the operators $\hat{I}_i^{\pm}$, are non-zero only when the suffixes p and p'', and s and s'', differ by ± 1 and ∓ 1 respectively. In addition it was considered that P_α is independent of α; this assumption is justifiable over a wide range of temperatures of the dipole lattice T_g up to $T_g \gg (\hbar/k)T_2^{-1}$.

The equation for $\left\langle \sum_j \hat{I}_{zj} \right\rangle$ differs from (21.14) only in the sign of the right-hand side. The equation obtained here for the mean value of the spin must be reduced to a more useful form.

We look upon the right-hand side of (21.14) as a formal function of ω_{12} for this purpose, rewriting (21.14) in the form

$$\frac{\partial}{\partial t} \langle \sum I_{zi} \rangle = -Ag(\omega_{12}). \tag{21.15}$$

If we then normalize $\int g(\omega_{12})\, d\omega_{12}$ to unity, we shall have for A (21.14), (21.15)

$$A = \frac{2\pi}{\hbar^2} \sum_{ij} B_{ij}^2(\langle I_i^+ I_i^- I_j^- I_j^+ \rangle - \langle I_i^- I_i^+ I_j^+ I_j^- \rangle). \tag{21.16}$$

We further use the fact that, outside the time intervals $\tau \gg T_2$ under discussion, each of the sub-systems can be considered to be in a state of thermal equilibrium; in this case we shall assume the temperatures to be sufficiently high that

$$\hbar\omega_1, \quad \hbar\omega_2 \ll kT. \tag{21.17}$$

It follows from inequality (21.17), which is satisfied in medium fields (thousands of gauss) right down to liquid-helium temperatures, that the mean differences of the numbers $\Delta N_{1(2)} = 2 \left\langle \sum_{i(j)} I_{zi(j)} \right\rangle$ of spins of each kind in the upper and lower states is much less than the total number of spins of a given kind $N_{1(2)}$.

Then, neglecting the quantities $\Delta n_{1(2)}/N_{1(2)}$ and also the difference between $\langle \hat{I}_i^+ \hat{I}_i^- \hat{I}_j^- \hat{I}_j^+ \rangle$ and $\langle \hat{I}_i^+ \hat{I}_i^- \rangle \langle \hat{I}_j^- \hat{I}_j^+ \rangle$, it is easy to obtain from (21.15) a similar expression for $\langle \sum \hat{I}_{zj} \rangle$, and from (21.16) the usual form of the cross-relaxation equations:

$$\begin{aligned} \frac{\partial}{\partial t} \left\langle \sum_i I_{zi} \right\rangle &= -\frac{1}{T_{12}} \left\langle \sum_i I_{zi} \right\rangle + \frac{1}{T_{21}} \left\langle \sum_j I_{zj} \right\rangle, \\ \frac{\partial}{\partial t} \left\langle \sum_j I_{zj} \right\rangle &= \frac{1}{T_{12}} \left\langle \sum_i I_{zi} \right\rangle - \frac{1}{T_{21}} \left\langle \sum_j I_{zj} \right\rangle, \end{aligned} \tag{21.18}$$

where

$$T_{12}^{-1} = \frac{2\pi}{\hbar^2 N_1} \sum_{ij} B_{ij}^2 g(\omega_{12});$$
$$T_{21}^{-1} = T_{12}^{-1} N_1 N_2^{-1}. \tag{21.19}$$

The value of the function $g(\omega_{12})$ must be taken, of course, when $\omega_{12} = \omega_1 - \omega_2$. This function has not yet been evaluated, and it may be time-dependent†.

To find $g(\omega_{12})$ we must use the method of moments, which has already been used in the preceding section for calculating the form of the absorption line.

Using the properties of the delta-function (see Appendix I) we have from (21.14) and (21.16)

$$g(\omega_{12}) = \frac{2\pi}{\hbar^2 A} \sum_{\alpha,\alpha',p,s,p'',s''} \int_{-\infty}^{\infty} dt' P(\alpha)\, \sigma_{ps}(t)\, V_{ps\alpha;p''s''\alpha'} \times V_{p''s''\alpha';ps\alpha}\, e^{j\omega_{\alpha\alpha'}t'}\, e^{i\omega_{12}t} \tag{21.20}$$

or

$$g(\omega_{12}) = \int_{-\infty}^{\infty} \eta_{12}(t')\, e^{i\omega t'} dt', \tag{21.21}$$

$$\eta(t') = \frac{2\pi}{\hbar^2 A} \langle \hat{V}(t')\, \hat{V}(0) \rangle. \tag{21.22}$$

Here the form of $\hat{V}(t')$ is determined wholly by $\hat{H}_{\text{dip}}$. We should further point out that it is essential, when obtaining (21.12), for the matrix elements of $\hat{V}^2$ to be independent of ω_{12}. Repeating exactly the procedure used in section 20 to find the moments $x(\omega)$, we find that the expression for the sth moment ω^s of the function $g(\omega_{12})$ is of the form

$$\omega^s = \frac{(i)^s\, \eta^s(0)}{\eta(0)},$$

where $\eta^s(0)$ is the sth derivative of $\eta(t)$ when $t = 0$. For calculating the derivatives η^s we make use, just as in section 20, of the quantum equation of

† Sometimes $g(\omega_{12})$ is taken to be of the form

$$g(\omega_{12}) = \int g_1(\omega_1')g_2(\omega_2')\, \delta(\omega_1' - \omega_2' + \omega_{12})\, d\omega_1'\, d\omega_2',$$

where $g_1(\omega'_1)$ and $g_2(\omega'_2)$ are the line-shape functions for the absorption lines of the first and second kind of spin respectively.

This choice of $g(\omega_{12})$ corresponds to neglecting the third term in (21.5) whose contribution in the general case is of the same order as that from the other terms.

motion for the operator

$$\dot{\hat{V}}(t) = \frac{i}{\hbar} [\hat{H}_{\text{dip}}, \hat{V}]. \tag{21.23}$$

In this case we obtain for the second and fourth moments of $g(\omega_{12})$ from (21.21)–(21.23)

$$\begin{aligned} \langle \omega^2 \rangle &= -\frac{1}{\hbar^2} \frac{\langle [\hat{H}_{\text{dip}} \hat{V}]^2 \rangle}{\langle \hat{V}^2 \rangle}, \\ \langle \omega^4 \rangle &= \frac{\langle [\hat{H}_{\text{dip}}, [\hat{H}_{\text{dip}}, \hat{V}]]^2 \rangle}{\hbar^4 \langle \hat{V}^2 \rangle}. \end{aligned} \tag{21.24}$$

Similar expressions hold for the higher moments. It can also be confirmed that, when condition (21.17) is satisfied, the moments, and therefore also the function $g(\omega_{12})$, are not time-dependent, and in it is sufficient to take the trace, instead of averaging in (21.24) and in similar expressions for the higher moments. The procedure for calculating the moments is exactly the same as the corresponding calculations of section 20 and we shall not do this here, especially as there are many examples of calculations of moments for various substances in the literature (see, e.g., Kiel, 1960; Kopvillem, 1960). After calculating a sufficient number of moments we can construct the function $g(\omega_{12})$ and, taking its value at the point $\omega_{12} = \omega_1 - \omega_2$, we obtain in accordance with (21.19) the values of the cross-relaxation times. The available estimates show that T_{12} is very strongly dependent on ω_{12} since $g(\omega_{12})$ has, roughly speaking, a Gaussian form when ω_{12} changes by an amount T_2^{-1}. For example (Bloembergen, Shapiro, Pershan and Artman, 1959)

$$T_2 < T_{12} < T_1 \quad \text{when} \quad 2T_2^{-1} > \omega_{12} > 10T_2^{-1}, \tag{21.25}$$

where T_1 is the spin–lattice relaxation time and is five to seven orders of magnitude greater than T_2.

Situations in which more than two spins take part in the simultaneous reversal are also of practical interest. Such cases can be treated by the above method (see Genkin, 1964). In future when we meet the phenomenon of cross-relaxation we shall treat it in a more qualitative manner. The purpose of this section was to show how the appropriate equations can be rigorously derived from those of section 8, and to indicate the approximations that were made in the derivations.

CHAPTER V

The Behaviour of Quantum Systems in Strong Fields

IN THIS chapter we shall continue our study of quantum systems in external fields. In the preceding chapter (sections 14–17) it was assumed that the external field was vanishingly small (only the first term of the expansion in the powers of the field was retained); we shall now take into consideration effects connected with the finite magnitude of the field. The solution of this problem is important in the theory of maser amplifiers and oscillators.

22. The Non-linear Properties of a Medium

The operating principles of masers depend essentially on the non-linear properties of a medium. Moreover, the development of lasers has led to the production in the optical region of high electromagnetic field strengths with a very narrow linewidth. It is essential to take into consideration the non-linear properties of the medium in fields of this kind†. It is therefore of interest to examine in detail the non-linear properties of a medium. The relationship between the polarization (dipole moment per unit volume) and the electric field can be described schematically (neglecting spatial and time dispersion and anisotropy) as follows:

$$P = \chi E + \chi_1 E^2 + \chi_2 E^3 + \dots, \tag{22.1}$$

where the coefficient χ is the usual susceptibility and the coefficients χ_1, χ_2 take account of the non-linear properties of the medium.

We can state a similar relation between the magnetization and the magnetic field.

Before finding the general phenomenological connection between the response and the force (allowing for the anisotropy and dispersion) we shall carry out a quantum-mechanical investigation using perturbation theory.

† See Chapter IX.

22.1. A General Approach using Perturbation Theory

In accordance with the general rules of quantum theory the external force acting on a system described by the Hamiltonian $\hat{H}$ can be taken into consideration by means of an interaction energy which is explicitly time-dependent:

$$\hat{V} = \hat{V}(t). \tag{22.2}$$

We shall solve the equation for the density matrix in the interaction representation (2.14), taking $\hat{V}$ to be a small quantity. To do this we write equation (2.14) in integral form:

$$\hat{\varrho}(t) = \hat{\varrho} + \frac{1}{i\hbar}\int_{-\infty}^{t} [\hat{V}(\tau'), \hat{\varrho}(\tau')]\, d\tau', \quad \hat{\varrho} = \hat{\varrho}(-\infty). \tag{22.3}$$

Solving this equation by iteration we obtain, as in (5.6)

$$\hat{\varrho}(t) = \hat{\varrho}^{(0)} + \hat{\varrho}^{(1)} + \hat{\varrho}^{(2)} + \hat{\varrho}^{(3)} + \dots + \hat{\varrho}^{(n)} + \dots, \tag{22.4}$$

where

$$\hat{\varrho}^{(0)} = \hat{\varrho};$$

$$\hat{\varrho}^{(1)}(t) = \frac{1}{i\hbar}\int_{-\infty}^{t} [\hat{V}(t_1), \hat{\varrho}]\, dt_1; \tag{22.5}$$

$$\hat{\varrho}^{(2)}(t) = \frac{1}{(i\hbar)^2}\int_{-\infty}^{t} dt_1 \int_{-\infty}^{t_1} dt_2 [\hat{V}(t_1), [\hat{V}(t_2), \hat{\varrho}]]; \tag{22.6}$$

$$\hat{\varrho}^{(3)}(t) = \frac{1}{(i\hbar)^3}\int_{-\infty}^{t} dt_1 \int_{-\infty}^{t_1} dt_2 \int_{-\infty}^{t_2} dt_3 [\hat{V}(t_1), [\hat{V}(t_2), [\hat{V}(t_3), \hat{\varrho}]]]; \tag{22.7}$$

. . . .

$$\hat{\varrho}^{(n)}(t) = \frac{1}{(i\hbar)^n}\int_{-\infty}^{t} dt_1 \int_{-\infty}^{t_1} dt_2 \dots \int_{-\infty}^{t_{n-1}} dt_n [\hat{V}(t_1), [\hat{V}(t_2) \dots [\hat{V}(t_n), \hat{\varrho}]] \dots]. \tag{22.8}$$

By using these expressions and the definition of the mean it is easy to obtain the expression for the response of the system in the form of the mean value of an operator $\hat{O}$:

$$\langle O \rangle = \langle O^{(1)} \rangle + \langle O^{(2)} \rangle + \langle O^{(3)} \rangle + \cdots \langle O^{(n)} \rangle + \cdots \tag{22.9}$$

(as before, we assume that when there is no external field the mean value $\langle O^{(0)}\rangle$ is zero). Here the quantities $\langle O^{(n)}\rangle$ are:

$$\langle O^{(1)}\rangle = \frac{1}{i\hbar}\int_{-\infty}^{t} \langle[\hat{O}(t), \hat{V}(t_1)]\rangle\, dt_1, \tag{22.10}$$

$$\langle O^{(2)}\rangle = \frac{1}{(i\hbar)^2}\int_{-\infty}^{t} dt_1 \int_{-\infty}^{t_1} dt_2 \langle[[\hat{O}(t), \hat{V}(t_1)], \hat{V}(t_2)]\rangle, \tag{22.11}$$

$$\langle O^{(3)}\rangle = \frac{1}{(i\hbar)^3}\int_{-\infty}^{t} dt_1 \int_{-\infty}^{t_1} dt_2 \int_{-\infty}^{t_2} dt_3 \langle[[[\hat{O}(t), \hat{V}(t_1)], \hat{V}(t_2)], \hat{V}(t_3)]\rangle, \tag{22.12}$$

. . . .

$$\langle O^{(n)}\rangle = \frac{1}{(i\hbar)^n}\int_{-\infty}^{t} dt_1 \int_{-\infty}^{t_1} dt_2 \ldots$$

$$\times \int_{-\infty}^{t_{n-1}} dt_n \langle[[[\ldots[\hat{O}(t), \hat{V}(t_1)], \hat{V}(t_2)]\ldots], \hat{V}(t_n)]\rangle. \tag{22.13}$$

We recall that the time-dependence of the operators $\hat{O}(t)$, $\hat{V}(t)$ is obtained without taking the interaction into consideration, i.e. it is determined by the unperturbed Hamiltonian $\hat{H}$. The $\langle\rangle$ brackets denote averaging using the density matrix $\hat{\varrho} = \hat{\varrho}(-\infty)$.

Let us examine the case where the interaction energy can be written in the form (14.1). Then the mean value of the quantity $\langle x_a\rangle$ becomes (in the following, we assume summation over repeated suffixes)

$$\langle x_a(t)\rangle = \int_{-\infty}^{t} dt_1 \varphi_{ab}(t, t_1) f_b(t_1) + \int_{-\infty}^{t} dt_1 \int_{-\infty}^{t_1} dt_2 \varphi_{abc}(t, t_1, t_2) f_b(t_1) f_c(t_2)$$

$$+ \int_{-\infty}^{t} dt_1 \int_{-\infty}^{t_1} dt_2 \int_{-\infty}^{t_2} dt_3 \varphi_{abcd}(t, t_1, t_2, t_3) f_b(t_1) f_c(t_2) f_d(t_3) + \ldots$$

$$+ \int_{-\infty}^{t} dt_1 \ldots \int_{-\infty}^{t_{n-1}} dt_n \varphi_{\underbrace{a,b,c\,\ldots\,u}_{n+1}}(t, t_1, t_2 \ldots t_n) f_b(t_1) f_c(t_2) \ldots$$

$$\ldots f_u(t_n) + \ldots, \tag{22.14}$$

where the response functions (or reaction functions) φ take the form (these are derived in Kubo, 1957; Bernard and Callen, 1959) (see also (14.5)),

$$\varphi_{ab}(t, t_1) = -\frac{1}{i\hbar}\langle[\hat{x}_a(t), \hat{x}_b(t_1)]\rangle \tag{22.15}$$

$$\varphi_{abc}(t, t_1, t_2) = \frac{1}{(i\hbar)^2} \langle[[\hat{x}_a(t), \hat{x}_b(t_1)], \hat{x}_c(t_2)]\rangle, \tag{22.16}$$

$$\varphi_{abcd}(t, t_1, t_2, t_3) = -\frac{1}{(i\hbar)^3} \langle[[[\hat{x}_a(t), \hat{x}_b(t_1)], \hat{x}_c(t_2)], \hat{x}_d(t_3)]\rangle \tag{22.17}$$

etc. When at $t = -\infty$ the system is in a stationary state $\hat{\varrho} = \hat{\varrho}(-\infty)$ the functions φ_{ab}, φ_{abc} and φ_{abcd} respectively depend only upon $t - t_1$; $t - t_1$, $t_1 - t_2$ and $t - t_1$, $t_1 - t_2$, $t_2 - t_3$.

In this case as well as φ the non-linear response of the system is determined by its frequency characteristics—*cross-susceptibilities.* These quantities are introduced as follows. We expand the external force $f_a(t)$ into a Fourier series (or integral):

$$f_a(t) = \sum_l f_a(\omega_l)\, e^{-i\omega_l t}; \quad f_a(-\omega_l) = f_a^*(\omega_l), \tag{22.18}$$

$$\omega_{-l} = -\omega_l.$$

Then, using (22.14)–(22.17) and allowing for the diagonality of $\hat{\varrho} = \hat{\varrho}(-\infty)$, we obtain (after symmetrization)

$$\langle x_a\rangle = \chi_{ab}(\omega_l) f_b(\omega_l)\, e^{-i\omega_l t} + \chi_{abc}(\omega_s, \omega_l) f_b(\omega_s) f_c(\omega_l)\, e^{-i(\omega_s+\omega_l)t}$$
$$+ \chi_{abcd}(\omega_s, \omega_l, \omega_r) f_b(\omega_s) f_c(\omega_l) f_d(\omega_r)\, e^{-i(\omega_s+\omega_l+\omega_r)t} + \dots, \tag{22.19}$$

where $\chi_{ab}(\omega_l)$ is the usual complex susceptibility (see (14.11));

$$\chi_{abc}(\omega_s, \omega_l) = \frac{1}{2!}\int_0^\infty d\tau_1 \int_0^\infty d\tau_2 \{\varphi_{abc}(\tau_1, \tau_2)\, e^{i(\omega_s+\omega_l)\tau_1 + i\omega_l\tau_2}$$
$$+ \varphi_{acb}(\tau_1, \tau_2)\, e^{i(\omega_s+\omega_l)\tau_1 + i\omega_s\tau_2}\}, \tag{22.20}$$

$$\chi_{abcd}(\omega_s, \omega_l, \omega_r) = \frac{1}{3!}\hat{P}_3 \int_0^\infty d\tau_1 \int_0^\infty d\tau_2$$
$$\times \int_0^\infty d\tau_3 \varphi_{abcd}(\tau_1, \tau_2, \tau_3)\, e^{i(\omega_s+\omega_l+\omega_r)\tau_1 + \omega_l\tau_2 + \omega_r\tau_3}. \tag{22.21}$$

Here P_3 is the operator of all the permutations of the quantities (ω_s, b), (ω_l, c) (ω_r, d), $\tau_1 = t - t_1$, $\tau_2 = t_1 - t_2$, $\tau_3 = t_2 - t_3$†. We notice that the symmetrization is carried out so that $f_b(\omega_s) f_c(\omega_l)$ and $f_c(\omega_l) f_b(\omega_s)$ have one and the same coefficient; during summation this coefficient is repeated twice

† Although we are using the same notation for the functions $\varphi_{abc}(t, t_1, t_2)$ and $\varphi_{abc}(\tau_1, \tau_2)$; $\varphi_{abcd}(t, t_1, t_2, t_3)$ and $\varphi_{abcd}(\tau, \tau_1, \tau_2)$ it must be remembered that these are different functions (for a number of reasons).

if (ω_s, b) and (ω_l, c) are different quantities. The tensors of the third- and fourth-rank susceptibilities $\chi_{abc}(\omega_s, \omega_l)$ and $\chi_{abcd}(\omega_s, \omega_l, \omega_r)$ therefore satisfy the obvious symmetry relations

$$\chi_{abc}(\omega_s, \omega_l) = \chi_{acb}(\omega_l, \omega_s), \quad \chi_{abcd}(\omega_s, \omega_l, \omega_r) = \chi_{abdc}(\omega_s, \omega_r, \omega_l) \text{ etc.} \tag{22.22}$$

The other symmetry relations will be discussed in section 22.3. We shall now express the coefficients χ_{abc} and χ_{abcd} in terms of the density matrix $\hat{\varrho}$ and the matrix elements $\hat{x}_a$. For this we can take (22.16), (22.17), (22.20) and (22.21) but it is more convenient to use the solutions (22.6) and (22.7), the definition of the mean value and (22.19). In a similar way to (14.12) we obtain

$$\chi_{abc}(\omega_s, \omega_l) = \tfrac{1}{2}[\varkappa_{abc}(\omega_s, \omega_l) + \varkappa_{acb}(\omega_l, \omega_s)],$$

$$\varkappa_{abc}(\omega_s, \omega_l) = \frac{1}{\hbar^2} x_{anm} x_{bmk} x_{ckn} [(\varrho_{nn} - \varrho_{kk})\, \zeta(\omega_l - \omega_{kn}) - (\varrho_{kk} - \varrho_{mm})$$
$$\times\, \zeta(\omega_s - \omega_{mk})]\, \zeta(\omega_l + \omega_s - \omega_{mn}), \tag{22.23}$$

$$\chi_{abcd}(\omega_s, \omega_l, \omega_r) = \frac{\hat{P}_3}{3!} \varkappa_{abcd}(\omega_s, \omega_l, \omega_r),$$

$$\varkappa_{abcd}(\omega_s, \omega_l, \omega_r) = -\frac{1}{\hbar^3} \zeta(\omega_s + \omega_l + \omega_r - \omega_{mn}) \{x_{dmp} x_{bpk} x_{ckn} x_{anm}$$
$$\times [(\varrho_{nn} - \varrho_{kk}) \zeta(\omega_l - \omega_{kn}) - (\varrho_{kk} - \varrho_{pp}) \zeta(\omega_s - \omega_{pk})]\, \zeta(\omega_l + \omega_s - \omega_{pn})$$
$$- x_{bmk} x_{ckp} x_{dpn} x_{anm} [(\varrho_{pp} - \varrho_{kk})\, \zeta(\omega_l - \omega_{kp}) - (\varrho_{kk} - \varrho_{mm})$$
$$\times\, \zeta(\omega_s - \omega_{mk})]\, \zeta(\omega_l + \omega_s - \omega_{mp})\} \tag{22.24}$$

(we recall that summation here is always carried out over repeated suffixes; in (22.23) and (22.24), however, there is no summation over the suffixes s, l, r).

22.2. *Phenomenological Treatment. Time and Spatial Dispersion*

In the phenomenological treatment we postulate a definite kind of relationship between the response and the force. Relation (22.1) is introduced in this way. A more general form of the connection between the polarization and the external electric field can be written as follows

$$P_a = \chi_{ab}(\omega_l)\, E_b(\omega_l)\, e^{-i\omega_l t} + \chi_{abc}(\omega_s, \omega_l)\, E_b(\omega_s)\, E_c(\omega_l)\, e^{-i(\omega_s + \omega_l)t}$$
$$+ \chi_{abcd}(\omega_s, \omega_l, \omega_r)\, E_b(\omega_s)\, E_c(\omega_l)\, E_d(\omega_r)\, e^{-i(\omega_s + \omega_l + \omega_r)t} + \ldots, \tag{22.25}$$

where it is taken that the electric field can be considered as the sum of time harmonics $\boldsymbol{E} = \boldsymbol{E}(\omega_l)\, e^{-i\omega_l t}$, $\boldsymbol{E}(-\omega_l) = \boldsymbol{E}^*(\omega_l)$, and the suffixes a, b, c, d take up the values x, y, z. We shall discuss the assumptions that are needed to obtain (22.25) from the general expressions (section 22.1). We shall first assume that the medium consists of particles (atoms, molecules) whose interaction with the field can be discussed in the dipole approximation. Then the interaction energy can be written in the form

$$\hat{V} = -\sum (\hat{\boldsymbol{d}}_j \cdot \boldsymbol{E}(\boldsymbol{x}_j, t)).$$

We can now make use of (22.19) (since the interaction energy takes the form of (14.1)). It follows from this that generally the mean value of the dipole moment at the point x_j is determined by the values of the field at all the other points. As has already been pointed out in section 14.3, this means that spatial dispersion is taken into consideration. In the following we shall not take it into consideration. Then (22.25) follows directly from (22.19), and we obtain the tensors $\chi_{abc}(\omega_s, \omega_i)$ and $\chi_{abcd}(\omega_s, \omega_l, \omega_r)$ from (22.20), (22.21), (22.23) and (22.24) by replacing x_a by d_a ($a = x, y, z$) and multiplying (22.23) and (22.24) by the number of particles per unit volume N (if all the particles are identical). For example $\chi_{abc}(\omega_s, \omega_i)$ in (22.25) becomes

$$\chi_{abc}(\omega_s, \omega_l) = \frac{N}{2!}\,[\varkappa_{abc}(\omega_s, \omega_l) + \varkappa_{acb}(\omega_l, \omega_s)],$$

$$\varkappa_{abc}(\omega_s, \omega_l) = \frac{1}{\hbar^2} d_{anm}\, d_{bmk}\, d_{ckn}[(\varrho_{nn} - \varrho_{kk})\, \zeta(\omega_l - \omega_{kn}) - (\varrho_{kk} - \varrho_{mm}) \times \zeta(\omega_s - \omega_{mk})]\, \zeta(\omega_l + \omega_s - \omega_{mn}). \tag{22.26}$$

We notice that, since the spatial dispersion is not taken into consideration in (22.25), $E_a(\omega_s)$ may be any function of the coordinates. Unlike (22.1) the relationship (22.25) allows for the time dispersion. The value of $P_a(t)$ at a given time is determined by the field E at preceding times. We can use (22.14) to write this dependence in the form

$$P_a(t) = \int_{-\infty}^{t} dt_1 \varphi_{ab}(t - t_1)\, E_b(t_1) + \int_{-\infty}^{t} dt_1 \int_{-\infty}^{t_1} dt_2 \varphi_{abc} \times (t - t_1, t_1 - t_2)\, E_b(t_1)\, E_c(t_2) + \int_{-\infty}^{t} dt_1 \int_{-\infty}^{t_1} dt_2 \int_{-\infty}^{t_2} dt_3 \varphi_{abcd} \times (t - t_1, t_1 - t_2, t_2 - t_3)\, E_b(t_1)\, E_c(t_2)\, E_d(t_3) + \ldots \tag{22.27}$$

Relations similar to (22.25) and (22.27) can be written for the connexion between the magnetization and the magnetic field, the polarization and the electric and magnetic fields, etc. In particular if we take into consideration the quadrupole interaction of an atom (molecule) with an electric field, then P_a will depend not only on E_a but also on the gradient ∇E_a; this essentially allows for spatial dispersion.

22.3. *Symmetry Relations*

Symmetry requirements impose limitations on the number of independent components of the tensors χ_{abc}, χ_{abcd} etc. It follows from the definition of these tensors that, for example, the components of the tensor in a new system of coordinates χ'_{abc} are connected with the components in the old system by the transformation

$$\chi'_{abc} = a_{ai}a_{bj}a_{ck}\chi_{ijk}, \tag{22.28}$$

where $\boldsymbol{a}$ is a transformation matrix containing an arbitrary combination of rotation and inversion.

If $\boldsymbol{a}$ is a symmetry transformation A, then all the properties of the medium should be described identically in both coordinate systems. This means in particular that

$$\chi_{abc} = A_{ai}A_{bi}A_{ck}\chi_{ijk}. \tag{22.29}$$

It is well known that there are 32 classes of crystals, each of which is characterized by its own combination of point symmetry transformations (see, e.g., Landau and Lifshitz, 1957; Nye, 1957). Each such transformation has a relation of the type (22.29). These relations limit the number of independent components in the different tensors. In particular if the class in question contains the inversion transformation $A_{ai} = -\delta_{ai}$, then it follows from (22.29) that

$$\chi_{abc} = -\chi_{abc} = 0. \tag{22.30}$$

From this we find that the components of χ_{abc} are non-zero only for those classes which have no centre of symmetry. There are altogether twenty of these, all of which show the phenomenon of piezoelectricity. Since the piezoelectric effect is also characterized by a third-rank tensor it is sufficient to use the solution of this problem for the tensor γ_{ikl}, which is related to the piezoelectric effect, in order to establish the non-zero independent components of χ_{abc}. Section 17 of the book by Landau and Lifshitz (1957) gives all the independent components of the third-rank tensor for the different classes of symmetry.

We note that the symmetry relations can also be derived from the quantum-mechanical expression (22.26). For example, if the wave functions Ψ_n used to find the matrix elements d_{ank} have a definite parity, then $\chi_{abc} = 0$ (and $\chi_{abcd} \neq 0$, generally speaking). In fact, the matrix elements of a polar vector such as the electric dipole moment $\boldsymbol{d}$ are non-vanishing for transitions with a change of parity (see, e.g., Landau and Lifshitz, 1963). In the triple product $d_{anm}d_{bmk}d_{ckn}$ at least one term should correspond to a transition with the same parity†. A product like $d_{amp}d_{bmk}d_{ckn}d_{dnm}$, however, may be non-vanishing, i.e. $\chi_{abcd} \neq 0$. It follows from this that for the third-rank tensor χ_{abc} to be finite the wave functions must have no centre of symmetry. This imposes a limitation on the symmetry of crystals in which non-linear effects connected with χ_{abc} are possible. When χ_{abc} is the magnetic susceptibility the conditions for χ_{abc} to be non-zero are the same as those above. For example, if the system possesses a centre of symmetry the components of χ_{abc} may be non-zero. At first glance this statement leads to a contradiction of the general proof given above that the tensor χ_{abc} is equal to zero. However, in the magnetic case χ_{abc} is not a tensor but a pseudo-tensor (and the magnetization is a pseudo-vector). In expression (22.26) $\boldsymbol{d}$ must be taken to be the magnetic dipole moment. Since the magnetic moment is an axial vector (pseudo-vector) its matrix elements for transitions without change of parity are non-zero. Therefore in the magnetic case χ_{abc} may be non-zero if the system has a centre of symmetry.

Let us examine in greater detail the properties of the quantity $\chi_{abc}(\omega_s, \omega_l)$. We can derive from the expression for $\varkappa$ (22.26) the relations

$$\varkappa_{abc}(\omega_s, \omega_l) = \varkappa^*_{acb}(-\omega_s, -\omega_l), \tag{22.31}$$

$$\chi_{abc}(-\omega_s, -\omega_l) = \chi^*_{abc}(\omega_s, \omega_l). \tag{22.32}$$

It follows from this that the real part of the quantity χ_{abc} is symmetric and the imaginary one is antisymmetric with respect to the simultaneous change in sign of all the frequency arguments. We notice that (22.32) can easily be obtained from the condition that the polarization P_a in (22.25) is real. The same symmetry conditions hold for higher-order tensors‡

$$\chi_{abcd\ldots}(-\omega_s, -\omega_l, -\omega_r, \ldots) = \chi^*_{abcd\ldots}(\omega_s, \omega_l, \omega_r \ldots). \tag{22.33}$$

We shall further examine the case when there is no absorption in the substance. This means that all the frequencies ω_s and $\omega_s + \omega_l$ are far from the eigenfrequencies of the system (there are no resonances). In this case we can replace the functions $\zeta(x)$ by $1/x$. In particular the expression for $\varkappa_{abc}(\omega_s, \omega_l)$ becomes

$$\varkappa_{abc}(\omega_s, \omega_l) = \frac{d_{anm}d_{bmk}d_{ckn}}{\hbar^2}\left(\frac{\varrho_{nn} - \varrho_{kk}}{\omega_l - \omega_{kn}} - \frac{\varrho_{kk} - \varrho_{mm}}{\omega_s - \omega_{mk}}\right) \times \frac{1}{\omega_s + \omega_l - \omega_{mn}}. \tag{22.34}$$

† Let, for example, n be even, m be odd and k be even; in this case d_{anm} and d_{bmk} are non-zero and $d_{ckn} = 0$.

‡ All the symmetry properties now being derived hold both for tensors and for pseudo-tensors.

We shall further assume that the matrix elements of $\boldsymbol{d}$ can be made real (see section 15). In this case $\varkappa_{abc}(\omega_s, \omega_l)$ is real and (22.32) becomes

$$\chi_{abc}(-\omega_s, -\omega_l) = \chi_{abc}(\omega_s, \omega_l). \tag{22.35}$$

An interesting situation in the field of non-linear optics (see Chapter IX) is the case where the frequencies of the fields acting on the system are related by

$$\omega_r + \omega_l + \omega_s = 0. \tag{22.36}$$

In this case we have from (22.34)

$$\varkappa_{abc}(-\omega_r, -\omega_l) = \varkappa_{cba}(\omega_r, \omega_s) = \varkappa_{bac}(\omega_s, \omega_l). \tag{22.37}$$

From this follows the relation between the cross-susceptibilities

$$\chi_{abc}(\omega_r, \omega_l) = \chi_{cba}(\omega_r, \omega_s) = \chi_{bac}(\omega_s, \omega_l). \tag{22.38}$$

These relations can be obtained phenomenologically from the condition for no absorption. In accordance with (17.7) and (22.25) the change in the internal energy of the system dependent on χ_{abc} is

$$\frac{dU}{dt} = -\dot{E}_a P_a = i\omega_r \chi_{abc}(\omega_s, \omega_l)\, E_a(\omega_r)\, E_b(\omega_s)\, E_c(\omega_l)\, e^{-i(\omega_r+\omega_l+\omega_s)t}. \tag{22.39}$$

The time-average of this quantity should be zero when there is no absorption. This average is not identically equal to zero for those terms for which condition (22.36) is satisfied. After regrouping these terms we obtain the condition for no absorption in the form

$$E_a(\omega_r)\, E_b(\omega_s)\, E_c(\omega_l)\, [\omega_r \chi_{abc}(\omega_s, \omega_l) + \omega_s \chi_{bac}(\omega_r, \omega_l) + \omega_l \chi_{cab}(\omega_r, \omega_s)] = 0.$$

Since this relation should be satisfied for any field amplitude (not only for small values)

$$\omega_r \chi_{abc}(\omega_s, \omega_l) + \omega_s \chi_{bac}(\omega_r, \omega_l) + \omega_l \chi_{cab}(\omega_r, \omega_s) = 0.$$

Remembering further that $\omega_r = -\omega_l - \omega_s$ we obtain†

$$\omega_s[\chi_{abc}(\omega_s, \omega_l) - \chi_{bac}(\omega_r, \omega_l)] + \omega_l\, [\chi_{abc}(\omega_s, \omega_l) - \chi_{cab}(\omega_r, \omega_l)] = 0. \tag{22.40}$$

Since this relation should hold for any arbitrary relation between ω_s and ω_l, we obtain (22.38) after permuting the suffixes. Similar relations can be derived for $\chi_{abcd}(\omega_s, \omega_l, \omega_r)$ if one or two conditions are imposed on the frequency ω_s. In Chapter IX we shall deal in greater detail with the properties of this tensor in connection with two-quantum processes.

† In (22.40) there is no summation over repeated suffixes.

22.4. Dispersion Relations

The cross-susceptibility tensors satisfy dispersion relations analogous to the Kramers–Kronig relations (section 16). We write χ_{abc} as a function of $\omega = \omega_s + \omega_l$ and ω_l:

$$\chi_{abc} = \chi_{abc}(\omega, \omega_l).$$

Then by using the analytical properties of $\chi_{abc}(\omega, \omega_l)$ as a function of ω it is easy to show (Bunkin, 1962; Kogan, 1962) that the dispersion relations

$$\begin{aligned} \operatorname{Re}\chi_{abc}(\omega, \omega_l) &= \frac{1}{\pi}\oint_{-\infty}^{\infty} \frac{d\omega'}{\omega' - \omega} \operatorname{Im}\chi_{abc}(\omega', \omega_l), \\ \operatorname{Im}\chi_{abc}(\omega, \omega_l) &= -\frac{1}{\pi}\oint_{-\infty}^{\infty} \frac{d\omega'}{\omega' - \omega} \operatorname{Re}\chi_{abc}(\omega', \omega_l) \end{aligned} \tag{22.41}$$

hold, as do also the analogous relations for $\chi_{abcd}(\omega, \omega_l, \omega_s)$ and the higher-order tensors. These relations are not difficult to obtain by using (22.20), (22.21), (22.23) and (22.24).

22.5. Estimates of the Terms in the Series Expansion in Powers of the Field

Let us now find the conditions under which we can take a limited number of terms of the expansions (22.9), (22.14), (22.15), etc. In order to do this we must make estimates of the individual terms of the expansion and, in particular, of the cross-susceptibilities.

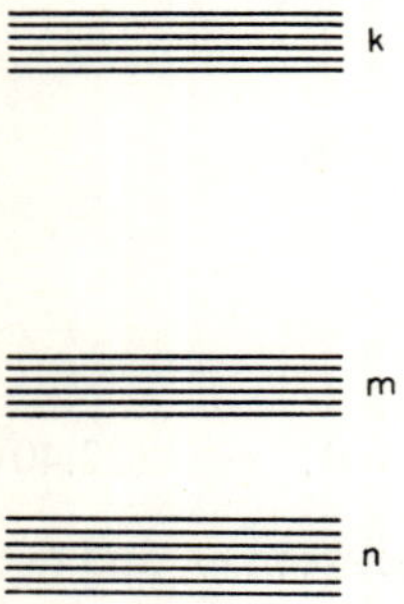

FIG. V.1. Three non-intersecting energy bands.

We shall discuss the following cases of practical interest:

(a) *The energy spectrum of the system consists of non-overlapping bands* (Fig. V.1). A spectrum of this kind is obtained if the energy levels *n*, *m*, *k* are initially degenerate and then the degeneracy is removed by a small enough perturbation so that the bands do not overlap. Let $f_m(E_m)$ be the density of states in the mth band:

$$dZ_m = f_m(E_m)\, dE_m, \tag{22.42}$$

where dZ_m is the number of states in the energy range dE_m.

Summation over m, n, k in (22.23) can now be replaced by integration with respect to E within the band, and summation over the bands (we shall denote the numbers of the bands by the same letters m, n, k):

$$\begin{aligned}\varkappa_{abc}(\omega_s, \omega_l) = \frac{1}{\hbar^2} \int dE_m \int dE_n \int dE_k f_m(E_m) f_n(E_n) f_k(E_k) \\ \times x_{anm} x_{bmk} x_{ckn} [(\varrho_{nn} - \varrho_{kk})\, \zeta(\omega_l - \omega_{kn}) - (\varrho_{kk} - \varrho_{mm}) \\ \times \zeta(\omega_s - \omega_{mk})]\, \zeta(\omega_l + \omega_s - \omega_{mn}).\end{aligned} \tag{22.43}$$

We can now estimate the orders of magnitude. Let us look at various cases. Let ω_l and ω_s be such that

$$\omega_l \approx \omega_{kn}, \quad \omega_s \approx \omega_{mk} \quad \text{and} \quad \omega_l + \omega_s \approx \omega_{mn}, \tag{22.44}$$

where the approximate equalities show that the range of values of the frequencies are determined by the widths of the bands. Further let x_{anm}, x_{bmk}, x_{ckn} and ϱ_{nn} vary little within a band. The terms with the δ-functions play the major rôle, as can easily be seen, in satisfying condition (22.44). We write explicitly the first term of (22.43) as the product of two δ-functions:

$$\begin{aligned}-\frac{\pi^2}{\hbar^2} \int dE_m \int dE_n \int dE_k x_{anm} x_{bmk} x_{ckn} \varrho_{nn} \delta(\omega_l - \omega_{kn})\, \delta(\omega_l + \omega_s - \omega_{mn}) \\ \times f_m(E_m) f_n(E_n) f_k(E_k) \approx -\pi^2 x_{anm} x_{bmk} x_{ckn} \\ \times \int \varrho_{nn} f_n(E_n) f_k(E_k^0 = E_n + \hbar\omega_l) f_m\, (E_m^0 = E_n + \hbar(\omega_l + \omega_s))\, dE_n \\ \approx -\pi^2 x_{anm} x_{bmk} x_{ckn} \bar{\varrho}_{nn} \frac{g_k}{\Delta E_k} \frac{g_m}{\Delta E_m},\end{aligned} \tag{22.45}$$

where $\bar{\varrho}_{nn} = \int f_n(E_n)\, \varrho_{nn}\, dE$ is the probability that the system is within the limits of band n;

g_k and g_m are the statistical weights of the bands;

ΔE_k and ΔE_m are the energy widths of these bands defined by the relations

$$\int dZ_m = g_m = \int f_m(E_m)\, dE_m = f(E_m^0)\, \Delta E_m,$$

where $f(E_m^0)$ is a mean value of f_m for the band m.

If one of the conditions (22.44) (or more than one) is not satisfied, for example $\omega_l \not\approx \omega_{kn}$, then instead of ΔE_k in (22.45) we must write the difference $\hbar(\omega_l - \omega_{kn})$. Collecting together all these expressions we obtain the following estimates for $\varrho_{mn}^{(2)}$ and $\langle x_a \rangle$:

$$\varrho_{mn}^{(2)} \sim \varrho_{nn} \frac{V_{mk}(\omega_s)\, V_{kn}(\omega_l)\, g_k g_m}{\hbar^2 \Delta\omega_k\, \Delta\omega_m},$$

$$\langle x_a \rangle \sim \int x_{anm} \frac{V_{mk}(\omega_s)\, V_{kn}(\omega_l)\, g_k g_m}{\hbar^2\, \Delta\omega_k\, \Delta\omega_m}\, \bar{\varrho}_{nn} f(E_n)\, dE_n, \tag{22.46}$$

where $V(\omega_s)$ is a Fourier component of the perturbation energy and the $\Delta\omega$ are either the differences of the frequencies or the widths of the terms, depending upon the case under discussion (non-resonance or resonance).

The estimates (22.46) immediately give the condition for applicability of second order perturbation theory

$$\frac{V_{mk}(\omega_s)\, V_{kn}(\omega_l)\, g_k g_m}{\hbar^2\, \Delta\omega_k\, \Delta\omega_m} \ll 1. \tag{22.47}$$

We notice that when we change to a continuous spectrum within the band the number of states $g_m \to \infty$. In this case, however, $V(\omega_s)\, g_k$ remains finite. This ensures that $\langle x_a \rangle$ and (22.47) are finite.

Let us move on now to another case.

(b) *The spectrum energy is discrete but there is attenuation in the system.* Strictly speaking, none of the preceding expressions can be used. The attenuation in the system can, however, sometimes be allowed for by introducing the complex eigenvalues of the energy (Landau and Lifshitz, 1963):

$$E_n = E_n^0 - i\Gamma_n, \quad \Gamma_n > 0.$$

This leads to the eigenfrequencies ω_{kn} becoming complex:

$$\omega_{kn} = \omega_{kn}^0 - i\gamma_{kn}, \quad \hbar\gamma_{kn} = \Gamma_k + \Gamma_n.$$

In this case we must substitute

$$\zeta(\omega_l - \omega_{kn}) \to \frac{1}{\omega_l - \omega_{kn}^0 + i\gamma_{kn}} \tag{22.48}$$

for the ζ-functions in (22.23) and (22.24). If we substitute this expression in the expression for the susceptibility (14.12) we obtain

$$\chi_{ab}(\omega) = \frac{1}{\hbar}(\varrho_{nn} - \varrho_{kk})\, x_{ank} x_{bkn} \frac{\omega - \omega_{nk}^0 - i\gamma_{nk}}{(\omega - \omega_{nk}^0)^2 + \gamma_{nk}^2}. \tag{22.49}$$

It is easy to show that the shape of line obtained is the same as (18.19). Here γ_{nk} plays the part of $1/\tau_{nk}$—the width of the absorption line. It can be shown that in second order perturbation theory expression (22.23), after substituting (22.48), agrees with the corresponding solution obtained by means of the transport equation (18.3) when $\gamma_{nk} = \tau_{nk}^{-1}$. The estimates of the orders of magnitude agree with (22.46) and (22.47), but in these expressions the part of $\Delta\omega$ is played by the line width (or the corresponding frequency differences). It is not surprising, of course, that the results obtained in cases (a) and (b) are in close agreement. The point is that, in accordance with the uncertainty relation between the energy and the time, the quasi-stationary level is to a considerable extent equivalent to the energy band $\Delta E_n = \Gamma_n$.

We shall now find the condition for the upper limit of the fields that allows the use of the first terms in the expansion (22.25). Using the above estimates it is easy to find that the condition for the upper limit is of the form

$$E(\omega_l) \ll E_{\mathrm{cr}} \approx \frac{\hbar\Delta\omega}{d}, \tag{22.50}$$

where d is the magnitude of the matrix element of the dipole moment (or the magnetic moment, in which case $E(\omega_c)$ must be taken to be a magnetic field);

$\Delta\omega$ is the characteristic difference between the frequencies of the field and the eigenfrequencies of the system.

In the case when resonance occurs (one of the field frequencies is equal to the difference between the terms of the system) $\Delta\omega$ must be taken as the corresponding line width. The upper limit condition for an nth order term in perturbation theory is, as can easily be seen, of the form

$$\underbrace{E(\omega_l)\, E(\omega_r) \cdots E(\omega_s)}_{n \text{ terms}} \ll E_{\mathrm{cr}}^{(1)} E_{\mathrm{cr}}^{(2)} \cdots E_{\mathrm{cr}}^{(n)}, \tag{22.51}$$

where each $E_{cr}^{(i)}$ has its own d and $\Delta\omega$. In the case when $\Delta\omega$ is of the order of frequency of an optical wave and $d \approx 10^{-18}$ CGSe (one Debye unit) the field E_{cr} is of the order of an atomic field $\sim 10^6$ CGSe $= 3 \times 10^{10}$ V/m. It is easy to see that the estimates for the quantities $\chi_{abc}(\omega_s, \omega_c)$ and $\chi_{abcd}(\omega_s, \omega_l, \omega_r)$ can be written as

$$\chi_{abc} \sim \frac{\chi_{ab}}{E_{cr}}, \quad \chi_{abcd} \sim \frac{\chi_{ab}}{E_{cr}^{(1)} E_{cr}^{(2)}}, \tag{22.52}$$

where χ_{ab} is the usual susceptibility.

22.6. *Connection between Actual and Macroscopic Fields (Armstrong, Bloembergen, Ducuing and Pershan, 1962)*

Expression (22.25) gives the connection between the actual or local field and the polarization. In the case of rarefied gases the actual field is the same as the macroscopic field and (22.25) can be used directly in the Maxwell equations. We still have to establish in the general case the connection between the local field $\boldsymbol{E}_{\text{loc}}$ and the macroscopic field. We know that in a liquid or in a crystal of cubic symmetry

$$\boldsymbol{E}_{\text{loc}} = \boldsymbol{E} + \frac{4\pi}{3}\boldsymbol{P} = \boldsymbol{E} + \frac{4\pi}{3}\boldsymbol{P}^{L} + \frac{4\pi}{3}\boldsymbol{P}^{NL}, \tag{22.53}$$

where $\boldsymbol{P}^L$ and $\boldsymbol{P}^{NL}$ are the linear and non-linear parts of the polarization respectively.

The displacement vector $\boldsymbol{D}$ contained in the Maxwell equations is of the form

$$\boldsymbol{D} = \boldsymbol{E} + 4\pi\boldsymbol{P}^L + 4\pi\boldsymbol{P}^{NL}. \tag{22.54}$$

Using the susceptibility the linear part of the polarization can be expressed as

$$\boldsymbol{P}^L = \chi\boldsymbol{E}_{\text{loc}}. \tag{22.55}$$

It follows from (22.53) and (22.55) that

$$\begin{aligned}\boldsymbol{P}^L &= \frac{\chi}{1-\frac{4}{3}\pi\chi}\boldsymbol{E} + \frac{\chi}{1-\frac{4}{3}\pi\chi}\frac{4\pi}{3}\boldsymbol{P}^{NL}\\ &= \frac{\varepsilon-1}{4\pi}\boldsymbol{E} + \frac{\varepsilon-1}{3}\boldsymbol{P}^{NL},\end{aligned} \tag{22.56}$$

where the last equation is obtained from the definition of the dielectric constant: the part of $\boldsymbol{D}$ that is linear in $\boldsymbol{E}$ is of the form

$$\varepsilon\boldsymbol{E} = \boldsymbol{E} + 4\pi\boldsymbol{P}^L = \boldsymbol{E} + \frac{4\pi\chi}{1-\frac{4\pi}{3}\chi}\boldsymbol{E},$$

where $\boldsymbol{P}^L$ is that part of $\boldsymbol{P}$ that is linear in $\boldsymbol{E}$.
Substituting (22.56) in (22.54) we obtain

$$\boldsymbol{D} = \varepsilon\boldsymbol{E} + \frac{(\varepsilon+2)}{3}4\pi\boldsymbol{P}^{NL} = \varepsilon\boldsymbol{E} + 4\pi\boldsymbol{P}^{NLS}. \tag{22.57}$$

It follows from this equation that the effective non-linear polarization is

$$\boldsymbol{P}^{NLS} = \frac{\varepsilon+2}{3}\boldsymbol{P}^{NL} = \frac{\varepsilon+2}{3}\hat{\chi}\boldsymbol{E}_{\text{loc}}\boldsymbol{E}_{\text{loc}}\boldsymbol{E}_{\text{loc}}, \tag{22.58}$$

where we have introduced a symbolic notation for the third term in (22.25)†: $\hat{\chi}$ is a fourth-rank tensor acting on the third-rank tensor $E_{loc}E_{loc}E_{loc}$ and χ takes account of the time dispersion. Equation (22.58) and the definition of the local field

$$E_{loc} = \frac{\varepsilon + 2}{3} E$$

which applies to isotropic materials and crystals with cubic symmetry, lead to the following change in the cross-susceptibility:

$$\chi_{abcd}(\omega_1, \omega_2, \omega_3) \to \frac{\varepsilon(\omega_1 + \omega_2, \omega_3) + 2}{3} \frac{\varepsilon(\omega_1) + 2}{3} \frac{\varepsilon(\omega_2) + 2}{3} \times \frac{\varepsilon(\omega_3) + 2}{3} \chi_{abcd}(\omega_1, \omega_2, \omega_3). \quad (22.59)$$

There is a generalization to the case of crystals with non-cubic symmetry in the paper by Armstrong, Bloembergen, Ducuing and Pershan (1962). In this case there is a tensor relation between the actual and macroscopic fields.

22.7. *The Choice of the Energy of Interaction with an External Field*

In the present chapter we are taking the interaction energy in the form (4.24)

$$\hat{V} = -\sum (\hat{d}_j \cdot E(r_j, t)). \quad (22.60)$$

In the same dipole approximation we could use the interaction energy in the form (4.14)

$$\hat{V}' = -\sum \left(\frac{e_j \hat{p}_j}{m_j c} \cdot A(r_j, t) \right). \quad (22.61)$$

The Hamiltonians with these energies are connected by canonical transformations and are completely equivalent. In approximate calculations and estimates, however, when we are not considering all the levels of the system but only a part of them (the *n*-level approximation) the form (22.60) is the correct choice for the energy. We shall explain this statement with an example.

The linear susceptibility connecting the polarization with the electric field

$$E = -\frac{1}{c} \frac{\partial A}{\partial t}$$

† In both an isotropic medium and a crystal of cubic symmetry with a centre of symmetry the second term in (22.25) is missing: $\chi_{abc} = 0$.

when calculated by means of (22.61) can easily be found to be of the form

$$\tilde{\chi}_{ba}(\omega) = \lim_{\varepsilon \to +0} \sum_{n,k} (\varrho_{nn} - \varrho_{kk}) \frac{d_{ank}\, d_{bkn}}{\hbar(\omega - \omega_{nk} - i\varepsilon)} \frac{\omega_{nk}}{\omega}. \qquad (22.62)$$

Using the interaction energy (22.60) it is not difficult to obtain (formula (14.12))

$$\tilde{\chi}_{ba}(\omega) = \lim_{\varepsilon \to +0} \sum_{n,k} (\varrho_{nn} - \varrho_{kk}) \frac{d_{ank} d_{bkn}}{\hbar(\omega - \omega_{nk} - i\varepsilon)}. \qquad (22.63)$$

It can be shown by direct calculation that these two expressions are the same. Let us now take the case when it would appear to be necessary to consider only the n lower levels of the system (the rest of the levels are sufficiently high and the frequencies ω of the external fields are much less than the difference $\Delta = (E_n - E_{n+1})/\hbar$. In this case the problem contains the small parameter ω/Δ.

If we use the expression (22.63) (and the similar expressions for the cross-susceptibilities) it can in fact be shown that there is a small correction if we allow for all the neglected levels starting at the $n + 1$-st. However, if we use expression (22.62) all the levels must be taken into account since the neglected levels give a large correction (because of the factor ω_{nk}/ω).

Therefore it is generally necessary to carry out canonical transformations which make it possible to use the small parameter ω/Δ in order to be able to make use of the n-level approximation. Genkin and Mednis (1967) show how to make this canonical transformation in the general case. (For our purposes it is sufficient to use (22.60).)

We should stress once more, to avoid misunderstandings, that in precise calculations (taking all levels into account) the different expressions for the interaction energy are fully equivalent; they are not equivaent in approximate calculations (taking into account only n levels).

We also note that in resonance cases when the frequencies of the external fields are close to the corresponding differences $(E_n - E_k)/\hbar$ the calculations using the interaction energies (22.60) and (22.61) are also the same (the factors $\omega_{nk}/\omega \sim 1$). This case corresponds exactly to the treatment of two-level systems used in section 19 and later.

23. Two-level Systems in a Strong Field

23.1. The General Discussion of the Behaviour of Quantum Systems in an Oscillatory Field. The Transition to Two-level Systems

We shall next discuss the action of a strong field in a number of special cases, but without limiting ourselves to approximate perturbation theory. We shall discuss the action of a sinusoidally time-dependent external field on a quantum system that can be described by the transport equation (18.3). The energy of interaction with the external field can be written in the form

$$\hat{V} = \hat{V}^0 \cos \Omega t, \tag{23.1}$$

where $\Omega \approx \omega_{21}$;

ω_{21} is the frequency difference between the two states 1 and 2 of the system under discussion.

We shall further assume that no other term differences are close to Ω. We shall find the steady-state solution of equations (18.3) in the form

$$\sigma_{mn} = \sum_s \sigma_{mn}^{(s)} \mathrm{e}^{is\Omega t}, \; s = 0, \pm 1, \pm 2. \tag{23.2}$$

Substituting (23.1) and (23.2) in (18.3) and equating the coefficients with the same harmonics $\mathrm{e}^{is\Omega t}$, we obtain the set of equations

$$is\Omega\sigma_{mm}^{(s)} = \sum_k w_{km}\sigma_{kk}^{(s)} - w_{mk}\sigma_{mm}^{(s)} - \frac{i}{2\hbar} \sum_k [V_{mk}^0(\sigma_{km}^{(s-1)} + \sigma_{km}^{(s+1)}) - (\sigma_{mk}^{(s-1)} + \sigma_{mk}^{(s+1)}) V_{km}^0], \tag{23.3}$$

$$[i(s\Omega + \omega_{mn}) + \tau_{mn}^{-1}] \, \sigma_{mn}^{(s)} = -\frac{i}{2\hbar} \sum_k [V_{mk}^0(\sigma_{kn}^{(s-1)} + \sigma_{kn}^{(s+1)}) - (\sigma_{mk}^{(s-1)} + \sigma_{mk}^{(s+1)}) V_{kn}^0]. \tag{23.4}$$

We shall further assume that the following conditions are satisfied†:

$$|\omega_{mn}| \gg \tau_{mn}^{-1} \quad \text{and} \quad |\omega_{mn} - \omega_{21}| \gg \frac{1}{\tau_{mn}}, \tag{23.5a}$$

when $\omega_{mn} \neq \omega_{21}$. Using these conditions it is easy to show that all $\sigma_{mn}^{(s)}$ are small compared with

$$\sigma_{12}^{(1)}, \sigma_{21}^{(-1)} = \sigma_{12}^{(1)*} \quad \text{and} \quad \sigma_{mm}^{0}. \tag{23.5b}$$

Leaving only these terms in the first approximation we arrive at the following set of equations:

$$-\frac{i}{2\hbar} (V_{12}^0\sigma_{21}^{(-1)} - \sigma_{12}^{(1)}V_{21}^0) + w_{21}\sigma_{22}^{(0)} - \sigma_{11}^{(0)} \sum_{k>1} w_{1k} + \sum_{k>2} w_{k1}\sigma_{kk}^{(0)} = 0, \tag{23.5c}$$

† The first of these conditions is also the condition under which (18.3) applies, since it was assumed in the derivation of (18.3) that $|\omega_{mn}| \gg \omega^*$, and the last term should always be far greater than τ_{mn}. This is one of the basic conditions of the derivation of the transport equations (see Chapter II).

$$-\frac{i}{2\hbar}(V^0_{21}\sigma^{(1)}_{12} - V^0_{12}\sigma^{(-1)}_{21}) + w_{12}\sigma^{(0)}_{11} - \sigma^{(0)}_{22}\sum_{k\neq 2} w_{2k} + \sum_{k>2} w_{k2}\sigma^{(0)}_{kk} = 0, \tag{23.6}$$

$$[i(\Omega - \omega_{21}) + \tau^{-1}_{12}]\,\sigma^{(1)}_{12} = -\frac{i}{2\hbar}V^0_{12}(\sigma^{(0)}_{22} - \sigma^{(0)}_{11}),\ \sigma^{(-1)}_{21} = \sigma^{(1)*}_{12}, \tag{23.7}$$

$$\sum_k w_{km}\sigma^{(0)}_{kk} - \sigma^{(0)}_{mm}\sum_k w_{mk} = 0,\quad m > 2, \tag{23.8}$$

$$\sum_k \sigma^{(0)}_{kk} = 1. \tag{23.9}$$

The last equation follows from the normalization condition. It is explicit from equations (23.5)–(23.9) that a strong field with a frequency $\Omega = \omega_{21}$ causes a redistribution of the populations $\sigma^{(0)}_{kk}$ not only of levels 1 and 2 but generally of all the other levels of the system†. This is due to the fact that all the levels of the system, because of the presence of relaxation, are coupled with each other by the probabilities of the direct and reverse transitions w_{mk} and w_{km}. Cases are possible (see (19.1)), however, when we need not take into consideration levels other than 1 and 2. In these cases we can consider just an ideal two-level system and equations (18.3) become

$$\frac{d\Delta}{dt} = -\frac{2i}{\hbar}[V_{12}\sigma_{21} - \sigma_{12}V_{21}) - \frac{1}{T_1}(\Delta - \Delta_0). \tag{23.10}$$

$$\left(\frac{d}{dt} + i\omega_{21} + \frac{1}{T_2}\right)\sigma_{21} = -\frac{i}{\hbar}(V_{22}\sigma_{21} - \sigma_{21}V_{11}) - \frac{i}{\hbar}V_{21}\Delta, \tag{23.11}$$

where

$$\Delta = \sigma_{11} - \sigma_{22};\quad \sigma_{11} + \sigma_{22} = 1; \tag{23.12}$$

Δ_0 is the equilibrium value of $\Delta(\Delta_0 = \sigma_{011} - \sigma_{022})$;

$$T_2 = \tau_{12} = \tau_{21};\quad w_{12} = \frac{1}{T_1}\sigma_{022};\quad w_{21} = \frac{1}{T_1}\sigma_{011}; \tag{23.13}$$

T_1 and T_2 are the transverse and longitudinal relaxation times.

The equations for the energy spin in an external field (19.41) can easily be obtained from (23.10) and (23.11).

23.2. The Equations for the Polarization of a Two-level System in an Electric Field (Fain, 1957 c)

It is easy to obtain from equations (23.10)–(23.12) the equations that give the polarization

$$\boldsymbol{P} = \sigma_{12}\boldsymbol{d}_{21} + \sigma_{21}\boldsymbol{d}_{12} \tag{23.14}$$

† If the field is weak, and the change in the level populations can be neglected, then, as can be seen from (23.7), we can ignore the other levels of the system.

in terms of the electric field $\boldsymbol{E}$. Here we make the following assumptions: $\hat{\boldsymbol{d}}_{11} = \hat{\boldsymbol{d}}_{22} = 0$, the vectors $\hat{\boldsymbol{d}}_{12}$ and $\hat{\boldsymbol{d}}_{21}$ are collinear (the dipole moment of the molecule is linearly polarized):

$$\frac{\boldsymbol{d}_{12}}{|d_{12}|} = \frac{\boldsymbol{d}_{21}}{|d_{21}|}, \tag{23.15}$$

and the energy of interaction of the molecule with the field is of the form $\hat{V} = -(\hat{\boldsymbol{d}} \cdot \hat{\boldsymbol{E}})$. In addition we shall assume that $\hat{\sigma}$ refers to unit volume. Only in this case is (23.14) strictly the polarization. Using these assumptions and equations (23.10)–(23.12), (23.14) and (23.15) it is easy to obtain (assuming that $\omega_{21} = -\omega_{12} = \omega_0 > 0$)

$$\begin{gathered} \ddot{\boldsymbol{P}} + \frac{2}{T_2}\dot{\boldsymbol{P}} + (\omega_0^2 + T_2^{-2})\boldsymbol{P} = -2N(\boldsymbol{E}\cdot \boldsymbol{d}_{21})\,\boldsymbol{d}_{12}\omega_0/\hbar, \\ \dot{N} + T_1^{-1}N - \frac{2}{\hbar\omega_0}(\boldsymbol{E}\cdot[\dot{\boldsymbol{P}} + T_2^{-1}\boldsymbol{P}]) = T_1^{-1}N_0. \end{gathered} \tag{23.16}$$

These are the equations which give the required relation between $\boldsymbol{P}$ and $\boldsymbol{E}$; N in these equations is the number of active molcules ein the field $\boldsymbol{E}$:

$$N = N_2 - N_1,$$

where N_2 and N_1 are the numbers of molecules per unit volume in the upper and lower levels respectively.

When $\boldsymbol{E} = 0$, $N = N_0$; this is the definition of N_0, the number of active molecules per unit volume when there is no field.

23.3. *Saturation Effect*

Let us find the steady-state solution of (23.10) and (23.11) taking the resonance field (23.1) into account. Equations (23.5)–(23.9) become

$$-\frac{i}{\hbar}(V_{12}^0\sigma_{21}^{(-1)} - \sigma_{12}^{(1)}V_{21}^0) - \frac{1}{T_1}(\Delta^{(0)} - \Delta_0) = 0, \tag{23.10$'$}$$

$$[i(\omega_{21} - \Omega) + T_2^{-1}]\,\sigma_{21}^{(-1)} = -\frac{i}{2\hbar}V_{21}^0\Delta^{(0)}, \quad \sigma_{12}^{(1)} = \sigma_{21}^{(-1)*}. \tag{23.11$'$}$$

Solving these equations we find after making some simple transformations

$$\Delta^{(0)} = \frac{(\omega_{21} - \Omega)^2 + T_2^{-2}}{(\omega_{21} - \Omega)^2 + T_2^{-2} + \dfrac{|V_{12}|^2}{\hbar^2}T_1T_2^{-1}}\Delta_0, \tag{23.17}$$

$$\sigma_{21}^{(-1)} = -\frac{V_{21}^{(0)}}{2\hbar}\frac{(\omega_{21} - \Omega) + iT_2^{-1}}{(\omega_{21} - \Omega)^2 + T_2^{-2} + \dfrac{|V_{12}^0|^2}{\hbar^2}T_1T_2^{-1}}\Delta_0. \tag{23.18}$$

By using these solutions we can determine the susceptibility of the system in a strong field. Let the interaction energy be of the form

$$\hat{V} = -\hat{x}_b f_0 \cos \Omega t.$$

From the definitions of the susceptibility (14.10) and of the mean value of the quantity $\hat{x}_a$

$$\langle x_a \rangle = \mathrm{Tr}\, \hat{\sigma}\hat{x}_a,$$

follows the expression for the susceptibility:

$$\chi_{ab} = \frac{2\sigma_{21}^{(-1)} x_{a12}}{f_0} = -\frac{x_{b12}x_{a21}}{\hbar} \frac{(\omega_{21} - \Omega) + iT_2^{-1}}{(\omega_{21} - \Omega)^2 + T_2^{-2} + \dfrac{|x_{b12}|^2}{\hbar^2} f_0^2 T_1 T_2^{-1}} \Delta_0. \tag{23.19}$$

Let us examine the special case when x_b and x_a are the x-components of the magnetic dipole moment $\boldsymbol{\mu}$. In this case χ_{ab} becomes the usual high-frequency magnetic susceptibility (per molecule)

$$\chi_{xx} = \frac{|\mu_{12}|^2}{\hbar} \frac{e^{\hbar\omega_{21}/kT} - 1}{e^{\hbar\omega_{21}/kT} + 1} \frac{\omega_{21} - \Omega + iT_2^{-1}}{(\omega_{21} - \Omega)^2 + T_2^{-2} + \dfrac{|\mu_{12}|^2 f_0^2}{\hbar^2} T_1 T_2^{-1}}, \tag{23.20}$$

where f_0 is the amplitude of the magnetic field, and we have replaced Δ_0 by its steady-state value.

For small values of f_0 this expression becomes (18.9). As has already been pointed out, the imaginary part of the susceptibility determines the rate of absorption of energy from the external field. This is still valid in strong fields (as can easily be checked from the derivation of (17.9)). It follows from (17.9) and (23.17) that the absorbed power is

$$Q = \frac{\Omega}{2} \chi''_{xx} f_0^2 = \frac{|\mu_{12}|^2}{2\hbar} \frac{\Omega T_2^{-1} f_0^2 \Delta_0}{(\omega_{21} - \Omega)^2 + T_2^{-2} + \dfrac{|\mu_{12}|^2 f_0^2}{\hbar^2} T_1 T_2^{-1}}. \tag{23.21}$$

In weak fields the absorbed power is proportional to the incident energy ($\sim f_0^2$), and for strong enough fields

$$\frac{|\mu_{12}|^2 f_0^2}{\hbar^2} \gg T_2^{-1} T_1^{-1} \tag{23.22}$$

so-called saturation occurs, and the absorbed power takes up a value which is independent of the external field and equal to

$$Q_{\text{sat}} = \frac{\Delta_0 \hbar \Omega}{2T_1}. \tag{23.23}$$

The meaning of this expression becomes clear if we assume that T_1 is the mean time between collisions. In a sufficiently strong field f_0 a molecule has time to make many transitions between the levels E_1 and E_2 in the time between two collisions. It is therefore equally probable that a collision will occur when the molecule in in either the upper or the lower state. If the molecule is in the upper state at the time of the collision there is absorption of a quantum $\hbar\Omega$; if it is in the lower state there is no absorption. Therefore absorption occurs on the average $1/2T_1$ times per second. Of course not all molecules take part in the absorption, the difference between the numbers of molecules in the upper and lower levels, i.e. Δ_0. If the difference of the populations is zero, transitions to the upper and lower levels are equally probable and there is on the average no absorption. We notice that when condition (23.22) is satisfied the difference of the populations (23.17) and the susceptibility decrease as $1/f^2$ as the field amplitude increases†.

23.4. *The Effect of two Fields that are Close in Frequency*

Let us now discuss the question of the simultaneous action of two fields of frequency Ω and ω on a two-level system. We shall examine the case when the two frequencies are fairly close to the transition frequency:

$$|\Omega - \omega_{21}| \lesssim T_2^{-1}, \quad |\omega - \omega_{21}| \lesssim T_2^{-1}, \tag{23.24}$$

and we shall consider that the field of frequency ω is weak. We shall impose no limitations on the magnitude of the field of frequency Ω. The interaction energy with the external field can now be written in the form

$$\hat{V} = \hat{V}^0 \cos \Omega t + \hat{V}' \cos (\omega t + \varphi). \tag{23.25}$$

We shall look for a solution of equations (23.10) and (23.11) in the form

$$\Delta = \Delta^0 + \Delta_1^+ \mathrm{e}^{i(\Omega-\omega)t} + \Delta_1^- \mathrm{e}^{-i(\Omega-\omega)t}, \tag{23.26}$$

$$\sigma_{21} = P_{21}^- \mathrm{e}^{-i\Omega t} + R_{21}^- \mathrm{e}^{-i\omega t} + Q_{21}^- \mathrm{e}^{-i(2\Omega-\omega)t},$$

$$\sigma_{12} = P_{12}^+ \mathrm{e}^{i\Omega t} + R_{12}^+ \mathrm{e}^{i\omega t} + Q_{21}^+ \mathrm{e}^{-i(2\Omega-\omega)t}. \tag{23.27}$$

It is easy to check that these are the only significant resonance terms. Terms in P_{21}^+, R_{21}^+ etc., are small and have been omitted in the approximation used. The relations

$$P_{21}^- = (P_{12}^+)^*, \quad R_{21}^- = (R_{12}^+)^*, \quad Q_{21}^- = (Q_{12}^+)^*, \quad \Delta_1^- = (\Delta_1^+)^* \tag{23.28}$$

† For further detail on the saturation effect taking into account the degeneracy of the levels see Karplus and Schwinger (1948) and Strandberg (1954).

follow from the Hermitian nature of $\hat{\sigma}$. Substituting (23.26) and (23.27) in (23.10) and (23.11), using the fact that $\hat{V}'$ is small and equating the terms in the same harmonics, we obtain the set of equations:

$$\frac{i}{\hbar}(V_{12}^{0}P_{21}^{-} - P_{12}^{+}V_{21}^{0}) + \frac{1}{T_1}(\Delta^0 - \Delta_0) = 0, \tag{23.29}$$

$$P_{21}^{-}\left[i(\omega_{21} - \Omega) + \frac{1}{T_2}\right] = -\frac{i}{2\hbar}V_{21}^{0}\Delta^0, \tag{23.30}$$

$$\left[i(\Omega - \omega) + \frac{1}{T_1}\right]\Delta_1^{+} = -\frac{i}{\hbar}(V_{12}^{0}R_{21}^{-} - P_{12}^{+}V_{21}' \mathrm{e}^{-i\varphi} - Q_{21}^{-}V_{21}^{0}), \tag{23.31}$$

$$R_{21}^{-}\left[i(\omega_{21} - \omega) + \frac{1}{T_2}\right] = -\frac{i}{2\hbar}V_{21}'\Delta^0 \mathrm{e}^{-i\varphi} - \frac{i}{2\hbar}V_{21}^{0}\Delta_1^{+}. \tag{23.32}$$

Equations (23.29) and (23.30) are the same as the equations for Δ^0 and $\sigma_{21}^{(-1)}$ (23.10′) and (23.11′) which are valid when there is no field V'. Therefore

$$P_{21}^{-} = \sigma_{21}^{(-1)}, \quad \Delta^0 = \Delta^{(0)}, \tag{23.33}$$

where $\sigma_{21}^{(-1)}$ and $\Delta^{(0)}$ are defined by (23.17) and (23.18). Using (23.31) and (23.32) we can express Δ_1^{+}, Q_{12}^{+} and R_{21}^{-} in terms of $\sigma_{21}^{(-1)}$ and $\Delta^{(0)}$:

$$\Delta_1^{+} = \left\{2\left[i(\omega_{21} - \omega) + \frac{1}{T_2}\right](\sigma_{21}^{(-1)})^{*} + \frac{i}{\hbar}V_{12}^{0}\Delta^{(0)}\right\} \times \frac{i}{2\hbar}V_{21}'\mathrm{e}^{-i\varphi} \Bigg/ \left\{\left[i(\Omega - \omega) + \frac{1}{T_1} + \frac{|V_{12}^{0}|^2}{2\hbar^2}\frac{i}{(\omega_{21} - 2\Omega + \omega) + (i/T_2)}\right] \times \left[i(\omega_{21} - \omega) + \frac{1}{T_2}\right] + \frac{|V_{12}^{0}|^2}{2\hbar^2}\right\} \tag{23.34}$$

$$R_{21}^{-} = -\left\{\left[i(\Omega - \omega) + \frac{1}{T_1}\right]\Delta^{(0)}\right.$$

$$+ \frac{i|V_{12}^0|^2/_2\hbar^2}{\omega_{21} - 2\Omega + \omega + (i/T_2)} \frac{i}{2\hbar} V'_{21}\,e^{-i\varphi} \bigg/ \left\{\left[i(\Omega - \omega) + \frac{1}{T_1}\right.\right.$$

$$\left. + \frac{|V_{12}^0|^2}{2\hbar^2} \frac{i}{(\omega_{21} - 2\Omega + \omega) + (i/T_2)}\right]$$

$$\times \left[i(\omega_{21} - \omega) + \frac{1}{T_2}\right] + \left.\frac{|V_{12}^0|^2}{2\hbar^2}\right\} \tag{23.35}$$

We can introduce the susceptibility at a frequency ω; as in (23.19) we obtain

$$\chi_{ab}(\omega) = \frac{2R_{21}^{-}x_{a12}}{f_1}, \tag{23.36}$$

$$Q_{12}^{+} = \frac{V_{12}^0}{2\hbar} \frac{\Delta_1^{+}}{[(\omega_{21} - 2\Omega + \omega) + (i/T_2)]}. \tag{23.37}$$

where f_1 is the amplitude of the radiation field of frequency ω.

Taken together with (23.35) expression (23.36) gives the absorption line shape of a weak field of frequency ω in the presence of a field of frequency Ω. As can be seen from the above expressions, the shape of the line is quite complex. When $V_{12}^0 \to 0$ the shape of the line is the same as that given by (18.9), as it should be. When the field V_{12}^0 increases the absorption decreases as f^{-2}. At the same time the following effect occurs. Let x_a^0 be a certain quantity having diagonal matrix elements†; then the mean value of this quantity due to the action of the two fields of frequencies ω and Ω is of the form

$$\langle x_a \rangle = \mathrm{Tr}\,\hat{\sigma}\hat{x}_a = \sigma_{11}x_{a11} + \sigma_{22}x_{a22} + \sigma_{12}x_{a21} + \sigma_{21}x_{a12}$$

$$= \frac{x_{a11} + x_{a22}}{2} + \frac{1}{2}(x_{a11} - x_{a22})\Delta^0 + \frac{1}{2}(x_{a11} - x_{a22})$$

$$\times (\Delta_1^{+}\,e^{i(\Omega-\omega)t} + \Delta_1^{-}\,e^{-i(\Omega-\omega t)} + \sigma_{21}^{(-1)}\,x_{a12}\,e^{-i\Omega t} + \text{complex conj.}$$

$$+ R_{21}^{-}x_{a12}\,e^{-i\omega t} + \text{complex conj.} + Q_{21}^{-}x_{a12}\,e^{-i(2\Omega-\omega)t}$$

$$+ \text{complex conj.}$$

† For example the magnetic dipole moment, if it is proportional to the angular momentum, whether orbital or spin.

Therefore not only is there a constant part and components with frequencies Ω and ω, but also combination frequencies whose amplitude, generally speaking, is not small when condition (23.22) is satisfied. This was to be expected from the general analysis obtained by using approximate perturbation theory in the preceding section. The quantities $\Delta_{1}^{\pm}$ determine the cross-susceptibility. As can be seen from (23.24), in weak fields f_0 and f_1 are proportional to the product of these fields, as expected from the results of the general theory of the preceding section.

24. Three-level Systems

24.1. Initial Equations

We shall move on to find the effect of alternating fields on three-level systems. At least two fields of frequencies close to ω_{32} and ω_{31} (or ω_{32} and ω_{21}, or ω_{31} and ω_{21}) are necessary for resonance action on a three-level system, where $E_1 < E_2 < E_3$ are the three levels question and, as usual, $\omega_{mn} = (E_m - E_n)/\hbar$. In accordance with the general theory of section 22, oscillations of the density matrix appear at the combination frequencies, one of which is close to the third resonance frequency of the three-level system. This, in its turn, may lead to the appearance of a field at this frequency (see below). We shall therefore discuss first the case when the three-level system is acted upon by a field of three frequencies: Ω_{31}, Ω_{21} and Ω_{32}, each of which is close (within the limits of the line width) to the transition frequencies ω_{31}, ω_{21} and ω_{32}:

$$\hat{V} = \hat{V}^{(31)} \cos(\Omega_{31}t - \varphi_{31}) + V^{(21)} \cos(\Omega_{21}t - \varphi_{21})$$
$$+ V^{(32)} \cos(\Omega_{32}t - \varphi_{32}). \tag{24.1}$$

Further, since the case when one of the frequencies is the sum of the other two is of most interest, we shall take

$$\Omega_{32} + \Omega_{21} = \Omega_{31}. \tag{24.2}$$

In this case each of the frequencies is the same as the combination frequency of the two other frequencies. It is convenient to introduce the notation

$$\Omega_{mn} = -\Omega_{nm}, \quad \varphi_{mn} = -\varphi_{nm}, \quad \hat{V}^{(nm)} = \hat{V}^{(mn)}. \tag{24.3}$$

Considering an ideal three-level system we assume that relaxation has an effect only on the populations of the three levels in question. In this case we can put

$$\sigma_{11} + \sigma_{22} + \sigma_{33} = 1, \tag{24.4}$$

and equations (18.3) can be rewritten in the form

$$\left(\frac{\partial}{\partial t} + i\omega_{mn}\right)\sigma_{mn} + \frac{i}{\hbar}\sum_{k=1}^{3}(V_{mk}\sigma_{kn} - \sigma_{mk}V_{kn})$$

$$= \begin{cases} \sum\limits_{k=1}^{3} w_{km}\sigma_{kk} - w_{mk}\sigma_{mm} & \text{when} \quad m = n, \\ -\dfrac{1}{\tau_{mn}}\sigma_{mn} & \text{when} \quad m \neq n. \end{cases} \tag{24.5}$$

When solving this set of equations we shall keep only the resonance terms. Then an analysis similar to that made in section 23 leads to the following non-zero matrix elements of the density matrices:

$$\sigma_{nm} = \lambda_{nm}\,\mathrm{e}^{-i\Omega_{nm}t}; \quad \lambda_{mn} = \lambda_{nm}^{*}, \tag{24.6}$$

where in the stationary case λ_{nm} are not time-dependent and are subject to the set of algebraical equations (Clogston, 1958)

$$\left[-i(\Omega_{nm} - \omega_{nm}) + \frac{1}{\tau_{nm}}\right]\lambda_{nm} = i\sum_{k=1}^{3}(A_{nk}\lambda_{km} - \lambda_{nk}A_{km}) \quad (n \neq m), \tag{24.7}$$

$$\sum_{k=1}^{3}(w_{kn}\lambda_{kk} - w_{nk}\lambda_{nn}) + i\sum_{k=1}^{3}(A_{nk}\lambda_{kn} - \lambda_{nk}A_{kn}) = 0 \tag{24.8}$$

and

$$A_{nk} = -\frac{1}{2\hbar}V_{nk}^{(nk)}\,\mathrm{e}^{i\varphi_{nk}}. \tag{24.9}$$

24.2. The Density Matrix of a Three-level System in an External Field

We shall now find the solution of equations (24.7) and (24.8). We shall discuss the case when only one of the fields (of frequency Ω_{31}) is allowed to become strong, whilst the other two fields are weak, so we can neglect the change in the populations $\lambda_{nn} = \sigma_{nn}$ caused by these fields. (We shall keep only terms that are linearly dependent on A_{21}, A_{12}, A_{32} and A_{23}.) In this case only four of the equations (24.7) and (24.8) contain A_{13} and A_{31}:

$$-(w_{12} + w_{13})\,\sigma_{11} + w_{21}\sigma_{22} + w_{31}\sigma_{33} + i(A_{13}\lambda_{31} - \lambda_{13}A_{31}) = 0, \tag{24.10}$$

$$-(w_{21} + w_{23})\,\sigma_{22} + w_{12}\sigma_{11} + w_{32}\sigma_{33} = 0, \tag{24.11}$$

$$-(w_{31} + w_{32})\,\sigma_{33} + w_{13}\sigma_{11} + w_{23}\sigma_{22} - i(A_{13}\lambda_{31} - \lambda_{13}A_{31}) = 0, \tag{24.12}$$

$$\left[-i(\Omega_{31} - \omega_{31}) + \frac{1}{\tau_{31}}\right]\lambda_{31} = iA_{31}(\sigma_{11} - \sigma_{33}). \tag{24.13}$$

In view of the fact that equations (24.7) and (24.8) are governed by condition (24.4) we use this condition instead of equation (24.12). The relaxation times compatible with the condition of detailed balancing (8.12)† are:

$$w_{km} = \frac{1}{T_{km}}\,\sigma^0_{mm}, \quad w_{mk} = \frac{1}{T_{mk}}\,\sigma^0_{kk}, \quad T_{mk} = T_{km}, \tag{24.14}$$

where σ^0_{mm} is the equilibrium density matrix.

Solving equations (24.10), (24.11), (24.4) and (24.13) we find:

$$\sigma_{11} = \sigma^0_{11} - \frac{2\,|A_{31}|^2\,\tau_{31}^{-1}}{[(\Omega_{31} - \omega_{31})^2 + \tau_{31}^{-2}]} \times \left[\frac{1}{T_{32}}(\sigma^0_{22} + \sigma^0_{33}) + \frac{1}{T_{12}}\,\sigma^0_{11}\right]\Sigma, \tag{24.15}$$

$$\sigma_{22} = \sigma^0_{22}\left\{1 + \frac{2\,|A_{31}|^2\,\tau_{31}^{-1}}{[(\Omega_{31} - \omega_{31})^2 + \tau_{31}^{-2}]}\left(\frac{1}{T_{32}} - \frac{1}{T_{12}}\right)\Sigma\right\}, \tag{24.16}$$

$$\sigma_{33} = \sigma^0_{33} + \frac{2\,|A_{31}|^2\,\tau_{31}^{-1}}{[(\Omega_{31} - \omega_{31})^2 + \tau_{31}^{-2}]} \times \left[\frac{1}{T_{12}}(\sigma^0_{22} + \sigma^0_{11}) + \frac{1}{T_{23}}\,\sigma^0_{33}\right]\Sigma, \tag{24.17}$$

$$\sigma_{33} - \sigma_{22} = \frac{\sigma^0_{33} - \sigma^0_{22} + \dfrac{4\,|A_{31}|^2\,\tau_{31}^{-1}}{[(\Omega_{31} - \omega_{31})^2 + \tau_{31}^{-2}]}\,T_e}{1 + \dfrac{4\,|A_{31}|^2\,\tau_{31}^{-1}}{[(\Omega_{31} - \omega_{31})^2 + \tau_{31}^{-2}]}\,T}, \tag{24.18}$$

† A similar definition is given in section 23, (23.13).

$$\sigma_{22} - \sigma_{11} = \frac{\sigma^0_{22} - \sigma^0_{11} - \dfrac{4\,|A_{31}|^2\,\tau_{31}^{-1}}{[(\Omega_{31} - \omega_{31})^2 + \tau_{31}^{-2}]}\,T_e}{1 + \dfrac{4\,|A_{31}|^2\,\tau_{31}^{-1}}{[(\Omega_{31} - \omega_{31})^2 + \tau_{31}^{-2}]}\,T}; \quad (24.19)$$

$$\sigma_{33} - \sigma_{11} = \frac{\sigma^0_{33} - \sigma^0_{11}}{1 + \dfrac{4\,|A_{31}|^2\,\tau_{31}^{-1}}{[(\Omega_{31} - \omega_{31})^2 + \tau_{31}^{-2}]}\,T}, \quad (24.20)$$

$$\lambda_{31} = A_{31}\,\frac{[(\omega_{31} - \Omega_{31}) + i\tau_{31}^{-1}]\,(\sigma^0_{11} - \sigma^0_{33})}{(\Omega_{31} - \omega_{31})^2 + \tau_{31}^{-2} + 4\,|A_{31}|^2\,\tau_{31}^{-1}T}, \quad (24.21)$$

where

$$\sum = (\sigma^0_{11} - \sigma^0_{33})\left[D + \frac{4\,|A_{31}|^2\,\tau_{31}^{-1}}{[(\Omega_{31} - \omega_{31})^2 + \tau_{31}^{-2}]}\right.$$

$$\left.\times\left\{\frac{1}{T_{12}}\,\sigma^0_{11} + \frac{1}{2}\left(\frac{1}{T_{32}} + \frac{1}{T_{12}}\right)\sigma^0_{22} + \frac{1}{T_{32}}\,\sigma^0_{33}\right\}\right]^{-1};$$

$$D = \frac{1}{T_{13}T_{12}}\,\sigma^0_{11} + \frac{1}{T_{12}T_{23}}\,\sigma^0_{22} + \frac{1}{T_{13}T_{23}}\,\sigma^0_{33} \quad (24.22)$$

and we have used the effective relaxation times

$$T = \frac{\dfrac{1}{T_{12}}\,\sigma^0_{11} + \dfrac{1}{2}\left(\dfrac{1}{T_{23}} + \dfrac{1}{T_{12}}\right)\sigma^0_{22} + \dfrac{1}{T_{32}}\,\sigma^0_{33}}{\dfrac{1}{T_{13}T_{12}}\,\sigma^0_{11} + \dfrac{1}{T_{12}T_{23}}\,\sigma^0_{22} + \dfrac{1}{T_{13}T_{23}}\,\sigma^0_{33}}, \quad (24.23)$$

$$T_e = \frac{1}{2}\,\frac{\dfrac{1}{T_{12}}\,(\sigma^0_{11} - \sigma^0_{22}) + \dfrac{1}{T_{32}}\,(\sigma^0_{33} - \sigma^0_{22})}{\dfrac{1}{T_{13}T_{12}}\,\sigma^0_{11} + \dfrac{1}{T_{12}T_{23}}\,\sigma^0_{22} + \dfrac{1}{T_{13}T_{23}}\,\sigma^0_{33}}. \quad (24.24)$$

From the remaining equation (24.7) we can express λ_{21} and λ_{23} in terms of the population differences (24.18)–(24.20) and λ_{13}, λ_{31}:

$$\lambda_{21} = \{i[i(\Omega_{32} - \omega_{32}) + \tau_{32}^{-1}]\,[A_{21}(\sigma_{11} - \sigma_{22}) + A_{23}\lambda_{31}] + A_{31}[A_{23}(\sigma_{33} - \sigma_{22}) + A_{21}\lambda_{13}]\} \times \left\{\left[-i(\Omega_{21} - \omega_{21}) + \frac{1}{\tau_{21}}\right] \times \left[i(\Omega_{32} - \omega_{32}) + \frac{1}{\tau_{23}}\right] + |A_{13}|^2\right\}^{-1}, \tag{24.25}$$

$$\lambda_{23} = \{i[-i(\Omega_{21} - \omega_{21}) + \tau_{12}^{-1}]\,[A_{23}(\sigma_{33} - \sigma_{22}) + A_{21}\lambda_{13}] + A_{13}[A_{21}(\sigma_{11} - \sigma_{22}) + A_{23}\lambda_{31}]\} \times \left\{\left[-i(\Omega_{21} - \omega_{21}) + \frac{1}{\tau_{21}}\right] \times \left[i(\Omega_{32} - \omega_{32}) + \frac{1}{\tau_{23}}\right] + |A_{13}|^2\right\}^{-1} \tag{24.26}$$

Therefore in the approximation used we have found all the elements of the density matrix of a three-level system acted upon by three resonance fields (one of which may be strong). Equations (24.20) and (24.21) describe the saturation effect on two levels (1 and 3) of a three-level system (in the case when the effect of the rest of the fields can be neglected). These equations have the same form as the corresponding equations (23.17) and (23.18) in a two-level system. Allowance for level 2 (which is coupled with the rest of the levels by relaxation) is made in that (24.20) and (24.21) contain the other relaxation times. Instead of T_1 and T_2 in (23.14) and (23.15) equations (24.20) and (24.21) contain the times T and τ_{31}. The effective relaxation time T depends upon σ_{22}^0 and T_{12}, T_{23}. When $T_{12} = T_{23} = \infty$ there is no need to allow for level 2 and then T, as can easily be checked from (24.23), becomes T_{13}. In the general case, not only do the populations σ_{11} and σ_{33} change because of the field $\hat{V}^{(13)}$ of frequency $\Omega_{31} \approx \omega_{31}$, but also, as can be seen from (24.16), the population σ_{22} which is coupled with levels 1 and 3 by relaxation.

24.3. The Susceptibility and Cross-susceptibility of a Three-level System

We shall now calculate the susceptibility and cross-susceptibility of a three-level system. We first find the general expressions and then discuss some special cases. Since A_{21} and A_{23} are contained linearly (each of them separately) in the expression for λ_{21} and λ_{23} they can be written in the form

$$\lambda_{21} = \lambda_{21}^{(1)} A_{21} + \lambda_{21}^{(2)} A_{23} A_{31}, \tag{24.27}$$

$$\lambda_{32} = \lambda_{32}^{(1)} A_{32} + \lambda_{32}^{(2)} A_{12} A_{31}, \tag{24.28}$$

where $\lambda_{21}^{(1)}$, $\lambda_{21}^{(2)}$, $\lambda_{23}^{(1)}$ and $\lambda_{23}^{(2)}$ generally depend upon A_{13} and A_{31}.

Only when field A_{13} is weak enough can the dependence of these coefficients on A_{13} be neglected. We also introduce the forces f_a^{nm} by means of the relations†:

$$\hat{V}^{nm}(t) = -\hat{x}_a f_a^{(nm)} \cos(\Omega_{nm} t - \varphi_{nm}), \tag{24.29}$$

where the operator $\hat{x}_a$ may be, for example, the magnetic dipole moment operator.

Generalizing (23.19) and using (22.19) it is easy to find the susceptibility and cross-susceptibility in terms of $\lambda^{(1)}$ and $\lambda^{(2)}$:

$$\chi_{ba}(\Omega_{21}) = \lambda_{21}^{(1)} \frac{x_{b12} x_{a21}}{\hbar}, \quad \chi_{ba}(\Omega_{32}) = \lambda_{32}^{(1)} \frac{x_{b23} x_{a32}}{\hbar}, \tag{24.30}$$

$$\chi_{abc}(\Omega_{32}, \Omega_{31}) = \frac{1}{2} \lambda_{21}^{(2)} \frac{x_{a12} x_{b23} x_{c31}}{\hbar^2} e^{-i(\varphi_{32} - \varphi_{31})}, \tag{24.31}$$

$$\chi_{abc}(\Omega_{21}, \Omega_{31}) = \frac{1}{2} \lambda_{32}^{(2)} \frac{x_{a23} x_{b12} x_{c31}}{\hbar^2} e^{-i(\varphi_{31} - \varphi_{21})}. \tag{24.32}$$

(We recall that in the summation in (22.19) the term $\chi_{abc}(\Omega_{21}, \Omega_{31}) f_b(\Omega_{21}) \times f_c(\Omega_{31})$ occurs twice if (b, Ω_{21}) and (c, Ω_{31}) are different.) We find from (24.25) and (24.26):

$$\lambda_{21}^{(1)} = \frac{i[i(\Omega_{32} - \omega_{32}) + \tau_{32}^{-1}](\sigma_{11} - \sigma_{22}) + A_{31}\lambda_{13}}{[-i(\Omega_{21} - \omega_{21}) + \tau_{12}^{-1}][i(\Omega_{32} - \omega_{32}) + \tau_{32}^{-1}] + |A_{13}|^2}, \tag{24.33}$$

$$\begin{aligned} \lambda_{21}^{(2)} = {} & -\{[i(\Omega_{32} - \omega_{32}) + \tau_{32}^{-1}][-i(\Omega_{31} - \omega_{31}) + \tau_{31}^{-1}]^{-1} \\ & \times [(\sigma_{11} - \sigma_{33}) + (\sigma_{33} - \sigma_{22})]\} \\ & \times \{[-i(\Omega_{21} - \omega_{21}) + \tau_{12}^{-1}][i(\Omega_{32} - \omega_{32}) + \tau_{32}^{-1}] \\ & + |A_{13}|^2\}^{-1}, \end{aligned} \tag{24.34}$$

† Here there is no summation for repeated indices.

$$\lambda_{32}^{(1)} = \frac{-i[i(\Omega_{21} - \omega_{21}) + \tau_{12}^{-1}](\sigma_{33} - \sigma_{22}) + A_{31}\lambda_{13}}{[i(\Omega_{21} - \omega_{21}) + \tau_{12}^{-1}][-i(\Omega_{32} - \omega_{32}) + \tau_{23}^{-1}] + |A_{13}|^2}, \tag{24.35}$$

$$\begin{aligned}\lambda_{32}^{(2)} = {} & \{[i(\Omega_{21} - \omega_{21}) + \tau_{12}^{-1}][-i(\Omega_{31} - \omega_{31}) + \tau_{31}^{-1}]^{-1} \\ & \times [(\sigma_{11} - \sigma_{33}) + (\sigma_{11} - \sigma_{22})]\} \\ & \times \{[i(\Omega_{21} - \omega_{21}) + \tau_{12}^{-1}][-i(\Omega_{32} - \omega_{32}) + \tau_{23}^{-1}] \\ & + |A_{13}|^2\}.\end{aligned} \tag{24.36}$$

24.4. Special Cases

1. First we shall examine the conditions appropriate to a solid-state three-level microwave device (Clogston, 1958; Scovil, Feher and Seidel, 1957). We assume that the energy level differences are small compared with kT (T is the working temperature). Then we can say approximately

$$\sigma_{11}^0 - \sigma_{22}^0 \equiv D_{12}^0 = \frac{E_2 - E_1}{3kT}, \quad \sigma_{11}^0 - \sigma_{33}^0 \equiv D_{13}^0 = \frac{E_3 - E_1}{3kT},$$

$$\sigma_{22}^0 - \sigma_{33}^0 \equiv D_{23}^0 = \frac{E_3 - E_2}{3kT}. \tag{24.37}$$

Here, and in what follows, we shall use the notation

$$D_{mn} = \sigma_{mm} - \sigma_{nn}. \tag{24.38}$$

We shall further assume that (see Scovil, Feher and Seidel, 1957)

$$T_{12} \ll T_{13} \quad \text{and} \quad T_{12} \ll T_{23}. \tag{24.39}$$

In this case we find from (24.23) and (24.24)

$$T_e = \frac{1}{2}\frac{D_{12}^0}{\sigma_{11}^0 + \frac{1}{2}\sigma_{22}^0} T = \frac{D_{12}^0}{1 + D_{13}^0} T \approx D_{12}^0 T, \tag{24.40}$$

$$T = \frac{\sigma_{11}^0 + \frac{1}{2}\sigma_{22}^0}{\frac{1}{T_{13}}\sigma_{11}^0 + \frac{1}{T_{23}}\sigma_{22}^0} \approx \frac{1}{2}\frac{1}{\frac{1}{T_{13}}\sigma_{11}^0 + \frac{1}{T_{23}}\sigma_{22}^0}. \tag{24.41}$$

From (24.19) we have approximately

$$D_{12} \approx D_{12}^0. \tag{24.42}$$

We shall further examine the case when all the "off-diagonal" relaxation times are

$$\tau_{mn} = \tau \quad \text{and} \quad \Omega_{31} = \omega_{31}. \tag{24.43}$$

Introducing the notation

$$b = \frac{|V_{31}^{(31)}|^2}{4\hbar^2}\tau^2 = |A_{31}|^2\tau^2, \quad R = \frac{D_{13}^0}{D_{12}^0}\cdot\frac{1}{1 + \frac{4T}{\tau}b} \tag{24.44}$$

and using equations (24.18), (24.20), (24.33)–(24.36) we find:

$$D_{13} = \frac{D_{13}^0}{1 + 4b\frac{T}{\tau}}, \quad D_{32} = \frac{D_{12}^0}{1 - R}, \tag{24.45}$$

$$\lambda_{21}^{(1)} = i\tau D_{12}^0 \frac{1 + i\delta - bR}{(1 + i\delta)^2 + b}, \tag{24.46}$$

$$\lambda_{21}^{(2)} = \tau^2 D_{12}^0 \frac{1 - 2R - i\delta R}{(1 + i\delta)^2 + b}, \tag{24.47}$$

$$\lambda_{32}^{(1)} = -i\tau D_{12}^0 \frac{(1 - R)(1 - i\delta) + bR}{(1 - i\delta)^2 + b}, \tag{24.48}$$

$$\lambda_{32}^{(2)} = \tau^2 D_{12}^0 \frac{1 + R(1 - i\delta)}{(1 - i\delta)^2 + b}, \tag{24.49}$$

where $\sigma = (\Omega_{32} - \omega_{32})\tau = -(\Omega_{21} - \omega_{21})\tau$ (if $\Omega_{31} = \omega_{31}$, as we have assumed). Relations (24.46)–(24.49) and (24.30)–(24.32) make it possible to find the susceptibility and cross-susceptibility for different values of the parameters. In particular the susceptibility

$$\chi_{aa}(\Omega_{32}) = \chi' + i\chi''$$

provided that $|x_{a23}| = |x_{a12}| = |\mu_{23}| = |\mu_{12}|$ (where μ is the magnetic dipole moment) can be found from

$$\frac{\chi'}{\chi_{12}} = \frac{\delta[(1 - R) + (3R - 1)b + (1 - R)\delta^2]}{(1 + b - \delta^2)^2 + 4\delta^2},$$

$$\frac{\chi''}{\chi_{12}} = \frac{(1 + b)(1 - R + Rb) + (1 - R - Rb)\delta^2}{(1 + b - \delta^2)^2 + 4\delta^2}, \tag{24.50}$$

where

$$\chi_{12} = \frac{1}{\hbar} |\mu_{12}|^2 \tau D^0_{12}.$$

Figures V.2 and V.3 show a series of curves for the susceptibilities (24.50). (The curves are taken from the paper by Clogston (1958).) An interesting feature of these curves is the formation of a doublet (the absorption line splitting in two) when b is high enough. In order to estimate the magnitudes of the susceptibility and cross-susceptibility in the case under discussion we shall give the expressions for the quantities

$$\lambda^{(1)}_{21} \equiv i\tau D_{ef}, \quad \lambda^{(2)}_{21} \equiv -\tau^2 D_e, \quad \lambda^{(1)}_{32} \equiv i\tau \tilde{D}_{ef}, \quad \lambda^{(2)}_{32} \equiv -\tau^2 \tilde{D}_e \tag{24.51}$$

in the most interesting range of variation of $|A_{13}|^2$.

Table V.1 gives the values of the effective population differences D_{ef}, D_e, $\tilde{D}_{ef}$, $\tilde{D}_e$ (defined by the equations (24.51) in the case of strong resonance, $\delta = 0$).

The values given in the centre column of the table are valid for $\tau \ll T$. If $\tau \approx T$, then only the first and third columns can be used.

2. Let us now look at the case when all the times T_{ik} are equal:

$$T_{12} = T_{23} = T_{13} = T_1. \tag{24.52}$$

TABLE V.1

	$\lvert A_{13}\rvert^2 \ll (\tau T)^{-1}$	$(\tau T)^{-1} \ll \lvert A_{13}\rvert^2 \ll \tau^{-2}$ $\tau \ll T$	$\lvert A_{13}\rvert^2 \gg \tau^{-2}$
D_{ef}	D^0_{12}	D^0_{12}	$\left(D^0_{12} - \frac{\tau}{4T} D^0_{13}\right)\Big/ b$
D_e	$D^0_{23} + D^0_{13}$	$-D^0_{12}$	$-D^0_{12}/b$
$\tilde{D}_{ef}$	D^0_{23}	$-D^0_{12}$	$-\left(D^0_{12} + \frac{\tau}{4T} D^0_{13}\right)\Big/ b$
$\tilde{D}_e$	$-(D^0_{13} + D^0_{12})$	$-D^0_{12}$	$-D^0_{12}/b$

This can be the case in three-level gas devices (Kontorovich and Prokhorov, 1957; Javan, 1957). From (24.23) and (24.24) we find

$$T = T_1, \quad T_e = \tfrac{1}{2} T_1 (D^0_{12} + D^0_{32}). \tag{24.53}$$

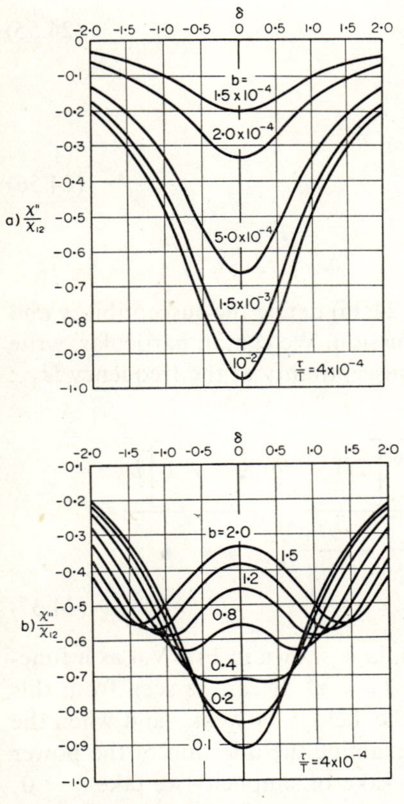

FIG. V.2. Imaginary part of the normalized susceptibility as a function of $\delta = (\Omega_{32} - \omega_{32})$ for $\tau/T = 4 \times 10^{-4}$ and different values of the parameter $b = (|V_{(31)}^{(3)}|/4\hbar^2)\tau^2$.

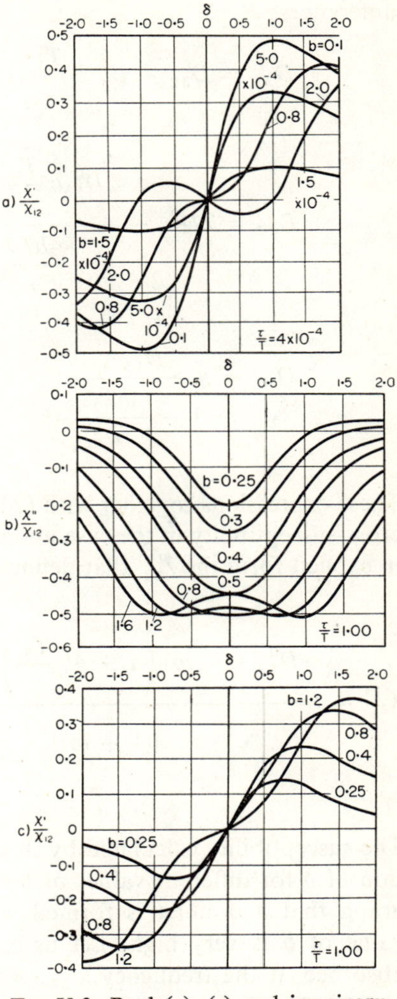

FIG. V.3. Real (*a*), (*c*) and imaginary (*b*) parts of normalized susceptibility as a function of $\delta = (\Omega_{32} - \omega_{32})$ for different values of the parameter b.

Substituting these values in (24.18)–(24.20) we obtain for the population differences:

$$D_{32} = D_{32}^0 + D_{13}^0 b \frac{T_1}{\tau} \Big/ (1 + 4bT_1/\tau), \tag{24.54}$$

$$D_{12} = 2D_{12}^0 - \frac{2D_{13}^0 b \dfrac{T_1}{\tau}}{1 + \dfrac{4bT_1}{\tau}}, \tag{24.55}$$

$$D_{13} = \frac{D_{13}^0}{1 + \dfrac{4bT_1}{\tau}}. \tag{24.56}$$

These expressions together with (24.30)–(24.36) define the susceptibility and cross-susceptibility in the case under discussion. We can, in particular, write in explicit form the $\lambda_{21}^{(1)}$ that defines the susceptibility at the frequency Ω_{21}:

$$\lambda_{21}^{(1)} = i \frac{D_{12}^0(1 + i\delta)\left(1 + 4b \dfrac{T_1}{\tau}\right) - D_{13}^0\left[2(1 + i\delta)\dfrac{T_1}{\tau} + 1\right] b}{[(1 + i\delta)^2 + b]\left(1 + 4\dfrac{bT_1}{\tau}\right)} \tau. \tag{24.57}$$

The susceptibility calculated by this formula is shown in Fig. V.4 as a function of δ for different values of b (when $T_1 = \tau$). It can be seen from this graph that a doublet is formed when the field b is weak, and when the value of b is very high. Let us briefly examine the question of the power absorbed at the frequency Ω_{21}. For the sake of simplicity we take $\delta = 0$, $T_1 = \tau$. In this case the imaginary part of the susceptibility, which determines the absorption, is

$$\chi'' = -\frac{|\mu_{12}|^2}{\hbar} D_{23}^0 \frac{3b - p(1 + b)}{(1 + b)(1 + 4b)} \tau, \quad p = \frac{D_{12}^0}{D_{23}^0}. \tag{24.58}$$

When $b \to 0$ this quantity, as expected, is positive. When

$$b(3 - p) > p \tag{24.59}$$

χ'' becomes negative, i.e. there is negative absorption or stimulated emission. The condition (24.59) can be satisfied only when $p < 3$ or $\sigma_{11}^0 - \sigma_{22}^0 < 3(\sigma_{22}^0 - \sigma_{33}^0)$. In this case when the level differences are much less than kT condition (24.59) can be satisfied if

$$E_2 - E_1 < 3(E_3 - E_2).$$

We notice that the condition that χ'' is negative is not the same as the condition for inversion of levels 1 and 2. It is not difficult to see that it is

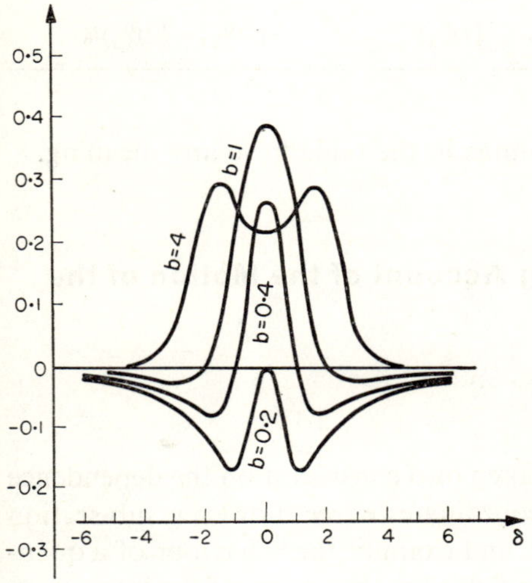

FIG. V.4. Imaginary part of susceptibility with opposite sign (in relative units) as a function of $\delta = (\Omega_{21} - \omega_{21})\tau$ for different values of the parameter b.

necessary only that $p > 1$ for $\sigma_{11} - \sigma_{22}$ to be negative. It is possible to select the field b so that the population difference is positive and χ'' is negative. This result cannot be obtained from the balance scheme in which the pump field is allowed for in the form of an additional transition probability, as is done for example by Bloembergen (1956). In the case when $\tau \ll T_1$ it is easy to show that the conditions for χ'' to be negative, and for the inversion of the populations $\sigma_{11} - \sigma_{22}$, are the same.

We have already pointed out that oscillations of dipole moment (or other appropriate quantities in a three-level system) occur at the combination frequency when it is subjected to two fields A_{13} and A_{23} (or A_{21}). The characteristic response of a system at the combination frequency is the cross-relaxation. We shall now give estimates of the cross-susceptibility and the susceptibility of a system at resonance when $\tau \ll T_1$.

Introducing definitions similar to (24.51) and using expressions (24.33)–(24.36) and (24.54)–(24.56) we arrive at the following table (Fain, Khanin and Yashchin, 1962, 1964).

TABLE V. 2

	$\lvert A_{13}\rvert^2 \ll (\tau T)^{-1}$	$(\tau T)^{-1} \ll \lvert A_{13}\rvert^2 \ll \tau^{-2}$, $\tau \ll T$	$\lvert A_{13}\rvert^2 \gg \tau^{-2}$
D_{ef}	D^0_{12}	$D^0_{12} - \frac{1}{2}D^0_{13}$	$(D^0_{12} - \frac{1}{2}D^0_{13})/b$
D_e	$D^0_{23} + D^0_{13}$	$D^0_{23} - \frac{1}{2}D^0_{13}$	$(D^0_{23} - \frac{1}{2}D^0_{13})/b$
$\tilde{D}_{ef}$	D^0_{23}	$D^0_{23} - \frac{1}{2}D^0_{13}$	$(D^0_{23} - \frac{1}{2}D^0_{13})/b$
$\tilde{D}_e$	$-(D^0_{13} + D^0_{12})$	$-(D^0_{12} - \frac{1}{2}D^0_{13})$	$-(D^0_{12} - \frac{1}{2}D^0_{13})/b$

If $\tau \approx T$ only the first and third columns in the table have any meaning.

25. Distributed Systems, taking Account of the Motion of the Molecules

25.1. Initial Equations

Up to now we have not explicitly taken into consideration the dependence of the external field on the spatial coordinates $\boldsymbol{r}$ (see, however, sub-section 14.3). We shall now take an example and examine the behaviour of a quantum system affected by a non-uniform field. Let the energy of the interaction with the external field be of the form

$$\hat{V} = \hat{V}(\boldsymbol{r}, t). \tag{25.1}$$

If the equations for the density matrix do not contain a coordinate, then $\boldsymbol{r}$ can be considered to be a given parameter and we may use all the earlier solutions in which A_{ik} and the phases φ_{mn} are functions of the coordinates. This situation occurs if a gas (or solid) consists of fixed molecules. In this case each molecule is acted upon by the field in its own locality and the density matrix depends on the magnitude of the field at this point. If, however, the molecule is moving in a non-uniform field the interaction of the molecule with the field will depend on the nature of its motion. In this case the density matrix of the system as a function of $\boldsymbol{r}$ cannot be obtained directly from the solutions of the earlier section.

We shall now derive the transport equation for the density matrix of a system of weakly interacting moving molecules. First of all let us take a look at a single molecule moving in a non-uniform field. We shall consider the motion of the molecules to be classical (taking place on a definite trajectory) and given by

$$\boldsymbol{r} = \boldsymbol{r}(t). \tag{25.2}$$

In this case the density matrix of the molecule is a complex time function

$$\hat{\sigma}(t) = \hat{\sigma}(t, \boldsymbol{r}(t)). \tag{25.3}$$

The change with time of this density matrix is given by the transport equation

$$\frac{d\hat{\sigma}}{dt} = \frac{\partial\hat{\sigma}}{\partial t} + (\boldsymbol{v}\cdot\nabla\hat{\sigma}) = -\frac{i}{\hbar}[\hat{E}_0 + \hat{V}(\boldsymbol{r}(t), t), \hat{\sigma}] + \hat{\Gamma}(\hat{\sigma}), \tag{25.4}$$

where $\boldsymbol{v} = \boldsymbol{r}$.

We note that the form of the relaxation term $\hat{\Gamma}(\hat{\sigma})$ may, generally speaking, alter because of the motion of the molecule. At low (non-relativistic) velocities, however, this effect may be neglected. Let us take, for example, the case where relaxation is due to radiation into free space (see below, section 32, formula (32.6a)). Then the transition probability for a molecule in uniform motion is

$$w_{mk} = 2\pi \sum_{\nu} |B_{\nu mk}|^2 \frac{(\bar{n}_\nu + 1)}{2[\omega_\nu - (\boldsymbol{k}_\nu\cdot\boldsymbol{v})]}\,\delta\,(E_m - E_k + \hbar[\omega_\nu - (\boldsymbol{k}_\nu\cdot\boldsymbol{v})]);$$

$$E_k > E_m. \tag{25.5}$$

This expression can be derived from (32.6a) if we take account of the Doppler effect for the moving molecule. It can be seen from (25.5) that for small enough $(\boldsymbol{k}_\nu\cdot\boldsymbol{v})$ compared with the transition frequency ω_0 the change in w_{mk} because of the motion can be neglected. Therefore when discussing a single moving molecule we shall use (25.4) in which $\hat{\Gamma}(\hat{\sigma})$ is of the form (see (18.3)):

$$\Gamma_{mn} = \begin{cases} \sum_k (w_{km}\sigma_{kk} - w_{mk}\sigma_{mm}) & \text{when} \quad m = n, \\[2ex] -\dfrac{1}{\tau_{mn}}\sigma_{mn} & \text{when} \quad m \neq n, \end{cases} \tag{25.6}$$

where the coefficients w_{km} and τ_{mn} have the same values as for stationary molecules.

We are generally interested in the density matrix as a function of the independent variables $\boldsymbol{r}$ and t (and not as a function of t for a single molecule). Let us first examine a uniform beam of molecules with identical velocities in order to obtain a density matrix of this type. Every molecule in this beam is subject to the same equation (25.4) with (25.6). Therefore this equation is valid at any point $\boldsymbol{r}$ of the beam. The equation for the density matrix $\sigma(t, \boldsymbol{r})$ (where t and $\boldsymbol{r}$ are independent variables) becomes (for a uniform molecular beam)

$$\frac{\partial\hat{\sigma}(t, \boldsymbol{r})}{\partial t} + (\boldsymbol{v}\cdot\nabla)\hat{\sigma}(t, \boldsymbol{r}) = -\frac{i}{\hbar}\,[\hat{E}_0 + \hat{V}(\boldsymbol{r}, t), \hat{\sigma}(t, \boldsymbol{r})] + \Gamma\,[\hat{\sigma}(t, \boldsymbol{r})]. \tag{25.7}$$

Let us now deal with a gas which has a certain velocity distribution (which is Maxwellian in the equilibrium case) and in which the molecules undergo collisions and change their velocity. Then $\boldsymbol{v}$ for each molecule is a complicated function of time, so in the general case we cannot use equation (25.7), even for a group of gas molecules which have at a given time a velocity $\boldsymbol{v}$ (at subsequent times each of the molecules in the group will be described by different equations since $\boldsymbol{v}$ will change in a different way for each molecule). There is a case, however, when (25.7) describes the gas as a whole or, to be more precise, a group of gas molecules having a velocity $\boldsymbol{v}$. This is the case when the time of free flight τ_{fr} of a molecule is greater than τ (the characteristic relaxation time):

$$\tau_{\mathrm{fr}} > \tau. \tag{25.8}$$

In this case we can consider that basically each molecule moves in a straight line with uniform velocity; at the same time there are collisions which may play no part in the relaxation process but which make the distribution of the gas velocities random. The solution of (25.7) will, of course, have to be averaged for all $\boldsymbol{v}$ (in particular for the Maxwellian velocity distribution).

Therefore the equation for the density matrix of moving molecules in a non-uniform field differs from (18.3) by the presence of the convective term $(\boldsymbol{v}\cdot\nabla)\hat{\sigma}$ (and the necessity for velocity averaging). If we examine the steady state solution, then

$$\nabla\hat{\sigma} \sim \boldsymbol{k}\hat{\sigma}, \tag{25.9}$$

where $\boldsymbol{k}$ is the wave vector which specifies the non-uniform field (i.e. $k \sim 2\pi/\lambda$, where λ is the characteristic dimension of an inhomogeneity in the external field).

Comparing the second terms in the left and right-hand sides we find that their ratio is of the order of

$$(\boldsymbol{v}\cdot\nabla)\sigma/\Gamma(\hat{\sigma}) \approx (\boldsymbol{k}\cdot\boldsymbol{v})\tau. \tag{25.10}$$

Therefore the convective term can be neglected when

$$(\boldsymbol{k} \cdot \boldsymbol{v})\tau \ll 1. \tag{25.11}$$

On the other hand, it is obvious that when the time of free flight is very small (and does not satisfy (25.8)) we need not allow for the motion of the molecules provided that

$$(\boldsymbol{k} \cdot \boldsymbol{v})\tau_{\text{fr}} \sim \frac{l}{\lambda} \ll 1, \tag{25.12}$$

where l is the mean free path.

In this case each molecule is in a quasi-uniform field. In the extreme case the molecules are stationary ($l = 0$). Condition (25.12) holds for the molecules of a solid which make small oscillations about their equilibrium positions. Condition (25.11) is generally satisfied for a gas at radiofrequencies because of the large wavelength. In optics, as a rule, the opposite case is true:

$$(\boldsymbol{k} \cdot \boldsymbol{v})\tau \gg 1. \tag{25.13}$$

25.2. *Steady-state Solutions*

Let us now find the steady-state solutions of equation (25.7). We shall examine first the case of a weak external field. In this case we can neglect the change in the populations of the levels, and the off-diagonal elements of the density matrix are subject to the equation

$$\begin{aligned}\left(\frac{\partial}{\partial t} + i\omega_{mn} + \frac{1}{\tau_{mn}} + (\boldsymbol{v}\cdot\nabla)\right)\sigma_{mn} &= -\frac{i}{\hbar}\sum_k (V_{mk}\sigma^0_{kn} - \sigma^0_{mk}V_{kn}) \\ &= -\frac{i}{\hbar} V_{mn}(\sigma^0_{nn} - \sigma^0_{mm}),\end{aligned} \tag{25.14}$$

where in the last equation it is assumed that the unperturbed density matrix σ^0_{mn} is diagonal.

We expand the energy of the interaction with the external field into a Fourier series (or integral):

$$V_{mn} = \sum_{k,\omega} v_{mn}(\boldsymbol{k}, \omega)\, \mathrm{e}^{i((\boldsymbol{k}\cdot\boldsymbol{r})-\omega t)}. \tag{25.15}$$

We look for a solution of (25.14) of the form

$$\sigma_{mn} = \sum_{k,\omega} \sigma_{mn}(\boldsymbol{k}, \omega)\, \mathrm{e}^{i((\boldsymbol{k}\cdot\boldsymbol{r})-\omega t)}. \tag{25.16}$$

Substituting (25.15) and (25.16) in (25.14) we find

$$\sigma_{mn}(\boldsymbol{k}, \omega) = -\frac{i}{\hbar} \frac{v_{mn}(\boldsymbol{k}, \omega)(\sigma^0_{nn} - \sigma^0_{mm})}{i[\omega_{mn} - \omega + (\boldsymbol{k}\cdot\boldsymbol{v})] + \tau^{-1}_{mn}}.$$

Writing the interaction energy in the form (14.1) it is not difficult by analogy with section 14 to introduce the susceptibility as a function of ω and $\boldsymbol{k}$:

$$\chi_{ba}(\boldsymbol{k}, \omega) = \frac{1}{\hbar} \sum_{m,n} x_{amn} x_{bnm} \frac{\sigma^0_{mm} - \sigma^0_{nn}}{[\omega_{mn} - \omega + (\boldsymbol{k}\cdot\boldsymbol{v})] - i\tau^{-1}_{mn}}. \qquad (25.17)$$

We are therefore dealing with the case of spatial dispersion when the susceptibility depends on the wave vector (see sub-section 14.3). Expression (25.17) holds for a monokinetic beam of molecules. In the general case we must average (25.17) over the velocity distribution.

We now go on to a discussion of the case when the external field is not weak. Let us take a two-level system. We write equations (25.7) in a form similar to (23.10) and (23.11):

$$\left(\frac{\partial}{\partial t} + (\boldsymbol{v}\cdot\nabla)\right)\Delta = -\frac{2i}{\hbar}(V_{12}\sigma_{21} - \sigma_{12}V_{21}) - \frac{1}{T_1}(\Delta - \Delta_0), \qquad (25.18)$$

$$\left(\frac{\partial}{\partial t} + i\omega_{21} + T_2^{-1} + (\boldsymbol{k}\cdot\nabla)\right)\sigma_{21} = -\frac{i}{\hbar} V_{21}\Delta, \qquad (25.19)$$

where Δ is the difference between the populations of the upper and lower levels of the system; we omit the terms containing the diagonal elements $\hat{V}$ which are non-resonant in nature and small. The external field is taken to be a standing wave

$$\hat{V} = \hat{V}^0 \cos(\boldsymbol{k}\cdot\boldsymbol{r})\cos(\omega t + \varphi). \qquad (25.20)$$

We obtain a set of equations similar to (23.10′) and (23.11′) for the Fourier components of the density matrix:

$$\left[i(\omega_0 - \omega) + \frac{1}{T_2} + (\boldsymbol{v}\cdot\nabla)\right]\bar{\sigma}_{21} = -\frac{i}{2\hbar} e^{-i\varphi} \Delta V^0_{21} \cos(\boldsymbol{k}\cdot\boldsymbol{r}); \qquad (25.21)$$

$$\left((\boldsymbol{v}\cdot\nabla) + \frac{1}{T_1}\right)\Delta^{(0)} = -\frac{i}{\hbar}[V^0_{12}\bar{\sigma}_{21} e^{i\varphi} - \overset{+}{\sigma}_{12} V^0_{21} e^{-i\varphi}] + \frac{1}{T_1}\Delta_0, \qquad (25.22)$$

where

$$\bar{\sigma}_{21} \equiv \sigma^{(-1)}_{21} = (\overset{+}{\sigma}_{12})^* \equiv (\sigma^{(1)}_{12})^* \quad \text{and} \quad \omega_0 = \omega_{21}.$$

We look for a solution of equations (25.21) and (25.22) in the form of a Fourier series of spatial harmonics:

$$\sigma_{21}^{-} = \sum \sigma_n^{-}\, e^{in(k\cdot r)}, \quad \sigma_{12}^{+} = \sum \sigma_n^{+}\, e^{in(k\cdot r)},$$

$$\sigma_{-n}^{+} = (\sigma_n^{-})^*, \qquad \Delta = \sum d_n\, e^{in(k\cdot r)}. \tag{25.23}$$

Substituting (25.23) in (25.21) and (25.22), and equating the terms with $e^{in(k\cdot r)}$, we obtain an infinite set of coupled equations:

$$\sigma_n^{-}\left[i(\omega_0-\omega)+\frac{1}{T_2}+in(\boldsymbol{k}\cdot\boldsymbol{v})\right] = -\frac{i}{4\hbar}V_{21}^{0}\,e^{-i\varphi}(d_{n-1}+d_{n+1}), \tag{25.24}$$

$$\left[\frac{1}{T_1}+n(\boldsymbol{k}\cdot\boldsymbol{v})\right]d_n = -\frac{i}{2\hbar}V_{21}^{0}\,e^{i\varphi}(\sigma_{n-1}^{-}+\sigma_{n+1}^{-})$$

$$+\frac{i}{2\hbar}V_{21}^{0}\,e^{-i\varphi}(\sigma_{n+1}^{+}+\sigma_{n-1}^{+})+\delta_{n0}\frac{1}{T_1}\Delta_0. \tag{25.25}$$

A simple analysis of these equations shows that when the condition

$$\frac{|V_{21}^{0}|^2}{\hbar^2} \gg \sqrt{\left(\frac{1}{T_2^2}+(\boldsymbol{k}\cdot\boldsymbol{v})^2\right)\left(\frac{1}{T_1^2}+(\boldsymbol{k}\cdot\boldsymbol{v})^2\right)} \tag{25.26}$$

is satisfied we can neglect the higher harmonics d_n and $\sigma_n^{(\pm)}$ compared with d_0, $\sigma_{\pm1}^{(-)}$ and $\sigma_{\pm1}^{(+)}$. In this case the solution of equations (25.24) and (25.25) is of the form

$$d_0 = \frac{[(\omega_0-(\boldsymbol{k}\cdot\boldsymbol{v})-\omega)^2+T_2^{-2}]\,[(\omega_0+(\boldsymbol{k}\cdot\boldsymbol{v})-\omega)^2+T_2^{-2}]\,\Delta_0}{[(\omega_0-(\boldsymbol{k}\cdot\boldsymbol{v})-\omega)^2+T_2^{-2}]\,[(\omega_0+(\boldsymbol{k}\cdot\boldsymbol{v})-\omega)^2+T_2^{-2}]+\dfrac{|V_{12}^{0}|^2}{2\hbar^2}\dfrac{T_1}{T_2}[(\omega-\omega_0)^2+T_2^{-2}+(\boldsymbol{k}\cdot\boldsymbol{v})^2]}, \tag{25.27}$$

$$\sigma_1^{-} = -\frac{i}{4\hbar}V_{21}^{0}\,e^{-i\varphi}\,d_0\,\frac{1}{i[\omega_0-\omega+(\boldsymbol{k}\cdot\boldsymbol{v})]+\dfrac{1}{T_2}}; \quad \sigma_{-1}^{+} = (\sigma_1^{-})^*,$$

$$\sigma_{-1}^{-} = -\frac{i}{4h}V_{21}^{0}\,e^{-i\varphi}\,d_0\,\frac{1}{i[\omega_0-\omega-(\boldsymbol{k}\cdot\boldsymbol{v})]+\dfrac{1}{T_2}}; \quad \sigma_1^{+} = (\sigma_{-1}^{-})^*. \tag{25.28}$$

These expressions define in particular the susceptibility of the system, allowing for spatial dispersion and the dependence on the external field (in the monokinetic approximation). We notice, however, that the external field is limited in magnitude by condition (25.26).

Let us further examine the case of resonance, $\omega = \omega_0$. In this case

$$d_0 = \frac{(\boldsymbol{k}\cdot\boldsymbol{v})^2 + T_2^{-2}}{(\boldsymbol{k}\cdot\boldsymbol{v})^2 + T_2^{-2} + \dfrac{|V_{12}^0|^2}{2\hbar^2}\dfrac{T_1}{T_2}}\,\Delta_0, \tag{25.29}$$

$$\sigma_1^- = -\frac{V_{21}^0}{2\hbar}\,\mathrm{e}^{-i\varphi}\,\frac{(\boldsymbol{k}\cdot\boldsymbol{v}) + iT_2^{-1}}{2[(\boldsymbol{k}\cdot\boldsymbol{v})^2 + T_2^{-2}] + \dfrac{|V_{12}^0|^2}{\hbar^2}\dfrac{T_1}{T_2}}, \tag{25.30}$$

$$\sigma_{-1}^- = -\frac{V_{21}^0}{2\hbar}\,\mathrm{e}^{-i\varphi}\,\frac{-(\boldsymbol{k}\cdot\boldsymbol{v}) + iT_2^{-1}}{2[(\boldsymbol{k}\cdot\boldsymbol{v})^2 + T_2^{-2}] + \dfrac{|V_{12}^0|^2}{\hbar^2}\dfrac{T_1}{T_2}}. \tag{25.31}$$

For a gas in which there is a Maxwellian velocity distribution it is necessary, as has already been pointed out, to carry out additional velocity averaging. Let v be a component of the velocity along the direction $\boldsymbol{k}$; it is then necessary, as can easily be seen, to carry out averaging over a uniform Maxwellian distribution (see, e.g., Landau and Lifshitz, 1951)

$$dw_v = \sqrt{\frac{m}{2\pi kT}}\,\mathrm{e}^{-\frac{mv^2}{2kT}}\,dv.$$

As a result of this averaging of (25.29)–(25.31) we obtain†

$$\bar{d_0} = \Delta_0\left\{1 - \frac{\sqrt{\pi}}{k\bar{v}}\left(\sqrt{T_2^{-2} + \frac{|V_{12}^0|^2}{2\hbar^2}} - \frac{T_2^{-2}}{\sqrt{T_2^{-2} + \dfrac{|V_{12}^0|^2}{2\hbar^2}}}\right) + \frac{|V_{12}^0|^2}{\hbar^2k^2\bar{v}^2} + \cdots\right\},$$

$$\overline{\sigma_1^-} = \overline{\sigma_{-1}^-} = -i\,\frac{V_{21}^0}{4\hbar}\,\frac{1}{T_1}\,\mathrm{e}^{-i\varphi}\left(\frac{\sqrt{\pi}}{k\bar{v}\sqrt{T_2^{-2} + \dfrac{|V_{12}^0|^2}{2\hbar^2}\dfrac{T_1}{T_2}}} - \frac{1}{k^2\bar{v}^2} + \ldots\right)\Delta_0, \tag{25.32}$$

where $\bar{v} = \sqrt{2kT/m}$ is the mean thermal velocity of the molecules.

When averaging we took account of condition (25.26).

It follows from the above expressions that when the motion of the molecules is allowed for in the case (25.26) saturation occurs not when$(|V_{12}^0|^2/\hbar^2)$

$(T_1/T_2) \sim 1/T_2^2$ but when $(|V_{12}^0|^2/\hbar^2)\,(T_1/T_2) \sim (\boldsymbol{k}\cdot\boldsymbol{v})^2$. This is understandable, since the Doppler effect due to the motion of the molecules leads to an effective line width (for randomly moving molecules) of the order of $(\boldsymbol{k}\cdot\boldsymbol{v})$ instead of $1/T_2$.

Using (25.32) it is not difficult to obtain a solution for a three-level system. This case is discussed in greater detail by Fain, Khanin and Yashchin (1962, 1964).

† The following integrals hold (Gradshtein and Ryzhik, 1962):

$$I = \int_{-\infty}^{\infty} \frac{e^{-q^2v^2}}{p^2+v^2}\,dv = [1-\Phi(pq)]\frac{\pi}{p}e^{p^2q^2} \approx \frac{\pi}{p} - 2\sqrt{\pi}q,$$

$$I_1 = -\frac{1}{2q}\frac{\partial I}{\partial q} = \int_{-\infty}^{\infty} \frac{v^2 e^{q^2v^2}}{p^2+v^2}\,dv$$

$$= \frac{\pi}{q}\left\{\frac{1}{\sqrt{\pi}} - pq[1-\Phi(pq)]\,e^{p^2q^2}\right\} \approx \frac{\sqrt{\pi}}{q} - \pi p,$$

where $\Phi(x)$ is the normalized error function $(2/\sqrt{\pi})\,\mathrm{erf}\,(x)$. (The approximate equalities are valid when $qp \ll 1$.)

CHAPTER VI

Spontaneous and Stimulated Emission

SPONTANEOUS and stimulated emission play an important part in quantum electronics. This chapter introduces these concepts and discusses a number of questions that are important to their understanding, including the connection with classical theory, phase relations, the part played by zero-point fluctuations, and so on.

26. The Concept of Spontaneous and Stimulated Emission

The concept of spontaneous and stimulated emission in quantum theory was introduced by Einstein when discussing black-body radiation (Einstein, 1917) long before quantum electrodynamics became a distinct subject.

Let Z_m and Z_n be two stationary states of a quantum system (it is assumed that these are "unperturbed" states, i.e. states in which interaction with radiation is not taken into consideration). Let E_m and E_n ($E_m > E_n$) be the energies of these states. Einstein postulated the following fundamental emission processes.

(a) *Spontaneous emission.* If the system is in an excited state E_m there is a probability of a transition from the state Z_m to the state Z_n in the time dt with the emission of a photon

$$dW_s = A_m^n \, dt, \tag{26.1}$$

where A_m^n does not depend on the time or the intensity of the radiation field.

Emission of this type, which also occurs when there is no external radiation, is called spontaneous emission. The coefficient A_m^n, which is the probability (per unit time) that the quantum system will emit spontaneously a photon in a transition $m \to n$, is called the Einstein A-coefficient. We can also introduce differential coefficients $a_m^n(\Omega)$ for the spontaneous emission of a photon with given properties (the direction of propagation in the solid angle $d\Omega$, and with a polarization α)

$$dA_m^n = a_{m\alpha}^n(\Omega) \, d\Omega, \tag{26.2}$$

where α is a suffix denoting the two independent directions of polarization of the photon.

According to Bohr's frequency rule the frequency of the emitted radiation is

$$\omega = \frac{E_m - E_n}{\hbar}.$$

We also note the obvious relation: $A^n_m = \sum_{\alpha=1}^{2} \int a^n_{m\alpha}\, d\Omega$.

(b) *Stimulated emission and absorption.* If a quantum system is in an electromagnetic field with an energy density

$$\varrho = \sum_{\alpha=1}^{2} \int d\omega \int d\Omega \varrho_\alpha(\omega, \Omega),$$

then this field may make the quantum system undergo a transition from the higher energy state E_m to a lower one with a probability

$$dW_i = b^n_{m\alpha}\varrho_\alpha(\omega, \Omega)\, d\Omega\, dt, \tag{26.3}$$

and at the same time emit a photon of frequency $\omega = (E_m - E_n)/\hbar$ with a polarization α and a direction of propagation within the solid angle $d\Omega$. This is called stimulated or induced emission. The total probability of stimulated emission from a transition $m \to n$ in a time dt is

$$dW_i = \sum_{\alpha=1}^{2} \int_{4\pi} d\Omega b^n_{m\alpha}\varrho_\alpha(\omega, \Omega)\, dt.$$

The total probability of the emission of a photon from a transition $m \to n$ is equal to the sum of the probabilities of spontaneous and stimulated emission

$$dW_s + dW_i.$$

If the quantum system was originally in a state n the presence of radiation makes possible a transition to a higher energy state m with the absorption of a photon of frequency $\omega = (E_m - E_n)/\hbar$. The probability of this process in a time dt is

$$dW_r = \sum_{\alpha=1}^{2} \int_{4\pi} b^m_{n\alpha}\varrho_\alpha(\omega, \Omega)\, d\Omega\, dt. \tag{26.4}$$

This process is called absorption. The coefficients $b^n_{m\alpha}$, $b^m_{n\alpha}$ are the differential coefficients for stimulated emission and absorption respectively†. Einstein's coefficients depend only on the characteristics of the quantum sys-

† The Einstein B-coefficient is

$$B^n_m = \sum_{\alpha=1}^{2} \int b^n_{m\alpha}\, d\Omega.$$

tem under discussion and not on the radiation field. There is a conncteion between the Einstein A and B coefficients which can be derived by considering a state of equilibrium (Einstein, 1917).

The form of the dependence of the emission processes on the field and the time (26.1), (26.3), (26.4) which were postulated by Einstein has since been confirmed by the quantum theory of radiation (Heitler, 1954) (see also section 32). Before moving on to the quantum theory of spontaneous and stimulated emission and absorption we give the results of the classical treatment of these processes. This will clarify their essential features and enable us to distinguish their classical and quantum aspects.

27. The Classical Discussion (Fain, 1963c)

27.1. The Classical Radiation Field of an Oscillating Charge

Let us first examine the interaction of a one-dimensional classical oscillator with an electromagnetic field. The classical Hamiltonian of a system consisting of an oscillator interacting with such a field is of the form (see (4.8))

$$H = \frac{1}{2m}\left[\boldsymbol{p} - \frac{e}{c}\boldsymbol{A}(\boldsymbol{r})\right]^2 + \frac{1}{2}m\omega_0^2x^2 + \frac{1}{2}\sum_\nu (p_\nu^2 + \omega_\nu^2 q_\nu^2),$$

$$\boldsymbol{A}(\boldsymbol{r}) = \sum_\nu q_\nu \boldsymbol{A}_\nu(\boldsymbol{r}), \tag{27.1}$$

where $\boldsymbol{p}$ and $\boldsymbol{x}$ are the momentum and coordinate of the oscillator; p_ν, q_ν are canonical variables describing the field.

We shall further consider the oscillator to be a point oscillator (its dimensions are much less than the wavelength of the radiation†—the dipole approximation). In this approximation we can neglect the dependence of $\boldsymbol{A}_\nu(\boldsymbol{r})$ on the coordinate $\boldsymbol{r}$ and replace $\boldsymbol{A}_\nu(\boldsymbol{r})$ by $\boldsymbol{A}_\nu(0)$ (assuming that the mean position of the oscillator is at the origin). The canonical Hamiltonian equations ($\dot{p} = -\partial H/\partial q$; $\dot{q} = \partial H/\partial p$), using the Hamiltonian (27.1), are of the form:

$$\dot{q}_\nu = p_\nu, \quad \dot{p}_\nu = -\omega_\nu^2 q_\nu + \frac{e}{mc}\left(\boldsymbol{p} - \frac{e}{c}\boldsymbol{A}\cdot\boldsymbol{A}_\nu(0)\right),$$

$$\dot{p}_x = -m\omega_0^2 x, \quad \dot{x} = \frac{1}{m}p_x - \frac{e}{mc}A_x, \quad \dot{p}_y = \dot{p}_z = 0, \tag{27.2}$$

$$\dot{y} = \frac{1}{m}p_y - \frac{e}{mc}A_y, \quad \dot{z} = \frac{1}{m}p_z - \frac{e}{mc}A_z.$$

† To be more precise, the oscillator's dimensions are much less than the wavelengths of all those modes of the field whose effect is significant.

We shall solve these equations by the method of successive approximations, taking terms containing the charge e as small. Equations (27.2) can be conveniently rewritten in integral form

$$
\begin{aligned}
q_\nu &= q_\nu^{(0)}(t) + \frac{e}{mc\omega_\nu}\int_0^t (\boldsymbol{A}_\nu(0)\cdot \boldsymbol{p}(t_1)) \sin\omega_\nu(t-t_1)\,dt_1 \\
&\quad - \frac{e^2}{mc^2\omega_\nu}\sum_\lambda (\boldsymbol{A}_\nu(0)\cdot \boldsymbol{A}_\lambda(0))\int_0^t q_\lambda(t_1)\sin\omega_\nu(t-t_1)\,dt_1, \\
p_\nu &= p_\nu^{(0)}(t) + \frac{e}{mc}\int_0^t (\boldsymbol{A}_\nu(0)\cdot \boldsymbol{p}(t_1)) \cos\omega_\nu(t-t_1)\,dt_1 \\
&\quad - \frac{e^2}{mc^2}\sum_\lambda (\boldsymbol{A}_\nu(0)\cdot \boldsymbol{A}_\lambda(0))\int_0^t q_\lambda(t_1)\cos\omega_\nu(t-t_1)\,dt_1, \\
p_x &= p_x^{(0)}(t) + \frac{\omega_0 e}{c}\int_0^t A_x(t_1)\sin\omega_0(t-t_1)\,dt_1, \\
p_y &= p_z = 0, \quad y = -\frac{e}{mc}\int_0^t A_y(t_1)\,dt_1, \quad z = -\frac{e}{mc}\int_0^t A_z(t_1)\,dt_1.
\end{aligned}
\tag{27.3}
$$

Here the index 0 denotes solutions when there is no interaction between the oscillator and the field:

$$
\begin{aligned}
q_\nu^{(0)} &= q_\nu(0)\cos\omega_\nu t + \frac{p_\nu(0)}{\omega_\nu}\sin\omega_\nu t, \\
p_\nu^{(0)} &= -\omega_\nu q_\nu(0)\sin\omega_\nu t + p_\nu(0)\cos\omega_\nu t, \\
p_x^{(0)} &= p(0)\cos\omega_0 t - \omega_0 m x(0)\sin\omega_0 t, \\
x^{(0)} &= x(0)\cos\omega_0 t + \frac{p(0)}{m\omega_0}\sin\omega_0 t,
\end{aligned}
\tag{27.4}
$$

where $q_\nu(0), p_\nu(0), p(0), x(0)$ are the values of the field at the oscillator and its coordinates at $t = 0$.

The solutions of equations (27.3), including terms in e and e^2 only, are

$$
\begin{aligned}
p_\nu^{(1)} &= \frac{e}{mc}A_{\nu x}\int_0^t p_x^{(0)}(t_1)\cos\omega_\nu(t-t_1)\,dt_1 \\
&\approx \frac{e}{mc}A_{\nu x}\frac{\sin(\omega_0-\omega_\nu)t}{2(\omega_0-\omega_\nu)}[p(0)\cos\omega_\nu t - \omega_0 m x(0)\sin\omega_\nu t],
\end{aligned}
$$

$$q_\nu^{(1)} = \frac{eA_{\nu x}}{\omega_\nu mc}\int_0^t p_x^{(0)}(t_1)\sin\omega_\nu(t-t_1)\,dt_1$$

$$\approx \frac{e}{\omega_\nu mc}A_{\nu x}\frac{\sin(\omega_0-\omega_\nu)\,t}{2(\omega_0-\omega_\nu)}[p(0)\sin\omega_\nu t+\omega_0 mx(0)\cos\omega_\nu t],$$

$$(27.5)$$

$$p_x^{(1)} = \frac{e}{c}\,\omega_0\sum_\lambda A_{\lambda x}\int_0^t q_\lambda^{(0)}(t_1)\sin\omega_0(t-t_1)\,dt_1$$

$$\approx \frac{e\omega_0}{c}\sum_\lambda A_{\lambda x}\left[q_\lambda(0)\sin\omega_0 t-\frac{p_\lambda(0)}{\omega_\lambda}\cos\omega_0 t\right]\frac{\sin(\omega_0-\omega_\lambda)\,t}{2(\omega_0-\omega_\lambda)},$$

$$q_\nu^{(2)} = \frac{eA_{\nu x}}{mc\omega_\nu}\int_0^t p_x^{(1)}(t_1)\sin\omega_\nu(t-t_1)\,dt_1$$

$$-\frac{e^2}{mc^2\omega_\nu}\sum_\lambda(A_\nu\cdot A_\lambda)\int_0^t q_\lambda^{(0)}(t_1)\sin\omega_\nu(t-t_1)\,dt_1;$$

$$p_\nu^{(2)} = \dot q_\nu^{(2)}.$$

(The appropriate equalities are written on the assumption that $\omega_0 t \gg 1$.) Using these expressions we can find the field with a similar accuracy:

$$E = -\frac{1}{c}\dot{A} = -\frac{1}{c}\sum_\nu p_\nu(t)\,A_\nu(x);\quad H = \operatorname{curl} A = \sum_\nu q_\nu(t)\operatorname{curl} A_\nu(x).$$

Therefore p_ν determines the electric field and q_ν the magnetic field.

It is not difficult to see that $q_\nu^{(1)}, p_\nu^{(1)}$ gives the spontaneous emission field, and $q_\nu^{(2)}, p_\nu^{(2)}$ the field caused by the applied field $q_\nu^{(0)}, p_\nu^{(0)}$. In fact $q_\nu^{(1)}$ and $p_\nu^{(1)}$ do not depend on $q_\nu^{(0)}$ or $p_\nu^{(0)}$ and are finite even when there is no field at $t = 0$. On the other hand $q_\nu^{(2)}$ and $p_\nu^{(2)}$ depend on the amplitudes $q_\lambda(0)$ and $p_\lambda(0)$ and become zero when they are zero. Therefore $q_\nu^{(2)}$ and $p_\nu^{(2)}$ represent the stimulated emission and absorption fields.

This last field combined with the fields $q_\nu^{(0)} + q_\nu^{(1)}, p_\nu^{(0)} + p_\nu^{(1)}$ leads either to an increase in the energy of the radiation (stimulated emission) or a decrease (absorption). It can be seen from the expressions (27.5) that the stimulated emission and absorption field $q_\nu^{(2)}$, $p_\nu^{(2)}$ (we shall call it the stimulated field, since it is stimulated by the applied field $q_\nu^{(0)}, p_\nu^{(0)}$ and is proportional to it) is close in frequency to the field $q_\nu^{(0)}, p_\nu^{(0)}$ but has a different direction of propagation and polarization. Or, generally speaking, a given mode of the applied field stimulates different modes of the emission field (including the given mode). This effect is related to the dimensions of the system (we have assumed so far the dimensions of the radiating system are infinitely small). In fact let the system have dimensions of the order of a. Then the electromagnetic field stimulates in the system a certain charge (or dipole moment) distribution—the term $p_x^{(1)}$—which in its turn causes the electromagnetic field $q_\nu^{(2)}, p_\nu^{(2)}$. This charge distribution is characterized by a wave vector scatter of the order of

$$|\Delta \boldsymbol{k}| \gtrsim \frac{1}{a}.$$

This causes the appearance of different wave vectors $\boldsymbol{k}$ in the emission field (with the same modulus k which is determined by the frequency of the radiation). As we shall see below, the energy of the total emission field given by the quantities $q_\nu^{(0)} + q_\nu^{(1)} + q_\nu^{(2)}$ and $p_\nu^{(0)} + p_\nu^{(1)} + p_\nu^{(2)}$ may increase due to spontaneous and stimulated emission.

Therefore the stimulated field could have a different direction of propagation even in the case when the applied field is a plane wave with a definite wave vector.

We notice that an expansion of $\boldsymbol{A}$ in plane (travelling or standing) waves is not generally valid in a resonator of arbitrary form and in such a case the wave vectors cannot be defined. However, the general conclusion that a field in a given mode λ stimulates fields in different modes ν remains valid. In this case when these other modes are far enough apart in frequency (as occurs near the fundamental frequency of the resonator) we need take only the given mode ν, of frequency ω_ν, into consideration. If here ω_ν is the same as the eigenfrequency ω_0 of the oscillator the factor $\sin(\omega_0 - \omega_\nu)t/(\omega_0 - \omega_\nu)$ becomes t, i.e. the rate of spontaneous emission rises proportionally to the time.

The stimulated field intensity will rise as t^2 if the frequency of the applied field is the same as the oscillator frequency ω_0. In the case when the frequencies ω_ν form a continuous spectrum it is necessary to carry out summation over all modes; here the summation of $\sin(\omega_\nu - \omega_0)\, t/\pi(\omega_\nu - \omega_0)$ becomes the δ-function $\delta(\omega_\nu - \omega_0)$ if $t \gg 1/\delta\omega$ where $\delta\omega$ is the characteristic scale of a non-uniformity in the continuous spectrum. In this case the rate of spontaneous emission is independent of time and the stimulated field intensity rises linearly with time.

27.2. The Change in the Field Energy

Let us now examine the important question of how the energy of the field changes. We shall use canonical formalism. Let H_ν be the energy in the νth mode of the field

$$H_\nu = \tfrac{1}{2}(p_\nu^2 + \omega_\nu^2 q_\nu^2),$$

then the time derivative of H_ν is

$$\frac{dH_\nu}{dt} = \{H, H_\nu\} \equiv \sum_\lambda \left(\frac{\partial H}{\partial p_\lambda} \frac{\partial H_\nu}{\partial q_\lambda} - \frac{\partial H}{\partial q_\lambda} \frac{\partial H_\nu}{\partial p_\lambda} \right),$$

where $\{H, H_\nu\}$ is the classical Poisson bracket (see, e.g., Landau and Lifshitz, 1958).

After simple transformations we find

$$\frac{dH_\nu}{dt} = \frac{e}{mc} (\boldsymbol{p} \cdot \boldsymbol{A}_\nu)\, p_\nu - \frac{e^2}{mc^2} (\boldsymbol{A} \cdot \boldsymbol{A}_\nu)\, p_\nu$$

$$= \frac{e}{c} (\dot{\boldsymbol{r}} \cdot \boldsymbol{A}_\nu)\, p_\nu = -e(\dot{\boldsymbol{r}} \cdot \boldsymbol{E}(\nu)), \tag{27.6}$$

where $\dot{\boldsymbol{r}} = (\boldsymbol{p}/m) - (e/mc)\,\boldsymbol{A}$;
$\boldsymbol{E}(\nu) = -(1/c)\, \boldsymbol{A}_\nu p_\nu$ is the electric field in the νth mode.

We substitute in (27.6) the solutions (27.4) and (27.5), omitting terms higher than the second order in e:

$$\frac{dH_\nu}{dt} = \frac{e}{mc} A_{\nu x} p_x^{(0)} p_\nu^{(0)} + \frac{eA_{\nu x}}{mc} p_x^{(1)} p_\nu^{(0)}$$

$$- \frac{e^2}{mc^2} \left(\boldsymbol{A}_\nu \cdot \sum_\lambda \boldsymbol{A}_\lambda q_\lambda^{(0)} p_\nu^{(0)} \right) + \frac{eA_{\nu x}}{mc} p_x^{(0)} p_\nu^{(1)}. \tag{27.7}$$

Expression (27.7) gives the intensity of spontaneous and stimulated emission and absorption. We are generally interested in the time-averaged value of this quantity for a time far greater than the period of the radiation, i.e. for $\omega_0 t \gg 1$. The first three terms depend on the amplitude of the applied field $p_\nu(0)$, $q_\nu(0)$, and after time-averaging may take up positive (stimulated emission) or negative (absorption) values.

Let us first examine stimulated emission and absorption. Here two cases can be distinguished: (a) the field and the oscillator have random phases, and (b) the field and the oscillator have a definite phase relationship.

In the latter case the time-averaged value ($t \gg 2\pi/\omega_0$) of the first term (which is essentially finite only when $\omega_\nu = \omega_0$) is

$$\left(\frac{d\bar{H}_\nu}{dt} \right)_1 = \frac{1}{2} \frac{e}{mc} A_{\nu x} [p(0)\, p_\nu(0) + \omega_0^2 m x(0)\, q_\nu(0)]. \tag{27.8}$$

In the case when (27.8) is finite the intensity of the emitted radiation is determined by this term since the remaining terms in the right-hand side of (27.7) are quadratic in e. This term depends on the relation between the phases of the oscillator and the field and may take up both positive and negative values, i.e. produce stimulated emission and absorption respectively. If the phases of the oscillator or the field are random the phase-distribution-averaged value of the emission intensity is of interest in a number of situations.

If the oscillator phase is completely random, i.e. $\overline{p(0)} = \overline{x(0)} = 0$, then the mean value of (27.8) is zero. A similar situation occurs if the phase of the field is completely random, i.e. $\overline{p_\nu(0)} = \overline{q_\nu(0)}$. Let us now proceed to evaluate the rest of the terms in the right-hand side of (27.7). When averaging (27.7) we shall use the relations:

$$\overline{p_\nu^{(0)} q_\nu^{(0)}} = 0, \quad \overline{p_\nu(0)\, p_\lambda(0)} = \delta_{\nu\lambda}\overline{p_\nu^2(0)}, \quad \overline{q_\nu(0) q_\lambda(0)} = \delta_{\lambda\nu}\overline{q_\nu^2(0)}, \tag{27.9}$$

where the bar denotes double averaging, over a time $t \gg 1/\omega_0$, and over the phases.

Equation (27.9) holds if the mean values of $\overline{q_\nu(0)}$ and $\overline{p_\nu(0)}$ are zero, or if the mean values of $\overline{p_\nu(0)}$ and $\overline{q_\nu(0)}$ are zero for all modes except one. After averaging (27.7) we find†

$$\left(\frac{d\bar{H}_\nu}{dt}\right)_2 = -\frac{e^2}{2mc^2} A_{\nu x}^2 \overline{H_\nu(0)} \frac{\sin(\omega_0 - \omega_\nu)t}{(\omega_0 - \omega_\nu)}$$

$$+ \frac{e^2}{2mc^2} A_{\nu x}^2 \frac{1}{2}\left\{\frac{\overline{p^2(0)}}{m} + m\omega^2\overline{x^2(0)}\right\} \frac{\sin(\omega_0 - \omega_\nu)t}{(\omega_0 - \omega_\nu)}. \tag{27.10}$$

The first term in this expression is negative and denotes absorption. Therefore after averaging over completely random field phases for a classical linear oscillator only absorption is obtained and there is no stimulated emission. We shall see below that negative absorption, averaged over the phase of the field (or the negative absorption that is not dependent on the relation between the phases of the field and the oscillator), is possible only when there is a non-linearity in the system, i.e. in the case when the oscillator is non-linear or when the force acting on the oscillator (the electric field) depends on a coordinate of the oscillator (an approximation which involves higher multipoles than the dipole moment). There is a similar situation (as will be shown below) in quantum theory. The second term in (27.10) gives the intensity of the spontaneous emission. This intensity is proportional to the mean value of the oscillator energy and is equal to zero if the oscillator is not excited. We notice that in the case when the oscillator phase is random and the mean values are $\overline{p(0)} = \overline{x(0)} = 0$ the spontaneous emission intensity is finite and proportional to the mean square fluctuations of $p(0)$ and $x(0)$. In this case the spontaneous emission field is on the average zero, and the spontaneous emission has the nature of noise. The same can be said of the stimulated field which on the average is zero when $\overline{p_\nu(0)} = \overline{q_\nu(0)} = 0$, and the intensity of the

† As a result of the averaging, the third term on the right-hand side of (27.7) becomes zero. This allows us to discard the term $e^2 A_\nu^2/mc^2$ in the interaction energy between the oscillator and the field when discussing the processes of spontaneous and stimulated emission and absorption.

field absorption is proportional to the mean square fluctuations of the field or the mean energy of the field.

In the case when the oscillator interacts with one mode of the radiation field $\omega_\nu = \omega_0$ the factor

$$\frac{\sin(\omega_0 - \omega_\nu)\, t}{(\omega_0 - \omega_\nu)} = t$$

and the field energy varies as t^2. If the radiation field has a continuous spectrum $\sin(\omega_0 - \omega_\nu)\, t/(\omega_0 - \omega_\nu)$ can be replaced by $\pi\delta(\omega_0 - \omega_\nu)$ and we sum over all modes. The summation is thus reduced to integration. For emission into free space this integration is over all frequencies and angles of propagation, and the summation is over the two polarizations. The number of modes in a given range $d\omega_\nu$ with a given polarization α and direction of propagation within the solid angle $d\Omega$ is (see (3.55))

$$dZ = \frac{\omega_\nu^2\, d\omega_\nu\, d\Omega}{(2\pi c)^3} L^3.$$

The total intensity of the spontaneous emission will be†

$$I_{sp} = \frac{e^2}{2mc^2} \int_{4\pi} \frac{d\Omega A_{\nu x}^2\, \omega_0^2 L^3}{(2\pi c)^3} \left\{ \frac{\overline{p^2(0)}}{2m} + \frac{m\omega_0^2}{2} \overline{x^2(0)} \right\}$$

$$= \frac{e^2 \omega_0^4 \left\{ \overline{x^2(0)} + \dfrac{\overline{p^2(0)}}{m^2 \omega_0^2} \right\}}{3c^3}. \tag{27.11}$$

Expression (27.11) is the same as the usual expression for the intensity of spontaneous emission

$$\bar{\boldsymbol{I}} = \frac{2}{3c^3} (\ddot{\boldsymbol{d}})^2$$

averaged with respect to time. In fact for an oscillator

$$\boldsymbol{d} = \boldsymbol{d}_0 \sin(\omega_0 t + \theta), \quad \bar{\boldsymbol{I}} = \frac{2\omega_0^4}{3c^3} d_0^2 \overline{\sin^2(\omega_0 t + \theta)} = \frac{\omega_0^4}{3c^3} d_0^2,$$

† It is necessary to replace $A_{x\nu}$ by $A_\nu \cos\theta$, where θ is the angle between the direction of the oscillator (the x-axis) and $\boldsymbol{A}_\nu$, and then sum with respect to the polarization directions, which involves replacing $\cos^2\theta$ by $\sin^2\vartheta$, where ϑ is the angle between the oscillator and the direction of propagation. An element of solid angle is of the form $d\Omega = \sin\vartheta\, d\vartheta\, d\varphi$. After integration with respect to half the directions of propagation, ϑ from 0 to $\pi/2$, φ from 0 to 2π (standing waves), and substituting $A_\nu = \sqrt{8\pi/L^3}\, c$ we arrive at (27.11).

and in (27.11)

$$e^2\left[\overline{x^2(0)} + \frac{\overline{p^2(0)}}{m^2\omega_0^2}\right]$$

is just the mean square of the oscillator dipole amplitude.

27.3. Non-linear Systems

As has already been pointed out above, in the case of a harmonic oscillator in a uniform field any negative absorption must be dependent on the phase relation between the field and the oscillator. This can be understood from general arguments (Gaponov, 1960). In accordance with (27.6) the change in the energy of the νth mode of the field is equal to the work done on the system by the field in the νth mode, but of opposite sign:

$$\frac{dH_\nu}{dt} = -e(\dot{\boldsymbol{r}} \cdot \boldsymbol{E}(\nu)).$$

Furthermore $\dot{\boldsymbol{r}}$ can be split into two, an undisturbed part $\dot{\boldsymbol{r}}_0$ and a part which is stimulated by the field, $\dot{\boldsymbol{r}}_1$:

$$\dot{\boldsymbol{r}} = \dot{\boldsymbol{r}}_0 + \dot{\boldsymbol{r}}_1.$$

The work done by the field on the first part $\dot{\boldsymbol{r}}_0$ depends on the phase relation. The part $\dot{\boldsymbol{r}}_1$ stimulated by the field by virtue of the linearity of the equation of motion of the oscillator will be the same for an excited or an unexcited oscillator. It is obvious, however, that in the latter case there can be no stimulated emission† so there cannot generally be any for a linear system. This is not the case in non-linear systems. Here, $\dot{\boldsymbol{r}}_1$ depends essentially on the degree of excitation of the system, and stimulated emission that is independent of the phase relation is possible. Gaponov (1960) gives a general treatment of the behaviour of non-linear oscillators acted upon by an external field and shows that the intensity of the stimulated emission from oscillators of this kind is proportional to

$$M = d \ln \omega_0(E)/dE,$$

where $\omega_0(E)$ is the oscillator frequency, which is a function of its energy E. In the case of a harmonic oscillator this quantity is equal to zero.

To conclude this section we would remark that a clear division into spontaneous and stimulated emission can generally be made in the approximation of perturbation theory for a short time only. After a long enough time the

† With the exception of the case when the velocity v is greater than the phase velocity of light in the medium c/n.

spontaneous emission field rises and, in its turn, stimulates radiation which cannot be clearly described either as spontaneous or stimulated emission. This situation occurs, for example, in the case of emission from an oscillator in a resonator tuned to the oscillator frequency. On the other hand, in free space (and in a large enough resonator) we can neglect interaction of this sort and consider stimulated and spontaneous emission separately at all values of the time.

28. The Quantum Theory of Spontaneous and Stimulated Emission in a System of Two-level Molecules

We now move on to the quantum theory of spontaneous and stimulated emission. In this section we shall discuss emission from a collection of quantum objects (which we shall call molecules) that do not interact with each other in the absence of a field. As has been pointed out (section 19), if the levels of the molecules are not equidistant it is possible in a number of problems connected with emission to take into consideration only two energy levels of a molecule. We shall describe the two-level molecules by the energy spin operators r_i (see section 19). We shall further consider that all the molecules have the same energy level difference $\hbar\omega_0$. In the case when the dimensions of the system of molecules is much less than a wavelength (for the sake of simplicity we use the dipole approximation) the Hamiltonian of the system can be written in the form (see (19.23))†

$$\hat{H} = \hbar\omega_0 \hat{R}_3 + \tfrac{1}{2}\sum_{\nu} (\hat{p}_\nu^2 + \omega_\nu^2 \hat{q}_\nu^2) - (\boldsymbol{A}\cdot\boldsymbol{e}_1)\,\hat{R}_1, \tag{28.1}$$

where the operator $\hat{\boldsymbol{R}}$ is the total energy spin

$$\hat{R}_1 = \sum_j \hat{r}_{1j}, \quad \hat{R}_2 = \sum_j \hat{r}_{2j}, \quad \hat{R}_3 = \sum_j \hat{r}_{3j}.$$

The quantum equations of motion following from this Hamiltonian are of the form:

$$\begin{gathered}\dot{\hat{p}}_\nu = -\omega_\nu^2 \hat{q}_\nu + (\boldsymbol{e}_1\cdot\boldsymbol{A}_\nu)\,\hat{R}_1, \quad \dot{\hat{q}}_\nu = \hat{p}_\nu, \\ \dot{\hat{R}}_1 = -\omega_0 \hat{R}_2, \quad \dot{\hat{R}}_2 = \omega_0 \hat{R}_1 + \frac{(\boldsymbol{A}\cdot\boldsymbol{e}_1)}{\hbar}\hat{R}_3, \quad \dot{\hat{R}}_3 = -\frac{1}{\hbar}(\boldsymbol{A}\cdot\boldsymbol{e}_1)\,\hat{R}_2.\end{gathered} \tag{28.2}$$

Just as in the classical case (section 27), we shall solve these operator equations in the perturbation theory approximation, taking $\boldsymbol{e}_1$ to be a small

† Here no allowance is made for the part of the interaction energy proportional to $\boldsymbol{A}^2$ (see the footnote to equation (27.10)). In addition, we have made $\boldsymbol{e}_2 = 0$, which corresponds to linear polarization of the molecular dipole moment.

parameter. We change to the appropriate integral equations to do this:

$$\hat{p}_\nu = \hat{p}_\nu^{(0)} + (A_\nu \cdot e_\nu) \int_0^t \hat{R}_1(t_1) \cos \omega_\nu(t - t_1)\, dt_1,$$
$$\hat{q} = \hat{q}_\nu^{(0)} + \frac{1}{\omega_\nu}(A_\nu \cdot e_\nu) \int_0^t \hat{R}_1(t_1) \sin \omega_\nu(t - t_1)\, dt_1, \tag{28.3}$$

$$\hat{R}_1 = \hat{R}_1^{(0)} - \left(\frac{e_1}{\hbar} \cdot \int_0^t \hat{A}(t_1)\right) \hat{R}_3(t_1) \sin \omega_0(t - t_1)\, dt_1,$$
$$\hat{R}_2 = \hat{R}_2^{(0)} + \left(\frac{e_1}{\hbar} \cdot \int_0^t \hat{A}(t_1)\right) \hat{R}_3(t) \cos \omega_0(t - t_1)\, dt_1. \tag{28.4}$$

As in the classical case it is easy to check by direct substitution that equations (28.3), (28.4) satisfy the initial equations (28.2). In the zero order approximation, when there is no interaction, the solutions of equations (28.2) are of the form (this can also be checked by direct substitution):

$$\begin{aligned}
\hat{q}_\nu^{(0)} &= \hat{q}_\nu(0) \cos \omega_\nu t + \frac{\hat{p}_\nu(0)}{\omega_\nu} \sin \omega_\nu t,\\
\hat{p}_\nu^{(0)} &= \hat{p}_\nu(0) \cos \omega_\nu t - \omega_\nu \hat{q}_\nu(0) \sin \omega_\nu t,\\
\hat{R}_1^{(0)} &= \hat{R}_1(0) \cos \omega_0 t - \hat{R}_2(0) \sin \omega_0 t,\\
\hat{R}_2^{(0)} &= \hat{R}_1(0) \sin \omega_0 t + \hat{R}_2(0) \cos \omega_0 t,\\
\hat{R}_3^{(0)} &= \hat{R}_3(0).
\end{aligned} \tag{28.5}$$

In this approximation

$$\begin{aligned}
\hat{q}_\nu^{(1)} &= \frac{(A_\nu \cdot e_1)}{\omega_\nu} \int_0^t \hat{R}_1^{(0)}(t_1) \sin \omega_\nu(t - t_1)\, dt_1\\
&\approx \frac{(A_\nu \cdot e_1)}{2\omega_\nu} \frac{\sin(\omega_\nu - \omega_0)\, t}{(\omega_\nu - \omega_0)} [\hat{R}_1(0) \sin \omega_\nu t + \hat{R}_2(0) \cos \omega_\nu t],\\
\hat{p}_\nu^{(1)} &= (A_\nu \cdot e_1) \int_0^t \hat{R}_1^{(0)}(t_1) \cos \omega_\nu(t - t_1)\, dt_1\\
&\approx \frac{(A_\nu \cdot e_1)}{2} \frac{\sin(\omega_\nu - \omega_0)\, t}{(\omega_\nu - \omega_0)} [\hat{R}_1(0) \cos \omega_\nu t - \hat{R}_2(0) \sin \omega_\nu t],\\
\hat{R}_1^{(1)} &= -\frac{1}{\hbar} \hat{R}_3(0) \sum_\lambda (A_\lambda \cdot e_1) \int_0^t q_\lambda^{(0)}(t_1) \sin \omega_0(t - t_1)\, dt_1,\\
\hat{R}_2^{(1)} &= -\frac{1}{\omega_0} \dot{\hat{R}}_1^{(1)}.
\end{aligned} \tag{28.6}$$

The electromagnetic field in the second approximation is of the form

$$\hat{q}_\nu^{(2)} = -\frac{(A_\nu \cdot e_1)}{\hbar\omega_\nu} \hat{R}_3(0) \sum_\lambda (e_1 \cdot A_\lambda) \int_0^t dt_1 \sin\omega_\nu(t - t_1) \times \int_0^{t_1} dt_2 q_\lambda^{(0)}(t_2) \sin\omega_0(t_2 - t_1),$$

$$\hat{p}_\nu^{(2)} = -\frac{(A_\nu \cdot e_1)}{\hbar} \hat{R}_3(0) \sum_\lambda (e_1 \cdot A_\lambda) \int_0^t dt_1 \cos\omega_\nu(t - t_1) \times \int_0^{t_1} dt_2 \hat{q}_\lambda^{(0)}(t_2) \sin\omega_0(t_2 - t_1). \quad (28.7)$$

Before going on to discuss the solutions obtained we shall find the change with time of the operator for the electromagnetic field energy in the νth mode

$$\hat{H}_\nu = \tfrac{1}{2}(\hat{p}_\nu^2 + \omega_\nu^2 \hat{q}_\nu^2).$$

The derivative of this quantity can be found in the usual way by means of the quantum Poisson brackets

$$\frac{d\hat{H}_\nu}{dt} = \frac{i}{\hbar}[\hat{H}, \hat{H}_\nu] = \frac{i}{\hbar}\hat{R}_1\left[-\sum_\lambda (e_1 \cdot A_\lambda)\hat{q}_\lambda, \hat{H}_\nu\right]$$

$$= -\frac{i}{\hbar}(e_1 \cdot A_\nu)\hat{R}_1 \frac{1}{2}[\hat{q}_\nu, \hat{p}_\nu^2] = (e_1 \cdot A_\nu)\hat{R}_1 \hat{p}_\nu. \quad (28.8)$$

Including terms of the second order of smallness we find

$$\frac{d\hat{H}_\nu}{dt} = (e_1 \cdot A_\nu)\hat{R}_1^{(0)}\hat{p}_\nu^{(0)} + (e_1 \cdot A_\nu)\hat{R}_1^{(1)}\hat{p}_\nu^{(0)} + (e_1 \cdot A_\nu)\hat{R}_1^{(0)}\hat{p}_\nu^{(1)}. \quad (28.9)$$

We have thus found the change with time of the field operators $\hat{p}_\nu$ and $\hat{q}_\nu$ and the field energy operators $\hat{H}_\nu$. These operators have no immediate physical meaning, but a meaning can be given to their mean values taken in the appropriate quantum ensemble. The initial quantum states of the radiation field and the molecules must be stated in order to find the time-dependence of these mean values. It is therefore necessary to state the wave function (or density matrix) of the initial state.

We shall assume that the wave function (or density matrix) of the initial state is known. Part of the solutions (28.5)–(28.7) is linearly dependent on the field and molecule operators. The averaging of these solutions involves replacing the operators by their mean values. For example, the mean value of the initial field is

$$\overline{q_\nu^{(0)}}(t) = \overline{q_\nu(0)} \cos \omega_\nu t + \frac{\overline{p_\nu(0)}}{\omega_\nu} \sin \omega_\nu t,$$

$$\overline{p_\nu^{(0)}}(t) = \overline{p_\nu(0)} \cos \omega_0 t - \omega_\nu \overline{q_\nu(0)} \sin \omega_0 t \quad \text{etc.}$$

In order to determine the mean value of quantities of the $\hat{R}_3(0)\,\hat{q}_\nu(0)$ type contained in the right-hand side of (28.6) and (28.7) we shall make the additional assumption that at time $t = 0$ the wave function of the system (or its density matrix) can be split into the products of the molecular and the field wave functions (or density matrices). In other words, we must make the assumption of the statistical independence of the behaviour of the molecules and the field at $t = 0$. Then

$$\overline{R_3(0)\,q_\nu(0)} = \overline{R_3(0)}\;\overline{q_\nu(0)}, \quad \overline{R_3(0)\,p_\nu(0)} = \overline{R_3(0)}\;\overline{p_\nu(0)}. \tag{28.10}$$

Using these relations we can express all the mean quantities in terms of their initial values. If the mean values $\overline{R_1(0)}$ and $\overline{R_2(0)}$ are finite the mean values $\bar{q}_\nu^{(1)}$ and $\bar{p}_\nu^{(1)}$ that determine the spontaneous emission field are also finite. Further, if the mean values $\overline{p_\nu(0)}$ and $\overline{q_\nu(0)}$ are finite the mean stimulated field $\overline{q_\nu^{(2)}}$ and $\overline{p_\nu^{(2)}}$ is finite. In this case (when the mean radiation field is finite) everything that has been said of the spontaneous and stimulated emission field in the classical case holds good (section 27). On the other hand, if the molecules and the field were at $t = 0$ in a stationary state (corresponding to an unperturbed Hamiltonian), then the mean values are

$$\overline{R_1(0)} = \overline{R_2(0)} = 0, \quad \overline{p_\nu(0)} = \overline{q_\nu(0)} = 0;$$

which will also be so if the system can be described by a density matrix that is diagonal in the energy representation†. In this case the mean spontaneous and stimulated emission field is zero. Of course here the squares of the field, and the field energy, are finite. A field of this kind is pure noise and is fluctuating in nature. The classical analogue is a field with a completely random phase, discussed in section 27. In the quantum treatment of spontaneous and stimulated emission (Heitler, 1954) it is generally assumed that initially the atom (molecule) is in a definite energy state and there is a definite number

† This can easily be checked directly if we remember that in the stationary state (with a definite energy) the diagonal matrix elements of the quantities $\hat{R}_1$, $\hat{R}_2$, $\hat{p}_\nu$ and $\hat{q}_\nu$ are equal to zero (these diagonal elements are equal to the mean values of these quantities). This statement is still valid for a state with a density matrix $\hat{\varrho}$ which is diagonal in the energy representation:

$$\bar{q}_\nu = \sum_n \varrho_{nn} q_{\nu nn} = 0, \; \bar{p}_\nu = \sum_n \varrho_{nn} p_{\nu nn} = 0 \quad \text{etc.}$$

of photons in the radiation field. It is obvious that this is just the case when the mean field is zero.

Let us find the change in energy of the field. The mean value of the term of the first order of smallness in the right-hand side of (28.9) is finite if the mean values $\overline{R_i^{(0)}}$ and $\overline{p_\nu^{(0)}}$ are finite. Carrying out the above-mentioned time-averaging (see section 27) we obtain an expression that is analogous to the classical expression (27.8):

$$\left(\frac{d\bar{H}}{dt}\right)_1 = (\boldsymbol{e}_1 \cdot \boldsymbol{A}_\nu)\,(\overline{R_1(0)}\,\overline{p_\nu(0)} + \omega_\nu \overline{R_2(0)}\,\overline{q_\nu(0)}). \tag{28.11}$$

As in classical theory, this expression may take up positive and negative values, depending on the relation between the phases $\bar{R}_i^{(0)}(t)$ and $\bar{p}_\nu^{(0)}(t)$ (or, in other words, on the mean values of the initial amplitudes $\overline{R_1(0)}$, $\overline{R_2(0)}$, $\overline{p_\nu(0)}$ and $\overline{q_\nu(0)}$.

If the mean value of the field $\overline{q_\nu(0)}$, $\overline{p_\nu(0)}$ is zero (or the mean amplitudes of all the modes except one are zero), then

$$\overline{p_\nu(0)\,p_\lambda(0)} = \delta_{\nu\lambda}\overline{p_\nu^2(0)}; \quad \overline{q_\nu(0)\,q_\lambda(0)} = \delta_{\nu\lambda}\overline{q_\nu^2(0)}. \tag{28.12}$$

We must average expression (28.9) over both the quantum ensemble and time in order to obtain the field intensity in the νth mode. Using (28.12) we obtain after averaging (taking the second and third terms of (28.9) into consideration)

$$\left(\frac{d\bar{H}_\nu}{dt}\right)_2 = \frac{\sin(\omega_0 - \omega_\nu)\,t}{(\omega_0 - \omega_\nu)}\,\frac{(\boldsymbol{e}_1 \cdot \boldsymbol{A}_\nu)^2}{4} \times \left[\overline{R_1^2(0)} + \overline{R_2^2(0)} + \frac{\overline{R_3(0)}}{\hbar\omega_\nu}\,(\overline{p_\nu^2(0)} + \omega_\nu^2\overline{q_\nu^2(0)})\right]. \tag{28.13}$$

Let us discuss this expression. Just as in the classical case, if the molecular system interacts with only one of the resonator modes (and the others are far enough away in frequency), then if $\omega_\nu = \omega_0$

$$\frac{\sin(\omega_0 - \omega_\nu)\,t}{(\omega_0 - \omega_\nu)} = t$$

and the rate of emission is proportional to t, whilst the field energy rises as t^2. If the radiation field has a continuous spectrum the total intensity is obtained by summing $d\bar{H}_\nu/dt$ over all modes; this involves integration with respect to frequency. Here $\sin(\omega_0 - \omega_\nu)\,t/(\omega_0 - \omega_\nu)$ can be replaced by $\pi\delta(\omega_0 - \omega_\nu)$†. As a result the total rate of emission turns out to be independent of time and the field energy rises in proportion to t.

† After a time $t \gg 1/\delta\omega$, where $\delta\omega$ is a quantity of the order of the inhomogeneity of the continuous spectrum.

Consider only the term concerned with spontaneous emission in expression (28.13), which can be evaluated if the radiation field was initially in a vacuum. This does not imply that the energy of the radiation field

$$\bar{H}_\nu(0) = \tfrac{1}{2}[\overline{p_\nu^2(0)} + \omega_\nu^2 \overline{q_\nu^2(0)}] \tag{28.14}$$

is zero. In fact, the mean energy of the radiation field in the νth mode is of the form

$$\bar{H}_\nu(0) = (\bar{n}_\nu + \tfrac{1}{2})\,\hbar\omega_\nu, \tag{28.15}$$

where $\bar{n}_\nu$ is the mean number of photons in the νth mode.

The term $\tfrac{1}{2}\hbar\omega_\nu$ is the so-called zero-point energy of the field. For a given choice of Hamiltonian the zero-point energy serves as a measure of the zero-point fluctuations of the field, i.e. the quantities $\overline{p_\nu^2(0)}$ and $\overline{q_\nu^2(0)}$. Generally speaking, however, the zero-point energy is not coupled with fluctuations of the vacuum. In fact we can carry out a transformation of the Hamiltonian of the field

$$\hat{H}'_\nu = \hat{H}_\nu - \tfrac{1}{2}\hbar\omega_\nu.$$

As a result of this kind of transformation the equations of motion remain unchanged and the zero-point energy disappears. Using (28.13), (28.14) and (28.15) we can find for the intensity of the spontaneous emission the expression

$$I_\nu^s = \frac{\sin(\omega_0 - \omega_\nu)\,t}{(\omega_0 - \omega_\nu)} \frac{(e_1 \cdot A_\nu)^2}{4} [\overline{R_1^2(0)} + \overline{R_2^2(0)} + \overline{R_3(0)}]. \tag{28.16}$$

The remainder of the emission intensity in (28.13) is due to stimulated emission or absorption

$$I_\nu^i = \frac{\sin(\omega_0 - \omega_\nu)\,t}{(\omega_0 - \omega_\nu)} \frac{(e_1 \cdot A_\nu)^2}{2} \bar{n}_\nu \overline{R_3(0)}. \tag{28.17}$$

Depending upon the sign of $\overline{R_3(0)}$ (which is half the difference of the populations of the upper and lower levels of the molecule), either stimulated emission or absorption will occur. We shall leave the more detailed discussion of the spontaneous and stimulated emission formulae and the connection between these expressions and the corresponding classical relations to the next section.

29. The Correspondence Principle

It is well known that a change from quantum to classical theory can be made for high quantum numbers. Generally speaking, however, motion described by a wave function does not change in this case into motion along a definite trajectory. The change to a classical trajectory is made only for a special form of wave function, namely a wave packet whose dimensions approach zero in the limit $\hbar \to 0$. In the general case we can only in state that the classical limit the probability distribution of the possible values of a quantity will change in accordance with the laws of classical mechanics† (see, e.g., Landau and Lifshitz, 1963); the same is true, of course, of the change from quantum to classical field theory since the latter can be looked upon as a mechanical system with, generally speaking, an infinite number of degrees of freedom. It has already been pointed out in section 3 that the stationary state of a field with an energy $(n_\nu + \frac{1}{2})\,\hbar\omega_\nu$, which by virtue of the uncertainty relation (3.44) has no phase, corresponds to an ensemble of oscillators with randomly distributed phases. Averaging by means of the wave function (or density matrix) corresponds to averaging over this classical ensemble.

After these remarks we can make a comparison of the classical and quantum expressions for the intensity of emitted radiation, i.e. establish the correspondence principle. First we shall divide the total emission intensity (28.9), (28.11) and (28.13) into fluctuations (noise) and a steady part. We shall call the noise that part which is connected with the fluctuations of the emission field and the dipole moment, and which is proportional to the magnitude of the mean square fluctuations of the field and the dipole moment. The remaining part of the emission intensity connected with the finite mean values of the dipole moment and the field will be called the regular part of the emission intensity. It is obvious that this division of the emission intensity can be carried out in both quantum and classical theory if we remember the remarks made above. Using the definitions we have introduced we can rewrite (28.11) and (28.13) in the form

$$
\begin{aligned}
I_\nu &= I_\nu^{\text{noise}} + I_\nu^{\text{reg}}, \\
I_\nu^{\text{noise}} &= \frac{\sin(\omega_0 - \omega_\nu)\,t}{(\omega_0 - \omega_\nu)} \frac{(e_1 \cdot A_\nu)^2}{4} \\
&\quad \times \left[\overline{\Delta R_1^2} + \overline{\Delta R_2^2} + \frac{\overline{R_3(0)}}{\hbar\omega_\nu}\left(\overline{\Delta p_\nu^2(0)} + \omega_\nu^2\,\overline{\Delta q_\nu^2(0)}\right)\right], \\
I_\nu^{\text{reg}} &= (e_1 \cdot A_\nu)\left(\overline{R_1(0)}\,\overline{p_\nu(0)} + \omega_\nu \overline{R_2(0)}\,\overline{q_\nu(0)}\right) \\
&\quad + \frac{\sin(\omega_0 - \omega_\nu)\,t}{(\omega_0 - \omega_\nu)} \frac{(e_1 A_\nu)^2}{4} \\
&\quad \times \left[\overline{R_1(0)}^2 + \overline{R_2(0)}^2 + \frac{\overline{R_3(0)}}{\hbar\omega_\nu}\left(\bar{p}_\nu^2(0) + \omega_\nu^2 \bar{q}_\nu^2(0)\right)\right],
\end{aligned}
\tag{29.1}
$$

where $\overline{\Delta R_1^2} = \overline{R_1^2} - \overline{R_1}^2$ etc.

† In classical mechanics the probability can be defined as the density of points in phase space. Each point moves in accordance with the equations of classical dynamics.

When the molecules in the field are initially in unperturbed stationary states the regular part of the emission is zero. The total emission becomes zero if the molecules and the field are in the ground state. It can easily be checked that the intensity of regular spontaneous emission

$$I_\nu^{s\,\mathrm{reg}} = \frac{\sin(\omega_0 - \omega_\nu)\,t}{(\omega_0 - \omega_\nu)} \frac{(\boldsymbol{e}_1 \cdot \boldsymbol{A}_\nu)^2}{4} \left(\overline{R_1(0)}^2 + \overline{R_2(0)}^2\right) \tag{29.2}$$

is the same as the classical intensity of the spontaneous emission of an oscillating dipole moment. For this we use the relation between R_1 and R_2 and the dipole moment (see (19.12) $\boldsymbol{a} = \boldsymbol{e}_2 = \boldsymbol{e}_3 = 0$)

$$\frac{1}{c}\,\dot{\boldsymbol{d}} = \boldsymbol{e}_1 \bar{R}_1 = \boldsymbol{e}_1(\overline{R_1(0)} \cos \omega_0 t - \overline{R_2(0)} \sin \omega_0 t). \tag{29.3}$$

After substituting this in the classical expression for the intensity of the spontaneous emission in the νth mode we obtain (29.2). This complete correspondence of the classical and quantum expressions is connected with the linearity of the Maxwell equations (4.10), (4.11) and (28.2).

Let us examine the case of a single two-level molecule

$$\hat{R}_1^2 = \tfrac{1}{4}, \quad \hat{R}_2^2 = \tfrac{1}{4}, \quad \hat{R}_3 = \begin{pmatrix} \frac{1}{2} & 0 \\ 0 & -\frac{1}{2} \end{pmatrix}.$$

If the molecule is in the upper state, then $R_3 = \frac{1}{2}$, all the spontaneous emission is noise emission, $I_\nu^{s\,\mathrm{reg}} = 0$, and the noise part from (29.1) becomes

$$I_\nu^{s\,\mathrm{noise}} = I_{\nu 0}^s \left[\frac{1}{2} + R_3 \frac{\overline{p_\nu^2(0)} + \omega_\nu^2 \overline{q_\nu^2(0)}}{\hbar\omega_\nu}\right] = I_{\nu 0}^s .$$

In this case the zero-point fluctuations of the field $\frac{1}{2}\hbar\omega_\nu$ stimulate just half the emission, the other half being connected with the zero-point fluctuations of the dipole moment $R_1^2 + R_2^2 = \frac{1}{2}$. If the molecule is in the lower state $R_3 = -\frac{1}{2}$ and the zero-point fluctuations of the field cause absorption which just compensates for the emission connected with the dipole moment zero-point fluctuations.

It must be stressed that sometimes spontaneous emission is ascribed entirely to the zero-point fluctuations of the field (see, e.g., Weisskopf, 1935), but Fain (1963c) criticizes this statement (see also Ginzburg, 1939).

Let us continue the discussion of the correspondence principle.

We now compare the expressions for the total emission intensity of a harmonic oscillator in the quantum and classical case. In order to obtain the emission from a quantum oscillator we notice that the quantum equations

(28.2) change into the corresponding equations for a quantum oscillator interacting with radiation if we make the following substitution:

$$\hat{R}_3 \to \overline{R_3(0)}, \quad \hat{R}_1 \to \sqrt{\frac{-\overline{R_3(0)}}{m\hbar\omega_0}}\,\hat{p}_x, \quad \hat{R}_2 \to \sqrt{-\frac{\overline{R_3(0)}\,m\omega_0}{\hbar}}\,\hat{x} \tag{29.4}$$

and

$$e_1 \to \frac{e}{c}\sqrt{\frac{\hbar\omega_0}{-m\overline{R_3(0)}}}\,i.$$

For the operators $\hat{R}_1$, $\hat{R}_2$, $\hat{p}_x$ and $\hat{x}$ to be Hermitian $\overline{R_3(0)}$ must be negative.

This substitution is essentially a linearization of equations (28.2) with respect to the state with $\overline{R_3(0)} < 0$. All the subsequent relations follow from these linearized equations, in particular the expression for the emission intensity (28.13). Making the same substitution (29.4) in this expression we find

$$\overline{\frac{dH_\nu}{dt}} = \frac{e^2}{2mc^2}A_{\nu x}^2 \frac{\sin(\omega_0 - \omega_\nu)\,t}{(\omega_0 - \omega_\nu)}\left\{\left[\frac{\overline{p_x^2(0)}}{2m} + \frac{m\omega_0^2}{2}\overline{x^2(0)}\right] - \frac{1}{2}\left[\overline{p_\nu^2(0)} + \omega_\nu^2\overline{q_\nu^2(0)}\right]\right\}, \tag{29.5}$$

which is the same as the classical expression (27.10) within the accuracy of the quantum-mechanical averaging, denoted by the bar in (29.5). This averaging automatically takes into consideration the zero-point fluctuations of the field and of the oscillator mentioned above. The classical expression for the intensity of spontaneous emission (27.11) does not take into consideration the zero-point fluctuations of the field and is therefore not applicable for weakly excited states of the oscillator, when its energy is comparable with $\hbar\omega_0$. This is connected with the factor of $\frac{1}{2}$ that appears in the usual formulation of the correspondence principle (Heitler, 1954) (the other half of the contribution to the spontaneous emission—the zero-point fluctuations of the field—is not considered). We notice that if we use the classical expression for the emission intensity (27.10), but allow for the zero-point fluctuations of the field and the dipole moment, the correct quantum expression is obtained from a semi-classical derivation.

It follows from the quantum expression (29.5) that stimulated emission from a linear oscillator in a field with random phase $\overline{p_\nu(0)} = \overline{q_\nu(0)} = 0$ is impossible†. This is because the linear oscillator corresponds to the case

† Or, in other words, in the case of a non-linear oscillator it is impossible to have negative absorption which is independent of the phase relations between the oscillator and the field.

$\overline{R_3(0)} < 0$. Negative absorption in a system that can be described by the energy spin (when $\overline{R_3(0)} > 0$) depends upon the non-linear properties of such a system. Such a system has a limited number $(2R + 1)$ of equidistant levels, unlike a linear oscillator which has an equidistant spectrum with an unlimited number of levels; in particular when $R = \frac{1}{2}$ the molecule has only two levels. This leads to a limitation of the amplitude (the amplitude of a harmonic oscillator is not limited). If a harmonic oscillator is in a given energy level there is always a higher level. At the same time the probability of a transition to a higher level because of the presence of radiation is greater than the probability of a transition to a lower level. On the other hand, if a two-level system, for example, is in the upper state there is only a probability of transition to the lower state, i.e. negative absorption, for any phase relation.

Let us now briefly examine the question of the phase relations between stimulated and stimulating fields. In the classical case, as can be seen from (27.5), there are definite phase relations between the stimulating ("initial") field and the stimulated emission field—$q_\nu^{(2)}$ is connected with $q_\nu^{(0)}$. In quantum theory the situation is somewhat more complex since the initial field may have no definite phase. With a high enough amplitude, however, the phase can be considered to be well defined and in this case (28.7) leads to relations analogous to the classical relations (27.5). There may be a difference in the sign of $R_3(0)$.

To conclude the present section we should like to point out that there is a far greater correspondence between the classical and quantum theories of spontaneous and stimulated emission than is generally admitted. In particular the regular parts of the emission are subject to the same relations in both cases. On the other hand, an oscillator which is in a definite phase relation with the field is often taken as an example of the classical analogue of stimulated emission (see, e.g., Heitler, 1954). This emission is then compared with the quantum expression which has been phase-averaged. Of course, there is phase-independent stimulated emission in classical theory but, as in the quantum case, it appears only when there is non-linearity.

30. General Expressions for the Intensities of Spontaneous and Stimulated Emission

30.1. The Derivation of the Emitted Intensity

Up until now we have been discussing in the dipole approximation radiation from systems with a discrete energy spectrum. We shall now remove these restrictions and discuss emission from an arbitrary system (which we shall continue for the sake of the argument to call a molecule). The Hamiltonian of a molecule interacting with radiation can be written in the form (see (4.13)–(4.16))

$$\hat{H} = \hat{H}_0 + \tfrac{1}{2}\sum_{\nu} (\hat{p}_\nu^2 + \omega_\nu^2 \hat{q}_\nu^2) - \sum_{\nu} \hat{B}_\nu \hat{q}_\nu, \tag{30.1}$$

where $\hat{H}_0$ is the Hamiltonian of the unperturbed molecule,

$$\hat{B}_\nu = \sum_{s=1} \frac{e_s}{m_s c} (\hat{p}_s \cdot A_\nu(s)),$$

e_s, m_s, p_s are respectively the charge, mass and momentum of the sth particle of the molecule;

$A_\nu(s)$ is the value of the eigenvector A_ν of the νth mode at the position of the sth particle.

Just as in the preceding section we find the operator of the energy derivative of the νth mode of the field

$$\frac{d\hat{H}_\nu}{dt} = \frac{i}{\hbar} [\hat{H}, \hat{H}_\nu] = \hat{B}_\nu \hat{p}_\nu. \tag{30.2}$$

We then find the mean value of this quantity. For this we use the equation for the density matrix in the interaction representation

$$i\hbar \frac{\partial \hat{\varrho}}{\partial t} = [\hat{V}, \hat{\varrho}].$$

The solution of this equation from first order perturbation theory is of the form†

$$\hat{\varrho}(t) = \hat{\varrho}(0) - \frac{i}{\hbar} \int_0^t [\hat{V}(t_1), \hat{\varrho}(0)]\, dt_1. \tag{30.3}$$

The mean value of the intensity of emission, including terms of the second order of smallness, is≠

$$\frac{d\bar{H}_\nu}{dt} = \overline{B_\nu(t)}\; \overline{p_\nu(t)} - \frac{i}{\hbar} \int_0^t \mathrm{Tr}\, \{[\hat{V}(t_1), \hat{\varrho}(0)]\, \hat{B}_\nu(t)\, \hat{p}_\nu(t)\}\, dt_1. \tag{30.4}$$

† Here, just as in the preceding section, we find the change in energy with an accuracy up to the second approximation of perturbation theory, and for this it is sufficient to know the density matrix in the first approximation.

≠ We assume, just as before, that at $t = 0$ the density matrix has the form

$$\varrho(0) = \varrho_m^{(0)} \varrho_f^{(0)},$$

where ϱ_m and ϱ_f are the density matrices of the molecule and the field respectively.

Using the property of the trace of the product of operators

$$\mathrm{Tr}\,(\hat{A}\hat{B}) = \mathrm{Tr}\,(\hat{B}\hat{A}),$$

and also the fact that in our case

$$\hat{V} = -\sum_{\lambda} \hat{B}_\lambda \hat{q}_\lambda,$$

after simple transformations we arrive at the following expression for the second term on the right-hand side of (30.4)

$$\overline{\left(\frac{dH_\nu}{dt}\right)_2} = \frac{i}{\hbar}\sum_{\lambda}\int_0^t \{\overline{B_\nu(t)\,B_\lambda(t_1)}\;\overline{p_\nu(t)\,q_\lambda(t_1)}$$

$$- \overline{B_\lambda(t_1)\,B_\nu(t)}\;\overline{q_\lambda(t_1)\,p_\nu(t)}\}\,dt_1, \tag{30.5}$$

where the bar denotes averaging by using the density matrix $\hat{\varrho}(0)$. In the interaction representation, which we are using here, the time-dependence is defined by the Hamilton operator of the unperturbed system. Therefore

$$\hat{p}_\nu(t) = \hat{p}_\nu(0)\cos\omega_\nu t - \omega_\nu \hat{q}_\nu(0)\sin\omega_\nu t,$$

$$\hat{q}_\nu(t) = \hat{q}_\nu(0)\cos\omega_\nu t + \frac{\hat{p}_\nu(0)}{\omega_\nu}\sin\omega_\nu t,$$

$$\hat{B}_\nu(t) = \mathrm{e}^{\frac{i}{\hbar}\hat{H}_0 t}\,\hat{B}_\nu(0)\,\mathrm{e}^{-\frac{i}{\hbar}\hat{H}_0 t}. \tag{30.6}$$

Hence we find (in accordance with the definition of the mean)

$$\overline{B_\nu(t)\,B_\lambda(t_1)} = \sum_n \{\mathrm{e}^{\frac{i}{\hbar}\hat{H}_0 t}\,\hat{B}_\nu(0)\,\mathrm{e}^{-\frac{i}{\hbar}\hat{H}_0(t-t_1)}\,\hat{B}_\nu(0)\,\mathrm{e}^{-\frac{i}{\hbar}\hat{H}_0 t_1}\,\varrho(0)\}_{nn}$$

$$= \sum_{nkl} B_{\nu nk} B_{\lambda kl} \varrho_{ln}\,\mathrm{e}^{i(\omega_{nk}t + \omega_{kl}t_1)} \tag{30.7}$$

and likewise

$$\overline{B_\lambda(t_1)\,B_\nu(t)} = \sum_{nkl} B_{\lambda nk} B_{\nu kl} \varrho_{ln}\,\mathrm{e}^{i(\omega_{nk}t_1 + \omega_{kl}t)} \tag{30.8}$$

(for the sake of simplicity we have omitted the argument $t = 0$).

In expressions (30.7) and (30.8) we must take $\hat{\varrho}$ to be the density matrix of the molecule. The total change in the energy of the νth mode of the field is

$$\overline{\frac{dH_\nu}{dt}} = \overline{B_\nu(t)}\;\overline{p_\nu(t)} + \frac{i}{\hbar}\sum_{\lambda}\int_0^t \{\overline{B_\nu(t)\,B_\lambda(t_1)}\;\overline{p_\nu(t)\,q_\lambda(t_1)}$$

$$- \overline{B_\lambda(t_1)\,B_\nu(t)}\;\overline{q_\lambda(t_1)\,p_\nu(t)}\}\,dt_1. \tag{30.9}$$

This expression gives, in terms of the mean values of the unperturbed field and of the molecular quantities, the change in the field energy as a result of the interaction. The first term in this expression is analogous to (28.11) and depends upon the phase relation between the dipole moment (or a more general characteristic B_ν of the molecule) and the νth mode of the emission field. The second term gives the change in the field energy caused by the fluctuation characteristics of the molecule and the field—the correlations $\overline{B_\nu(t)\, B_\lambda(t_1)}$, $\overline{q_\lambda(t_1)\, p_\nu(t)}$, etc.

Let us look at the special case when the density matrix of the molecule (at $t = 0$) is diagonal and the fluctuations of the field satisfy the conditions (see (27.9) and (28.12))

$$\overline{p_\nu(t)\, q_\lambda(t_1)} = \delta_{\nu\lambda} \overline{p_\nu(t)\, q_\nu(t_1)}; \quad \overline{q_\lambda(t_1)\, p_\nu(t)} = \delta_{\nu\lambda} \overline{q_\nu(t_1)\, p_\nu(t)} \tag{30.10}$$

(the latter hold if the means of all modes of the field, with the exception perhaps of one, are zero). In this case, carrying out transformations similar to those made in the preceding sections, after time-averaging over a period much greater than the period of the radiation $2\pi/\omega_\nu$, we obtain

$$\begin{aligned} I_\nu = \frac{d\bar{H}_\nu}{dt} &= \pi \sum_{n,k} |B_{\nu nk}|^2\, \delta(\omega_{nk} - \omega_\nu) \\ &\quad \times \left\{ \frac{1}{2} \frac{\overline{p_\nu^2} + \omega_\nu^2 \overline{q_\nu^2}}{\hbar\omega_\nu} (\varrho_n - \varrho_k) + \frac{1}{2} (\varrho_n + \varrho_k) \right\} \\ &= \pi \sum_{nk} |B_{\nu nk}|^2\, \delta(\omega_{nk} - \omega_\nu) \left\{ \bar{n}_\nu (\varrho_n - \varrho_k) + \varrho_n \right\}. \end{aligned} \tag{30.11}$$

Here $\varrho_n = \varrho_{nn}(0)$ is the population of level n;
$\bar{n}_\nu$ is the mean number of photons of frequency ω_ν.

If the frequencies ω_ν or the energy levels of the molecule have a continuous spectrum, then we must take $\delta(\omega_{nk} - \omega_\nu)$ to be a δ-function; in the other case (when the frequency spectrum of the field and the levels of the molecule are discrete) we must replace

$$\delta(\omega_{nk} - \omega_\nu) \quad \text{by} \quad \frac{\sin(\omega_{nk} - \omega_\nu)\, t}{\pi(\omega_{nk} - \omega_\nu)}.$$

In accordance with (30.11) the spontaneous emission intensity has the form

$$I_\nu^s = \pi \sum_{nk} |B_{\nu nk}|^2\, \delta(\omega_{nk} - \omega_\nu)\, \varrho_n. \tag{30.12}$$

This can easily be obtained by using the expression for the transition probability for the molecule + field system, assuming that the latter is in a vacum.

The intensity of the stimulated emission (or absorption) is of the form

$$I_\nu^i = \pi \sum_{nk} |B_{\nu nk}|^2 \, \delta(\omega_{nk} - \omega_\nu) \, \bar{n}_\nu(\varrho_n - \varrho_k). \qquad (30.13)$$

This intensity may, depending on the sign of $\varrho_n - \varrho_k$, take positive (stimulated emission) and negative (absorption) values. We can show that in the case of a linear harmonic oscillator only absorption is possible. To do this we rewrite (30.13) in the form

$$I_\nu^i = \pi \sum_{nk} \varrho_n \bar{n}_\nu \, |B_{\nu nk}|^2 \, [\delta(\omega_{nk} - \omega_\nu) - \delta(\omega_{nk} + \omega_\nu)].$$

The last expression for the case of a harmonic oscillator of frequency ω_0 can be rewritten in the form

$$I_\nu^i = \pi \sum_{n} \varrho_n \bar{n}_\nu \, \delta(\omega_0 - \omega_\nu) \, [|B_{\nu n, n-1}|^2 - |B_{\nu n, n+1}|^2].$$

It is easy to check that the expression in square brackets for a linear oscillator (in the dipole approximation) is proportional (with a positive coefficient) to $|x_{n,n-1}|^2 - |x_{n,n+1}|^2 = -\hbar/2m\omega$. Therefore in this case $I_\nu^i < 0$, and only absorption is possible.

30.2. The Connection between the Emitted Intensity I_ν, the Imaginary Part of the Susceptibility χ'', and Spontaneous Noise

In sections 14–17 of Chapter IV we discussed the behaviour of a quantum system under the influence of an alternating field. We now consider the same problem, but, in contrast with the previous treatment, we describe the electromagnetic field within the framework of quantum theory. It is interesting to compare the conclusions of the present section and Chapter IV. For this we first note that if the field q_ν were described classically (and is a given function of time), then, since the energy of interaction with this field has the form

$$\hat{V} = -\sum_\nu \hat{B}_\nu \hat{q}_\nu,$$

we could introduce the susceptibility according to (14.12)

$$\chi(\omega_\nu) = \frac{1}{\hbar} \sum_{n,k} (\varrho_n - \varrho_k) \, |B_{\nu nk}|^2 \, \zeta(\omega_{nk} - \omega_\nu). \qquad (30.14)$$

This quantity describes the response of the system in the form of the mean value $\langle B_\nu \rangle$ in a field

$$q_\nu(t) = \operatorname{Re} q_\nu^0 \, e^{-i\omega_\nu t}.$$

In accordance with the definition of the susceptibility (14.10)

$$\langle B_\nu \rangle = \mathrm{Re}\,\{\chi(\omega_\nu)\, q_\nu^0\, \mathrm{e}^{-i\omega_\nu t}\}.$$

We can connect the imaginary part of the susceptibility $\chi''(\omega_\nu)$ with the expression for the emitted intensity (30.11). Comparing (30.14) and (30.11), and using (17.6) we obtain

$$I_\nu = \pi (B_\nu^2)_{\omega\nu} - \tfrac{1}{2}\hbar\chi''(\omega_\nu) - \bar{n}_\nu \hbar \chi''(\omega_\nu). \tag{30.15}$$

The first two terms give the intensity of the spontaneous emission, the last term gives either the intensity of the stimulated emission or the absorbed power, depending on the sign of $\chi''(\omega_\nu)$. Let us first examine this term. We compare it with the mean power absorbed by the system (17.9)

$$Q = \frac{\omega_\nu}{2}\chi''(\omega_\nu)\, q_{\nu 0}^2. \tag{30.16}$$

It is easy to see that in the classical limit ($\bar{n}_\nu \gg 1$)

$$\bar{n}_\nu = \frac{\bar{p}_\nu^2 + \omega_\nu^2 \bar{q}_\nu^2}{2\hbar\omega_\nu} \approx \frac{\omega_\nu q_{\nu 0}^2}{2\hbar}$$

and (30.16) agrees with the last term in (30.15) except for the sign ((30.15) gives the emitted power and (30.16) the absorbed power). Therefore the stimulated emission and absorption at the classical limit $\bar{n}_\nu \gg 1$ is described by the imaginary part of the susceptibility. The case of stimulated emission corresponds to $\chi'' < 0$ and of absorption to $\chi'' > 0$. Stimulated emission can thus be looked upon as negative absorption.

Let us now look at the first two terms in (30.15), which describe spontaneous emission

$$I_\nu^s = \pi (B_\nu^2)_{\omega\nu} - \tfrac{1}{2}\hbar\chi''(\omega_\nu). \tag{30.17}$$

As has already been pointed out, the spontaneous emission intensity consists of two parts: the fluctuations of a property of the molecule $(B_\nu^2)_{\omega\nu}$ (in the dipole approximation B_ν is proportional to the dipole moment), and the absorption or emission connected with the vacuum fluctuations of the field. These two parts describe the first and second terms of (30.17) respectively. In a state of thermal equilibrium, when the fluctuation–dissipation theorem (17.12) holds, the spectral density $(B_\nu^2)_{\omega\nu}$ is of the form

$$(B_\nu^2)_{\omega\nu} = \frac{\hbar\chi''(\omega_\nu)}{\pi}\left\{\frac{1}{2} + \frac{1}{\mathrm{e}^{\hbar\omega_\nu/kT} - 1}\right\}. \tag{30.18}$$

Substituting this expression in (30.17) we obtain the intensity of the spontaneous emission in the equilibrium case

$$I_\nu^s = \frac{\hbar\chi''(\omega_\nu)}{e^{\hbar\omega_\nu/kT} - 1}. \tag{30.19}$$

We notice that (30.18) and (30.19) can also be obtained by making the assumption that the radiation is in a state of equilibrium

$$\bar{n}_\nu = \frac{1}{e^{\hbar\omega_\nu/kT} - 1}.$$

Substituting this expression in (30.15) and equating I_ν to zero, since we have assumed a state of thermal equilibrium, we obtain (30.18) and (30.19).

In conclusion we would remark that the agreement, in the classical limit, of the expressions obtained in this section and in Chapter IV is due to the fact that the change to the delta-functions ($t \gg 1/\delta\omega$) means in both cases that a steady state of the system is under consideration. The absence of terms such as $\overline{B_\nu(t)}$ and $\overline{p_\nu(t)}$ which depend on the phase relation between $B_\nu(t)$ and $p_\nu(t)$ is connected with the assumption that when there is no field the mean values of the quantities $\hat{x}_a$ are zero.

CHAPTER VII

Spontaneous and Stimulated Emission in Free Space

IN THIS chapter it is assumed that the radiation has a continuous spectrum, in, for example, free space and waveguides. The emission process is then irreversible and the behaviour of a dynamic system in such a field can be described by transport equations. We shall use the mathematical techniques developed in Chapter II to describe a dynamic system interacting with radiation in free space.

31. Coherence during Spontaneous Emission

31.1. The Intensity of Emission from a System of Two-level Molecules

Let us consider radiation from a system of two-level molecules in free space. We shall assume that the dimensions of the molecular system are much less than the length of the emitted wave. We have already carried out summation over the free space modes for this kind of system in deriving (27.11). By carrying out a similar summation in the (28.16) and (28.17) we obtain the intensity of the spontaneous and stimulated emission from a system of two-level molecules in free space†

$$I^s = I_s^0(\overline{R_1^2} + \overline{R_2^2} + \overline{R_3(0)}), \tag{31.1}$$

$$I^i = I_i^0\, 2\overline{R_3(0)}, \tag{31.2}$$

where I_s^0 and I_i^0 are the intensities of the spontaneous and induced emission from a single isolated molecule.

As was to be expected, the intensity of the stimulated emission is propor-

† It is a good thing to remind ourselves once again that here we are not discussing that part of the stimulated emission (or absorption) that depends on the phase relations.

tional to the difference of the populations of the upper and lower levels (if $2R_3(0) < 0$ absorption occurs).

It can be seen from (31.1) that the spontaneous emission intensity from a system of molecules whose dimensions are much less than the length of the emitted wave is not equal to the sum of the intensities from the individual molecules. Let us find, for example, the intensity radiated from a system which starts by being in a pure state characterized by the quantum numbers R and M (see section 19). In this state

$$\overline{R_1^2} + \overline{R_2^2} + M^2 = R(R+1). \tag{31.3}$$

By using this equation we find that

$$\overline{R_1^2} + \overline{R_2^2} + \overline{R_3(0)} = R(R+1) - M^2 + M. \tag{31.4}$$

We thus arrive at the expression

$$I^s = I_s^0(R+M)(R-M+1), \tag{31.5}$$

which was first obtained by Dicke (1954) (see also Fain, 1958). We can obtain the same expression by using the fact that the energy of interaction with the radiation is proportional to $\hat{R}_1$ (see (28.1)) and therefore the probability of a transition with the emission of a photon is proportional to the square of the modulus of the matrix element $\hat{R}_1$, i.e. in accordance with (19.9), $(R+M)(R-M+1)$†. Therefore coupled states of a system of molecules are possible which give spontaneous emission with an intensity proportional to the square of the number of molecules. Such, for example, is the state with $R = \frac{1}{2}N$, $M = 0$ (where N is the total number of molecules). The intensity from a system in this state is

$$I^s = \tfrac{1}{2}N(\tfrac{1}{2}N + 1)\, I_s^0 \approx \tfrac{1}{4}N^2 I_s^0. \tag{31.6}$$

On the other hand states are possible in which the system emits no energy at all. An example of this is the state with $R = M = 0$. We notice that in all the transitions which are accompanied by emission the quantum "cooperation number" R (as it is called by Dicke, 1954) does not change since the operator $\hat{R}^2$ is an integral of the motion. This operator commutes with the Hamiltonian (28.1).

31.2. The Classical Analogue

The molecules thus do not emit independently and the radiation from individual molecules is coherent. This coherence can be understood if we remember that each molecule is in the field radiated by the other molecules.

† This result is the same, of course, if $e_2 \neq 0$.

Let us take a system of oscillating charges occupying a region with linear dimensions much less than the length of the emitted wave. In the dipole approximation the radiation damping acting on each of the oscillators is (see, e.g., Landau and Lifshitz, 1960)

$$\boldsymbol{F}_{\text{rad}} = \frac{2}{3c^2} \sum_{k=1}^{N} e_k^2 \dddot{\boldsymbol{r}}_k, \tag{31.7}$$

where $\boldsymbol{r}_k$ is the displacement of the kth oscillator;
e_k is the charge.

When there is no interaction between the oscillators (apart from interaction due to the radiation), the coordinates $\boldsymbol{r}_k$ are subject to the equations

$$\ddot{\boldsymbol{r}}_k + \omega_k^2 \boldsymbol{r}_k = \frac{2}{3mc^3} \sum_{s=1}^{N} e_s^2 \dddot{\boldsymbol{r}}_s. \tag{31.8}$$

It follows from this that because of the interaction due to the radiation the motion of each oscillator is dependent on the motion of all the other oscillators. In the case when all the oscillators have the same frequency ω_0, and all the charges $e_k = e$, equation (31.8) can be rewritten in the form

$$\ddot{\boldsymbol{r}}_k + \omega_0^2 \boldsymbol{r}_k = -\gamma_0 \sum_{l=1}^{N} \dot{\boldsymbol{r}}_l. \tag{31.9}$$

Here we have replaced $\dddot{\boldsymbol{r}}_k$ by $-\omega_0^2 \dot{\boldsymbol{r}}_k$, assuming that the width of the emission line, given by $\gamma_0 = 2e^2\omega_0^2/3mc^3$, is much less than the frequency ω_0†. The strength of the radiation field from the system of oscillators is proportional to the second derivative of the dipole moment of the system

$$\boldsymbol{d} = e \sum_{k=1}^{N} \boldsymbol{r}_k.$$

† It should be remembered that the only solutions of the set of equations (31.8) which (have any physical meaning are those which approximately satisfy (31.9). The point is that 31.8) contains not only decreasing solutions, but also exponentially increasing solutions, which have no physical meaning. It is easy to check the existence of such solutions in the special case of a single oscillator. Equation (31.8) becomes

$$\ddot{\boldsymbol{r}} + \omega_0^2 \boldsymbol{r} - \frac{2e^2}{3mc^3} \dddot{\boldsymbol{r}} = 0.$$

This equation has three linearly independent solutions (for $e^2\omega_0/mc^3 \ll 1$):

$$C_{1,2} \exp\left[\left(\pm i\omega_0 - \frac{2e^2\omega_0^2}{3mc^3}\right) t\right] \quad \text{and} \quad C_3 \exp\left[\left(\frac{3mc^3}{2e^2} + \frac{2e^2\omega_0^2}{3mc^3}\right) t\right]$$

The last solution increases exponentially and has no physical meaning.

Summing equations (31.9) we obtain

$$\ddot{\boldsymbol{d}} + \omega_0^2 \boldsymbol{d} + N\gamma_0 \dot{\boldsymbol{d}} = 0. \tag{31.10}$$

The solution of (31.10) is approximately

$$\boldsymbol{d} = \boldsymbol{d}_0 \mathrm{e}^{-\frac{N\gamma_0 t}{2}} \mathrm{e}^{i\omega_0 t}.$$

Therefore the strength of the radiation field has the form

$$\boldsymbol{E} = \boldsymbol{E}_0 \, \mathrm{e}^{-\frac{N\gamma_0 t}{2}} \mathrm{e}^{i\omega_0 t}. \tag{31.11}$$

In order to obtain the spectral distribution of the field we expand (31.11) as a Fourier integral:

$$\boldsymbol{E} = \int_{-\infty}^{\infty} \boldsymbol{E}(\omega)\, \mathrm{e}^{i\omega t}\, d\omega,$$

$$\boldsymbol{E}(\omega) = \frac{1}{2\pi} \boldsymbol{E}_0 \int_0^{\infty} \mathrm{e}^{i(\omega_0 - \omega)t}\, \mathrm{e}^{-\frac{N\gamma_0 t}{2}}\, dt$$

$$= \frac{1}{2\pi} \boldsymbol{E}_0 \frac{1}{i(\omega_0 - \omega) - \dfrac{N\gamma_0}{2}}.$$

This gives for the intensity distribution

$$J(\omega) \sim |E(\omega)|^2 \sim \frac{1}{(\omega_0 - \omega)^2 + \dfrac{N^2\gamma_0^2}{4}}. \tag{31.12}$$

Therefore a system of N oscillators has a natural line width

$$\gamma = N\gamma_0, \tag{31.13}$$

where γ_0 is the natural line width of a single isolated oscillator.

This investigation of radiation from a system of classical oscillators confirms the conclusion obtained above from quantum theory that the intensity radiated from a system of oscillators

$$I = \frac{2}{3c^3} (\ddot{\boldsymbol{d}})^2$$

is not equal to the sum of the intensities from the individual oscillators and, secondly, leads us to the conclusion that the width of the emission line $\gamma = N\gamma_0$ is not equal to the line width of a single individual oscillator. A similar conclusion follows from the quantum treatment of a system of two-level molecules (see section 33).

31.3. Examples

We can now continue the discussion of the coherence of the spontaneous emission from a system of molecules. Let us take two simple examples which can be solved completely.

First example. System of two molecules. Initially molecule 1 is in an excited state E_+ and molecule 2 is in the ground state E_-. We assume as before that the molecules have only the two non-degenerate energy levels E_+ and E_-, and are physically separated by a distance which is small compared with the length of the emitted wave. It is also assumed that direct interaction of the molecules can be neglected, the wave functions of the molecules do not overlap and questions of the symmetry of the molecular states are irrelevant. We are interested in the behaviour of the radiation field and of the molecules at a time t. We use the method of Weisskopf and Wigner (1930*a* and *b*) to solve this problem (see also Heitler, 1954)†. We shall solve the Schrödinger equation in the interaction representation (4.33)

$$i\hbar\dot{b}_n = \sum_m V_{nm}\, e^{i\omega_{nm}t}\, b_m,$$

where the probability amplitudes b_n are the coefficients of the expansion of the wave function of the system (4.30) in eigenfunctions such as (4.31). We shall use the following notation: b_{+-0} is the probability that the first molecule is in a state E_+, the second molecule in a state E_- and there are no photons; b_{-+0} is the same but with the molecules in interchanged energy states; b_{--1_ν} is the probability that both molecules are in a state E_- and that a photon of energy $\hbar\omega_\nu$ is emitted. We also use

$$V_{+-0;\,--1_\nu},\quad V_{-+0;\,--1_\nu},\quad V_{--1_\nu;\,+-0},\quad V_{--1_\nu;\,-+0}$$

to denote the matrix elements of the operator for the energy of interaction of the molecules with the radiation corresponding to the transitions $+\,-0 \to -\,-1_\nu$, $-\,+0 \to -\,-1_\nu$ etc. Because of the assumption that the molecules are far less than a wavelength apart

$$V_{+-0;\,--1_\nu} = V_{-+0;\,--1_\nu} \equiv V_{+0;\,-1_\nu}. \tag{31.14}$$

Following Weisskopf and Wigner's method we consider only those transitions in which the energy is at least approximately conserved. In this case

† These problems can, of course, be solved by the transport equations derived in Chapter II. For the purposes of illustration, however, we have decided to use the Weisskopf–Wigner method.

equations (4.33) become

$$i\hbar\dot{b}_{+-0} = \sum_{\nu} V_{+-0;\,--1_\nu} b_{--1_\nu}\, e^{\frac{i}{\hbar}(E_+ - E_- - \hbar\omega_\nu)\,t},$$

$$i\hbar\dot{b}_{--1_\nu} = (V_{--1_\nu;\,+-0}\, b_{+-0} + V_{--1_\nu;\,-+0}\, b_{-+0})\, e^{\frac{i}{\hbar}(E_- - E_+ + \hbar\omega_\nu)t},$$

$$i\hbar\dot{b}_{-+0} = \sum_{\nu} V_{-+0;\,--1_\nu}\, b_{--1_\nu}\, e^{\frac{i}{\hbar}(E_+ - E_- - \hbar\omega\nu)\,t},$$

or, remembering (31.14),

$$i\hbar\dot{b}_{+-0} = \sum_{\nu} V_{+0;\,-1_\nu}\, b_{--1_\nu}\, e^{\frac{i}{\hbar}(E_+ - E_- - \hbar\omega_\nu)\,t},$$

$$i\hbar\dot{b}_{--1_\nu;} = V_{-1_\nu;\,+0}(b_{+-0} + b_{-+0})\, e^{\frac{i}{\hbar}(E_- - E_+ + \hbar\omega_\nu)\,t},$$

$$i\hbar\dot{b}_{-+0} = \sum_{\nu} V_{+0;\,-1_\nu}\, b_{--1_\nu}\, e^{\frac{i}{\hbar}(E_+ - E_- - h\omega_\nu)\,t}.$$

The initial conditions have the form

$$b_{+-0}(0) = 1, \quad b_{-+0}(0) = b_{--1_\nu}(0) = 0.$$

We make the substitution

$$b_1 = b_{+-0} + b_{-+0}, \quad b_2 = b_{+-0} - b_{-+0}, \quad b_3 = b_{--1_\nu}$$

and obtain the equations

$$i\hbar\dot{b}_1 = \sum_{\nu} 2V_{+0;\,-1_\nu}\, b_3\, e^{i(\omega_0 - \omega_\nu)\,t}, \tag{31.15}$$

$$i\hbar\dot{b}_3 = V_{-1_\nu;\,+0}\, b_1\, e^{-i(\omega_0 - \omega_\nu)\,t}, \tag{31.16}$$

$$\dot{b}_2 = 0, \tag{31.17}$$

where

$$\omega_0 = (E_+ - E_-)/\hbar.$$

The initial conditions can be rewritten in the form

$$b_1(0) = 1, \quad b_2(0) = 1, \quad b_3(0) = 0. \tag{31.18}$$

From (31.17) and (31.18) we at once obtain

$$b_2 = 1. \tag{31.19}$$

We shall look for a solution of the rest of the equations in the form†

$$b_1(t) = e^{-\frac{1}{2}\gamma t}. \tag{31.20}$$

Substituting (31.20) in (31.16) we obtain the differential equation

$$i\hbar\dot{b}_3 = V_{-1_\nu;+0}\, e^{-i(\omega_0-\omega_\nu)t-\frac{1}{2}\gamma t}$$

with the solution

$$b_3 = V_{-1_\nu;+0}\frac{e^{-i(\omega_0-\omega_\nu)t-\frac{1}{2}\gamma t}-1}{\hbar\left(\omega_0-\omega_\nu-\frac{i}{2}\gamma\right)}. \tag{31.21}$$

In order to satisfy condition (31.15) we substitute (31.21) in (31.15) and obtain

$$-\frac{i\hbar\gamma}{2} = \sum_\nu \frac{2|V_{-1_\nu;+0}|^2\,[1-e^{(i(\omega_0-\omega_\nu)+\frac{1}{2}\gamma)t}]}{\hbar\left(\omega_0-\omega_\nu-\frac{i}{2}\gamma\right)}. \tag{31.22}$$

Summation over the radiation oscillators in the right-hand side should, as usual, be replaced by integration with respect to the frequency. By performing this integration it is easy to find (Fain, 1958) that

$$\gamma = 2\gamma_0,$$

where $\gamma_0 = w_{+-}$ is the width of the line from an isolated molecule (equal to the value of the probability of the transition $+ \to -$ per unit time).

Using (31.19), (31.20) and the definition of b_1, b_2 and b_3 we find that at time $t \gg 1/\gamma$

$$|b_{-+0}(\infty)|^2 = \tfrac{1}{4}, \quad |b_{+-0}(0)|^2 = \tfrac{1}{4}.$$

Therefore there is a probability of $\frac{1}{4}$ of no change in the state of the system (molecule 1 is excited, molecule 2 is in the ground state). There is also a probability of $\frac{1}{4}$ that the second molecule becomes excited and the first makes a transition to the ground state (molecule 1 emits a photon, molecule 2 absorbs it); and a probability of $\frac{1}{2}$ that a photon is emitted. The line broadening is due to the probability of excitation of the second molecule. We should stress

† We notice that the exponential dependence is always approximate. In the first place we find the solution for a time $t \gg 2\pi/\omega_0$ (the period of the radiation) and, secondly, it can be shown (Höhler, 1958) that even with $\gamma t \gg 1$ a solution of the type (31.20) differs from an exponential form. Khalfin (1957) has shown in the general case that exponential decay does not occur. Over a large time interval, however, the exponential can be used approximately. This remark also holds good for the transport equations of Chapter II from which, in particular, exponential decay follows.

the fact that, although the line width doubles, the initial intensity radiated is equal to the intensity from an isolated molecule. This can easily be proved by perturbation theory. Although the time during which the system emits energy has been halved ($\tau = \frac{1}{2}\gamma_0$), the emitted energy has also been halved (a system of two molecules on the average emits an energy of $\frac{1}{2}\hbar\omega_0$ instead of $\hbar\omega_0$ as in the case of a single molecule).

Second example. A system of N molecules. Initially molecule 1 is excited and the rest of the molecules are in the ground state. The initial conditions can be written in the form

$$b_{+-\cdots-0} = 1, \quad b_{-+-\cdots-0} = \ldots = b_{--\cdots+0} = b_{-\cdots-1_\nu} = 0. \tag{31.23}$$

Using the same assumptions as in the case of two molecules we obtain the following set of equations

$$\begin{aligned}
i\hbar\dot{b}_{+-\cdots-0} &= \sum_\nu V_{+0;1_\nu} b_{-\cdots-1_\nu} \mathrm{e}^{-i(\omega_\nu-\omega_0)t}, \\
i\hbar\dot{b}_{-+-\cdots-0} &= \sum_\nu V_{+0;-1_\nu} b_{-\cdots-1_\nu} \mathrm{e}^{-i(\omega_\nu-\omega_0)t}, \\
&\cdots\cdots\cdots\cdots\cdots\cdots \\
i\hbar\dot{b}_{-\cdots+0} &= \sum_\nu V_{+0;-1_\nu} b_{-\cdots-1_\nu} \mathrm{e}^{-i(\omega_\nu-\omega_0)t}, \\
i\hbar\dot{b}_{-\cdots-1_\nu} &= V_{1_\nu;+0}(b_{+-\cdots-0} + \ldots + b_{-\cdots-+0})\, \mathrm{e}^{i(\omega_\nu-\omega_0)t}.
\end{aligned} \tag{31.24}$$

From the initial conditions and equations (31.24) we find

$$b_{-+-\cdots-0} = \ldots = b_{-\cdots-+0} \equiv b. \tag{31.25}$$

We can then make the substitution

$$\begin{aligned}
b_1 &= b_{+-\cdots-0} + \ldots + b_{-\cdots-+0} = b_{+-\cdots-0} + (N-1)\,b, \\
b_2 &= b_{-\cdots-1_\nu}.
\end{aligned} \tag{31.26}$$

Solving the equations obtained we find

$$b_1(t) = \mathrm{e}^{-\frac{N\gamma_0 t}{2}}, \tag{31.27}$$

$$b_2(t) = V_{-1_\nu+0} \frac{\mathrm{e}^{-i(\omega_0-\omega_\nu)t - \frac{N\gamma_0 t}{2}} - 1}{\hbar\left(\omega_0 - \omega_\nu - \dfrac{i}{2} N\gamma_0\right)}. \tag{31.28}$$

We find from this that the radiation line width is $\gamma = N\gamma_0$. The probability of emission of a photon after a long time interval is $1/N$, the probability that molecule 1 remains in the ground state is $(N-1)^2/N^2$ and the probability of the excitation of any of the molecules (apart from the first) is $1/N^2$. Therefore the probability that a photon is not emitted is $1 - (1/N)$. The line broadening is due to the possibility of the excitation of each of the originally unexcited molecules. It should be pointed out that, as in the previous case, the initial intensity of the emitted radiation is the same as that for an isolated molecule.

32. Balance Equations and Transport Equations

32.1. The Balance Equations

We shall use the mathematical techniques developed in Chapter II to describe the behaviour of quantum systems interacting with radiation. Radiation in free space (or in a waveguide) has a continuous spectrum and may be considered to be a dissipative system. The Hamiltonian of the whole system, which consists of charged particles (the dynamic sub-system) interacting with the radiation field, can be written in the form (see (4.13)–(4.16), (30.1))

$$\hat{H} = \hat{E} + \hat{F} + \hat{V}, \tag{32.1}$$

where $\hat{E}$ is the Hamiltonian of the dynamic sub-system;

$$\hat{F} = \tfrac{1}{2}\sum_\nu (\hat{p}_\nu^2 + \omega_\nu^2 \hat{q}_\nu^2) \tag{32.2}$$

is the Hamiltonian of the radiation field and†

$$\hat{V} = -\sum_\nu \hat{B}_\nu \hat{q}_\nu; \quad \hat{B}_\nu = \sum_k \frac{e_k}{m_k c}(\hat{p}_k \cdot A_\nu(k)). \tag{32.3}$$

Using (4.35) we find that the non-zero matrix elements of $\hat{V}$ are of the form

$$\langle m; n_1, n_2, \ldots, n_\nu, n_{\nu+1} \ldots |V| k; n_1, n_2, \ldots n_\nu \pm 1, n_{\nu+1} \ldots\rangle$$

$$= -B_{\nu mn}\sqrt{\frac{\hbar}{2\omega_\nu}}\begin{Bmatrix}\sqrt{n_\nu + 1}\\ \sqrt{n_\nu}\end{Bmatrix}. \tag{32.4}$$

Let us now find the relaxation coefficients Γ (see (8.4)–(8.7)). We shall discuss the case of a sufficiently weak external field ((8.6) and (8.7))

$$\Gamma_{mkln} = \frac{\pi}{\hbar}\sum_{\{n_\nu\}\{n'_\nu\}} V_{m\{n_\nu\};k\{n'_\nu\}}\, P(\{n'_\nu\})\, V_{l\{n'_\nu\};n\{n_\nu\}}$$

$$\times\ \delta(E_l - E_n + F_{\{n'_\nu\}} - F_{\{n_\nu\}})\,\Delta(\omega_{mk} + \omega_{ln}),$$

† Here we are neglecting terms in the interaction energy that are proportional to the square of the charge.

where $\{n_\nu\}$ denotes the combination of the quantum numbers $n_1, n_2, \ldots, n_\nu, \ldots$ Using (32.4) and the fact that $P(\{n'_\nu\})$ is the probability distribution of $\{n'_\nu\}$ we obtain

$$\Gamma_{mkln} = \frac{\pi}{\hbar} \sum_\nu B_{\nu mk} B_{\nu ln} \Delta(\omega_{mk} + \omega_{ln}) \frac{\hbar}{2\omega_\nu}$$

$$\times [(\bar{n}_\nu + 1)\, \delta(E_l - E_n - \hbar\omega_\nu) + \bar{n}_\nu \delta(E_l - E_n + \hbar\omega_\nu)]. \tag{32.5a}$$

Likewise we find

$$\Gamma_{mk} = -\sum_{l,\nu} \Delta(\omega_{mk}) \frac{B_{\nu ml} B_{\nu lk}}{E_l - E_k + \hbar\omega_\nu} \frac{(\bar{n}_\nu + 1)\,\hbar}{2\omega_\nu}$$

$$-\sum_{l,\nu} \Delta(\omega_{mk}) \frac{B_{\nu ml} B_{\nu lk}}{E_l - E_k - \hbar\omega_\nu} \frac{\bar{n}_\nu \hbar}{2\omega_\nu}. \tag{32.5b}$$

In these expressions $\bar{n}_\nu$ denotes the mean value of the number of photons in a state denoted by the suffix ν. In particular if the radiation is in equilibrium, then

$$\bar{n}_\nu = \frac{1}{e^{\hbar\omega_\nu/kT} - 1}.$$

Equations (8.3), (32.5a) and (32.5b) describe the relaxation of the dynamic system due to its interaction with the radiation. We shall now discuss some special cases of these equations. If the conditions for changing to the balance equation are satisfied†

$$\dot{\sigma}_{mm} = \sum_k (w_{km}\sigma_{kk} - w_{mk}\sigma_{mm}), \tag{8.10}$$

then the transition probabilities which appear in these equations become

$$w_{mk} = 2\Gamma_{kmmk} = \begin{cases} 2\pi \sum_\nu |B_{\nu mk}|^2 \dfrac{\bar{n}_\nu + 1}{2\omega_\nu} \delta(E_m - E_k - \hbar\omega_\nu), & E_m > E_k, \\ 2\pi \sum_\nu |B_{\nu mk}|^2 \dfrac{\bar{n}_\nu}{2\omega_\nu} \delta(E_k - E_m - \hbar\omega_\nu), & E_k > E_m. \end{cases} \tag{32.6a}$$

These expressions provide the quantum-mechanical foundation for Einstein's theory of spontaneous and stimulated emission (see section 26). In order to connect expressions (32.6a) with the Einstein coefficients we must change from summation over ν to integration with respect to the frequencies

† We recall that in the case of relaxation caused by interaction with radiation the part of the quantity ω^* is played by the frequency $\omega \approx |\omega_{nm}|$ of the radiation field (see section 7).

and directions of propagation. As a result we obtain

$$w_{mk} = \begin{cases} \sum_{\alpha=1}^{2} \int d\Omega (a_k^m + \varrho_\alpha(\omega_{mk}, \Omega)\, b_{k\alpha}^m), & E_m > E_k, \\ \sum_{\alpha=1}^{2} \int d\Omega \varrho_\alpha(\omega_{km}, \Omega)\, b_{m\alpha}^k, & E_k > E_m, \end{cases} \tag{32.6b}$$

where $a_{k\alpha}^m$; $b_{k\alpha}^m = b_{m\alpha}^k$ are the differential forms of the Einstein coefficients for spontaneous emission, stimulated emission and absorption respectively, and†

$$\varrho_\alpha(\omega_\nu \Omega)\, d\omega_\nu\, d\Omega = \bar{n}_\nu \hbar\omega_\nu \frac{\omega_\nu^2\, d\omega_\nu\, d\Omega}{(2\pi c)^3}$$

is the density of the radiation in the range of frequencies $d\omega_\nu$, the range of solid angles $d\Omega$ and with a polarization α. In particular for dipole radiation from a single (generally bound) particle we obtain the usual expressions

$$a_{n\alpha}^m = \frac{\omega_{mn}{}^3}{2\pi\hbar c^3} |d_{mn}|^2 \cos^2\theta_\alpha; \quad b_{n\alpha}^m = \frac{4\pi}{\hbar^2} |d_{mn}|^2 \cos^2\theta_\alpha = b_{m\alpha}^n, \tag{32.7}$$

where d_{mn} is a matrix element of the dipole moment;
θ_α is the angle between the dipole moment and the polarization vector.

32.2. The Equations for the Mean Energy Spins

We shall now move on to discuss a system of two-level molecules. A system of this kind can be described by the energy spin (see section 19). In accordance with (19.11) the Hamiltonian of the dynamic sub-system can be written in the form

$$\hat{E} = \hbar\omega_0 \sum_j \hat{r}_{3j} = \hbar\omega_0 \hat{R}_3. \tag{32.8}$$

The energy of the interaction with radiation is of the form (see (19.23))‡

$$\hat{V} = -\sum_j (\hat{A}(j) \cdot [e_1 \hat{r}_{1j} + e_2 \hat{r}_{2j}]) = \sum_j (\hat{r}_j^+ \hat{F}_j^- + \hat{r}_j^- \hat{F}_j^+), \tag{32.9}$$

where

$$\hat{r}_j^\pm = \hat{r}_{1j} \pm i\hat{r}_{2j}, \quad \hat{F}_j^\pm = -\tfrac{1}{2}(A(j) \cdot e^\pm) \tag{32.10}$$

and $e^\pm = e_1 \pm ie_2 = \pm(2i\omega_0/c)\, d_{21}$ is a matrix element of the molecular dipole moment. In order to find the mean values $\langle r_j^\pm \rangle$ and $\langle r_{3j} \rangle$ we make

† Here we have used (3.55) for the number of modes per unit volume in free space.
‡ For the sake of simplicity we omit the magnetic dipole terms.

use of (8.18a) where the operators $\hat{r}_j$ and $\hat{r}_{3j}$ play the part of the operators v_j^r (see (19.30)). As a result we obtain for the derivative of the mean value of the operator O

$$\frac{d\langle O\rangle}{dt} = \frac{i}{\hbar}\langle[\hat{E} + \hat{\Gamma}, \hat{O}]\rangle + \sum_{i,i'} \Phi_{ii'}^{+-} \langle \hat{r}_{i'}^{-}[\hat{O}, \hat{r}_i^{+}] + [\hat{r}_{i'}^{-}, \hat{O}]\hat{r}_i^{+}\rangle$$

$$+ \sum_{ii'} \Phi_{ii'}^{-+} \langle \hat{r}_{i'}^{+}[\hat{O}, \hat{r}_i^{-}] + [\hat{r}_{i'}^{+}, \hat{O}]\hat{r}_i^{-}\rangle, \qquad (32.11)$$

where

$$\Phi_{i'i}^{+-} = \frac{\pi}{8\hbar}\sum_{\nu} (A_\nu(i)\cdot e^{-})(A_\nu(i')\cdot e^{+})\frac{n_\nu}{\omega_0}\,\delta(\omega_0 - \omega_\nu);$$

$$\Phi_{i'i}^{-+} = \frac{\pi}{8h}\sum_{\nu} (A_\nu(i')\cdot e^{+})(A_\nu(i)\cdot e^{-})\frac{\bar{n}_\nu + 1}{\omega_0}\,\delta(\omega_0 - \omega_\nu); \qquad (32.12)$$

$$\hat{\Gamma} = \sum_{i,i'} \varphi_{ii'}^{+-}\hat{r}_i^{+}\hat{r}_{i'}^{-} + \varphi_{i'i}^{-+}\hat{r}_{i'}^{-}\hat{r}_i^{+}.$$

If the radiation is in thermal equilibrium, then

$$\Phi_{ii'}^{+-} = e^{-\hbar\omega_0/kT}\Phi_{i'i}^{-+}.$$

The coefficients Φ are a measure of the relaxation of the dynamic sub-system due to interaction with the radiation†. Let us calculate these coefficients in the special case when the molecules have linearly polarized dipole moments which are at right angles to the line joining their centres of gravity. We shall assume that the mean occupation numbers $\bar{n}_\nu$ depend only upon the frequency $\omega_\nu = \omega_0$ (as occurs in the case of thermal equilibrium).

Then

$$\Phi_{ii'}^{+-} = \bar{n}_\nu\frac{\gamma_{ii'}}{2}; \quad \Phi_{i'i}^{-+} = (\bar{n}_\nu + 1)\frac{\gamma_{i'i}}{2},$$

$$\gamma_{ii'} = \gamma_{i'i} = \frac{\pi}{2}\frac{\omega_0}{\hbar c^2}\sum_{\nu} (A_\nu(i')\cdot d_{12})(A_\nu(i)\cdot d_{21})\,\delta(\omega_0 - \omega_\nu)$$

$$= \frac{\omega_0^3 d^2}{\pi c^3\hbar}\int_0^{2\pi} d\varphi \int_0^{\pi/2} d\theta\,\sin^3\theta\,\cos(ka\sin\theta\cos\varphi). \qquad (32.13)$$

Finally we obtain

$$\gamma_{ii'} = \frac{3}{2}\gamma_0\left\{\frac{\sin ka}{ka} + \frac{\cos ka}{(ka)^2} - \frac{\sin ka}{(ka)^3}\right\}. \qquad (32.14)$$

† Here we do not give the coefficients φ. Their calculation is similar to that for the coefficients Φ.

Here $k = \omega_0/c$;

a is the distance between the ith and i'th molecules;

$d = |\boldsymbol{d}_{12}|$;

γ_0 is the natural line width of an isolated molecule;

$$\gamma_0 = \tfrac{4}{3}\,\omega_0^3 d^2/c^3\hbar. \tag{32.15}$$

We find from (32.11) and (32.13)†

$$\frac{d\langle r_{3i}\rangle}{dt} = -2\bar{n}_\nu\gamma_0\langle r_{3i}\rangle - \sum_{i'}\frac{1}{2}\gamma_{ii'}\langle \hat{r}_{i'}^+\hat{r}_i^- + \hat{r}_i^+\hat{r}_{i'}^-\rangle$$

$$= -\gamma_0(2\bar{n}_\nu + 1)\left(\langle r_{3i}\rangle + \frac{1}{2}\frac{1}{2\bar{n}_\nu + 1}\right) - \sum_{i'(\neq i)}\frac{\gamma_{ii'}}{2}\langle \hat{r}_{i'}^+\hat{r}_i^- + \hat{r}_i^+\hat{r}_{i'}^-\rangle \tag{32.16}$$

$$\frac{d}{dt}\langle r_i^+\rangle = i\omega_0\langle r_i^+\rangle - \left(\bar{n}_\nu + \frac{1}{2}\right)\gamma_0\langle r_i^+\rangle + \sum_{i'(\neq i)}\gamma_{ii'}\langle \hat{r}_{i'}^+\hat{r}_{3i}\rangle, \tag{32.17}$$

$$\frac{d}{dt}\langle r_i^+\rangle = -i\omega_0\langle r_i^-\rangle - \left(\bar{n}_\nu + \frac{1}{2}\right)\gamma_0\langle r_i^-\rangle + \sum_{i'(\neq i)}\gamma_{ii'}\langle \hat{r}_{3i}\hat{r}_{i'}^+\rangle. \tag{32.18}$$

Here we have used the commutation relations for the energy spin $[\hat{r}\wedge\hat{r}] = i\hat{r}$ and the series of relations $\hat{r}_2\hat{r}_3 = (i/2)\,\hat{r}_1$, $\hat{r}_3\hat{r}_1 = (i/2)\,\hat{r}_2$, $\hat{r}_1\hat{r}_2 + \hat{r}_2\hat{r}_1 = 0$, $\hat{r}_2^2 = \frac{1}{4}$, etc. that are valid for a spin of $\frac{1}{2}$. It can be seen from equations (32.16)–(32.18) that the relaxation of the ith molecule (which is described by the operators $\hat{r}_i^\pm, \hat{r}_{3i}$) is connected with the relaxation of the other molecules. This connection appears because each molecule is in the radiation field of the rest of the molecules.

Let us examine some of the consequences of equations (32.16)–(32.18). Let the states of the molecules at a certain time be statistically independent and the mean values be $\langle r_i^\pm\rangle = 0$. (This equation holds in particular for stationary states of a system). The intensity of the spontaneous emission from the ith molecule at this time is‡

$$-\frac{d\langle r_{3i}\rangle}{dt}\hbar\omega_0 = \gamma_0\hbar\omega_0\left(\frac{1}{2} + \langle r_{3i}\rangle\right) = n_+\gamma_0\hbar\omega_0 = n_+I_0$$

† In order not to complicate the discussion we are neglecting the shift of the frequency ω_0 which appears because of the presence of the operator Γ in the right-hand side of the equation (32.11).

‡ We recall that the mean value $2\langle r_3\rangle = n_+ - n_-$, where n_+ and n_- are the populations of the upper and lower levels of a single molecule

$$n_+ + n_- = 1,\quad \tfrac{1}{2} + \langle r_3\rangle = n_+.$$

and is independent of the states of the rest of the molecules. The total intensity of spontaneous emission from the whole system in this case is equal to the sum of the emission intensities from the individual molecules. As time proceeds the correlations $\langle \hat{r}_i^+ \hat{r}_{i'}^- + \hat{r}_i^+ \hat{r}_{i'} \rangle$ $(i' \neq i)$ become finite and the intensity of the spontaneous emission is no longer equal to the sum of the intensities from the individual molecules.

32.3. *The Classical Approximation*

The set of equations (32.17) and (32.18) is non-linear. It is not complete since the mean values $\langle \boldsymbol{r}_i \rangle$ are expressed in terms of the correlations $\langle \hat{r}_i^+ \hat{r}_{i'}^- \rangle$, $\langle \hat{r}_{3i} \hat{r}_{i'}^- \rangle$, etc. The equations for the latter can be obtained from (32.11). We can thus obtain a set of coupled equations. However, for states that are weakly excited from equilibrium the set (32.16)–(32.18) can be linearized. Let us examine the case where $\bar{n}_\nu = 0$ (spontaneous emission). In equilibrium $r_{3i} = -\frac{1}{2}$, and substituting these values in the right-hand sides of (32.17) and (32.18) we obtain

$$\frac{d}{dt} \langle r_i^\pm \rangle = \pm i\omega_0 \langle r_i^\pm \rangle - \frac{1}{2} \sum_{i'} \gamma_{ii'} \langle r_{i'}^\pm \rangle. \tag{32.19}$$

We notice that exactly this kind of equation can be obtained for the annihilation and creation operators $\langle a_i \rangle$ and $\langle a_i^+ \rangle$ of a system of charged oscillators†. At the same time the equations for the mean values of the harmonic oscillators are the same as the classical equations (Ehrenfest's theorem). Essentially, therefore, (32.19) is a set of classical equations. It changes into (31.9) when the dimensions of the system of oscillators are much less than the wavelength of the emitted wave.

Let us examine the case of two molecules

$$\frac{d\langle r_1^\pm \rangle}{dt} = \pm i\omega_0 \langle r_1^\pm \rangle - \frac{1}{2} \gamma_0 \langle r_1^\pm \rangle - \frac{1}{2} \gamma_{12} \langle r_2^\pm \rangle,$$

$$\frac{d\langle r_2^\pm \rangle}{dt} = \pm i\omega_0 \langle r_2^\pm \rangle - \frac{1}{2} \gamma_0 \langle r_2^\pm \rangle - \frac{1}{2} \gamma_{12} \langle r_1^\pm \rangle. \tag{32.20}$$

The solution of these equations is of the form

$$\langle r_{12}^\pm \rangle = A_{1,2}^\pm \, e^{\pm i\omega_0 t - \frac{1}{2}\gamma_1 t} + B_{1,2}^\pm \, e^{\pm i\omega_0 t - \frac{1}{2}\gamma_2 t}, \tag{32.21}$$

† A two-level system and a harmonic oscillator are equivalent if the probability of being in an excited state is sufficiently small and the probability that the system is in the ground state is close to unity. In substituting the value $r_{3i} = -\frac{1}{2}$ in the right-hand sides of (32.17) and (32.18) we are using this approximation.

where

$$\gamma_{1,2} = \gamma_0 \pm \gamma_{12} = \gamma_0 \left\{ 1 \pm \frac{3}{2} \frac{\sin ka}{ka} \pm \frac{3}{2} \frac{\cos ka}{(ka)^2} \mp \frac{3}{2} \frac{\sin ka}{(ka^3)} \right\}.$$

The problem of radiation from an excited classical oscillator in the presence of unexcited oscillators has been solved (Podgoretskii and Roizen, 1960; Muzikarzh, 1961) on the assumption that $ka \gg 1$†. In this case (32.21) changes into the corresponding expressions of Podgoretskii and Roizen (1960). (Podgoretskii and Roizen (1960) also allow for the change in the emitted frequencies in the case of two oscillators.) It can be seen from (32.21) that the emission process takes place exponentially (the sum of the two exponents) leading to a non-Lorentzian line shape.

33. The Natural Width and Shift of the Emission Line

33.1. The Width and Shape of the Emission Line

The width and shape of the spontaneous emission line are always different from those of a line from an individual isolated molecule: the width and shape of the emission line are determined by a radiation process which takes place not at a definite time, but which is spread over a fairly long time. Therefore, even if the states of the molecules are statistically independent initially, a connection appears between them in the course of time (this connection is given by the correlations $\langle \hat{r}_{i'}^+ \hat{r}_i^- + \hat{r}_i^+ \hat{r}_{i'}^- \rangle$, $\langle \hat{r}_{3i} \hat{r}_{i'}^+ \rangle$, $\langle \hat{r}_{i'}^+ \hat{r}_{3i} \rangle$ for $i \neq i'$ in equations (32.16)–(32.18)).

Here we shall examine the width and shape of the spontaneous emission line from a system of molecules whose dimensions are much smaller than a wavelength (Dicke, 1954; Fain, 1957a; Fain, 1958; Itkina and Fain, 1958).

In this case ($ka \to 0$)

$$\gamma_{ii'} = \gamma_0 \tag{33.1}$$

for any pair of molecules i and i'. The total intensity of the spontaneous emission is

$$-\hbar\omega_0 \sum_i \frac{d}{dt} \langle r_{3i} \rangle = \hbar\omega_0 \gamma_0 \sum_{ii'} \langle \hat{r}_{i'}^+ \hat{r}_i^- \rangle = \hbar\omega_0 \gamma_0 \langle \hat{R}^+ \hat{R}^- \rangle$$

$$= I_0 \langle \hat{R}_1^2 + \hat{R}_2^2 + \hat{R}_3 \rangle = I_0 \langle \hat{R}^2 - \hat{R}_3^2 + \hat{R}_3 \rangle. \tag{33.2}$$

† In these papers the system of oscillators consists of nuclei emitting gamma-quanta under conditions where the Mössbauer effect occurs and interaction with the emission field can be observed despite its small magnitude.

The sum of the correlations can be expressed in terms of $\hat{R}$ in the form

$$\sum_{i \neq i'}\sum \langle \hat{r}_{i'}^{+} \hat{r}_{i}^{-} \rangle = \langle \hat{R}_1^2 + \hat{R}_2^2 \rangle - \frac{N}{2} = \langle \hat{R}^2 - \hat{R}_3^2 \rangle - \frac{N}{2}. \qquad (33.3)$$

In the case under discussion $\hat{R}^2$ is conserved during spontaneous emission† and, as we shall see below, R_3 decreases (i.e. the energy of the dynamic subsystem decreases). Therefore if the correlation (33.3) is zero at the start (here the emitted intensity is $N_+ I_0$, where N_+ is the number of molecules at the E_+ level) it will increase after this, as can be seen from (33.3).

We must solve (33.2) in order to find the width and shape of the spontaneous emission line. We first notice that this equation can be solved simply for the case of a single molecule $R = \frac{1}{2}$. In this case

$$-\langle \dot{r}_3 \rangle = \gamma_0(\langle r_3 \rangle + \tfrac{1}{2}). \qquad (33.4)$$

The solution of this equation is of the form‡

$$\langle r_3 \rangle + \tfrac{1}{2} = n_+ = n_+(0)\, e^{-\gamma_0 t},$$

i.e. γ_0 is the natural line width of an isolated molecule. Let us now examine the situation when there is a large number of molecules; essentially this is a quasi-classical case. We can therefore neglect the mean square deviations of the components $\hat{R}_3$ and $\hat{R}^2$ from their mean values and assume that $R^2 \gg R_3$. Equation (32.2) becomes

$$-\dot{R}_3 = \gamma_0(R^2 - R_3^2), \quad R^2 = \text{const}. \qquad (33.5)$$

To solve this equation we shall introduce the angle θ between the vector R and direction 3 (in the special case of a magnetic moment in a magnetic field this is the direction of the field)

$$R_3 = R \cos\theta. \qquad (33.6)$$

After this substitution equation (33.5) becomes

$$\frac{d\theta}{dt} = \gamma_0 R \sin\theta. \qquad (33.7)$$

The solution of (33.7) that satisfies the initial condition

$$\theta(0) = \theta_0$$

† When the dimensions of the molecular system are much less than the length of the emitted wave $\hat{R}^2$ commutes with the complete Hamiltonian (32.1) and is an integral of the motion.

‡ This solution can be obtained by using the Weisskopf–Wigner method (Weisskopf and Wigner, 1930a and b). Here it has been obtained by using the formalism of Chapter II. It may therefore be said that the theory developed there is a generalization of the Weisskopf–Wigner method.

is of the form

$$\tan\frac{\theta}{2} = \tan\frac{\theta_0}{2}\,\mathrm{e}^{\gamma_0 R t}.$$

From this we can find for the spontaneous emission intensity

$$I = I_0(R^2 - R_3^2) = I_0 R^2 \sin^2\theta$$

$$= \frac{4R^2 I_0}{\left(\tan\dfrac{\theta_0}{2}\,\mathrm{e}^{\gamma_0 R t} + \cot\dfrac{\theta_0}{2}\,\mathrm{e}^{-\gamma_0 R t}\right)^2}. \tag{33.8}$$

The strength of the radiation field is of the form

$$A(t) \sim \begin{cases} \mathrm{e}^{i\omega_0 t}\sin\theta, & t > 0, \\ 0, & t < 0. \end{cases}$$

When θ changes from θ_0 to $\pi/2$ (if $\theta_0 < \pi/2$) the amplitude of the radiation rises, reaches a maximum when $\theta = \pi/2$ and then drops. The spectral intensity is proportional to the square of the modulus of the Fourier component

$$J(\omega) \sim \left| \int_0^\infty \frac{\mathrm{e}^{i(\omega_0 - \omega)t}\,dt}{\tan\dfrac{\theta_0}{2}\,\mathrm{e}^{\gamma_0 R t} + \cot\dfrac{\theta_0}{2}\,\mathrm{e}^{-\gamma_0 R t}} \right|^2. \tag{33.9}$$

This expression defines the shape of the emission line which, as we can see, is not Lorentzian. The values of the line width that follow from (33.9) are (Fain, 1958)

$$\Delta\omega = \alpha\gamma_0 R, \tag{33.10}$$

where, in the range of angles θ_0 from 0 to π, α takes up a value from 1 to 2. (With angles θ that are very close to π, when equation (33.5) is inapplicable, Fain (1958) has given an additional treatment which shows that $\alpha = 2$.)

33.2. *The Frequency Shift. Classical Calculations*

Up to now we have been discussing electromagnetic interaction only in terms of effects such as attenuation. As well as attenuation (or broadening of the emission line), however, electromagnetic interaction leads to a shift in the energy levels of the system, or to a displacement of the centre of the emission line (the term Γ in the right-hand side of (32.11)) (Fain, 1959). For a single atom this is the well-known Lamb shift (Bethe, 1947). We shall first carry out a classical treatment to display the essential features of the effect.

In a system of classical particles each particle is in a quasi-stationary field due to the particle itself and to all the rest of the particles. The classical interaction energy of the system of charges, including terms up to the order of v^2/c^2, is (Landau and Lifshitz, 1960)

$$V = \sum_{A>B} \frac{e_A e_B}{r_{AB}} - \sum_{A>B} \frac{e_A e_B}{2c^2 m_A m_B r_{AB}} [(\boldsymbol{p}_A \cdot \boldsymbol{p}_B) + (\boldsymbol{p}_A \cdot \boldsymbol{n})(\boldsymbol{p}_B \cdot \boldsymbol{n})]. \tag{33.11}$$

Here e_A is the charge,
m_k is the mass,
$\boldsymbol{p}_A$ is the momentum of the particle,
r_{AB} is the distance between particles,
$\boldsymbol{n} = \boldsymbol{r}_{AB}/r_{AB}$ is a unit vector in the direction joining e_A to e_B.

The first term in (33.11) is the Coulomb interaction energy and leads to dipole interaction between the molecules. We shall not take into consideration the frequency shift connected with these terms. It can be shown that for a uniform and isotropic system of molecules this term does not lead to a shift. It is essential to take the second term into consideration if the distances between the particles are less than the length of the emitted wave. If we consider the charge to be "blurred" over the volume of the system the second term can be written approximately in the form $(e^2/Mc^2R)\,\boldsymbol{P}^2$, where $\boldsymbol{P}$ is the momentum of the whole system, M is its mass, R is a quantity of the order of the radius of the system and e is the charge of the system†. This term is of the same form as the kinetic energy of the system. It is clear from this that the shift of the eigenfrequencies of the system is connected with this term. (In a system of oscillators a change in the kinetic or potential energy leads to a change in the frequencies.) The next term in the expansion in powers of v/c (not given here) leads to radiation damping.

Let us consider the motion of the magnetic moment in a magnetic field in the quasi-classical calculation of the frequency shift. The equations of motion are of the form of (19.27) and (19.28) where $\boldsymbol{e}_1 = \boldsymbol{e}_2 = 0$ and the $\boldsymbol{m}_i$ are defined by (19.14). Equation (19.27) can be rewritten in the form

$$\dot{\boldsymbol{M}} = \gamma[\boldsymbol{M} \wedge (\boldsymbol{H}_0 + \boldsymbol{h})], \tag{33.12}$$

where $\boldsymbol{H}_0$ is an external constant field parallel to the z-axis and the eigenfield $\boldsymbol{h}$ is defined by the expression

$$\boldsymbol{h} = -\sum_{\nu} \omega_\nu \boldsymbol{H}_\nu q_\nu, \tag{33.13}$$

where q_ν satisfies equation (19.28) (in other words $\boldsymbol{h}$ satisfies the Maxwell equations). Elimination of the magnetic field $\boldsymbol{h}$ from equations (33.12) and

† We must, of course, remember the doubly qualitative nature of this estimate.

(19.28) for a system whose dimensions are much less than a wavelength leads to the following equation for the magnetic moment (Ginzburg, 1943):

$$\dot{\boldsymbol{M}} = \gamma[\boldsymbol{M} \wedge \boldsymbol{H}_0] - \frac{4\omega_m \gamma}{3\pi c^3} [\boldsymbol{M} \wedge \ddot{\boldsymbol{M}}] + \frac{2\gamma}{3c^3} [\boldsymbol{M} \wedge \dddot{\boldsymbol{M}}], \quad (33.14)$$

where $\omega_m \equiv c/l$;

l is a quantity of the order of the radius of the system.

The third term in the right-hand side of equation (33.14) is dissipative and corresponds to radiation damping of the magnetic moment. The natural width of the emission line (33.10) is given by this term. The second term in the right-hand side of (33.14) is conservative. It is not difficult to see that for $\omega_0 = \gamma H_0 \gg \Delta_1\omega$, where $\Delta_1\omega$ is the frequency shift, the second term in (33.14) leads to a frequency shift

$$\Delta_1\omega = \frac{4\omega_m \omega_0^2}{3\pi c^3} \gamma M_z. \quad (33.15)$$

In fact, under the assumptions that have been made, $\ddot{\boldsymbol{M}}$ in the second term can be found from

$$\dot{\boldsymbol{M}} = \gamma[\boldsymbol{M} \wedge \boldsymbol{H}_0].$$

In this case

$$\ddot{\boldsymbol{M}} = \gamma[\dot{\boldsymbol{M}} \wedge \boldsymbol{H}_0] = -\gamma^2\{\boldsymbol{M}H_0^2 - \boldsymbol{H}_0 M_z H_0\}$$

and

$$-\frac{4\omega_m \gamma}{3\pi c^3} [\boldsymbol{M} \wedge \ddot{\boldsymbol{M}}] = -\gamma \frac{4\omega_m \omega_0^2}{3\pi c^3} \frac{M_z}{H_0} [\boldsymbol{M} \wedge \boldsymbol{H}_0],$$

and (33.15) follows from this.

33.3. The Frequency Shift. Quantum Calculations

We now move on to the quantum calculations for the shift in the levels (and the corresponding frequencies) of a system of molecules. The energy levels of a system of molecules are given by (19.15)

$$E_{RM} = M\hbar\omega_0.$$

We shall find the displacement of these levels caused by interaction with radiation. The non-zero second-order correction from perturbation theory (see, e.g., the term Γ_{mm} in Landau and Lifshitz, 1963, and in (8.6)) is

$$\Delta E_{RM} = -\sum_{R'M'\nu} \frac{|V_{RM0;\,R'M'1_\nu}|^2}{E_{R'M'} - E_{RM} + \hbar\omega_\nu}, \quad (33.16)$$

where

$$\hat{V} = -\frac{1}{c}(A \cdot \dot{d})$$

and d is the dipole moment of the whole system.

Changing from summation to integration with respect to the frequencies and directions of propagation we obtain (see Bethe, 1947)

$$\Delta E_{RM} = -\frac{2}{3\pi\hbar c^3} \int_0^K k\, dk \sum_{R'M'} \frac{|\langle RM\,|\dot{d}|\,R'M'\rangle|^2}{E_{R'M'} - E_{RM} + k}, \tag{33.17}$$

where $\langle RM\,|\dot{d}|\,R'M'\rangle$ is a matrix element of the derivative of the dipole moment (for the sake of simplicity we are discussing the case of linear polarization of the dipole moment, when $e_1 || e_2$), $K = \hbar\omega_m$. For frequencies $\omega < \omega_M$ it can be assumed that the field is uniform over the system (the dipole approximation) and (33.17) is derived in this approximation. The contribution made by higher frequencies, $\omega > \omega_M$, must be calculated with an allowance for any non-uniformity of the field within the system. It can, however, be shown (see, e.g., Ginzburg, 1943) that the contribution made by fields of these frequencies can be neglected. Therefore cut-off occurs at wavelengths $\lambda = 2\pi c/\omega_m$ of the order of the dimensions of the system. When calculating the energy displacement of a single bound electron, which is considered to be a point charge in present-day theory, the correction to the energy (33.17) diverges. However if we carry out renormalization (subtract from the energy of the electron the electromagnetic energy of a free electron, which is also infinite) we obtain the finite displacement of the levels of atomic electrons discovered by Lamb (Bethe, 1947). With a certain amount of justification we can say that (33.15) and (33.17) give the Lamb shift for the whole system (by analogy with the natural line width of the whole system), since both the natural line width and the Lamb shift are the result of taking into consideration the interaction of the atom with the eigenfield of the radiation. We can use (33.17) to calculate the displacement of the levels of a molecular system. The matrix elements of the dipole moment take the form (see (19.12), (19.8) and (19.9))

$$|\langle RM\,|d|\,RM \mp 1\rangle|^2 = |d_{12}|^2 (R \pm M)(R \mp M + 1), \tag{33.18}$$

where d_{12} is a matrix element of the transition $1 \to 2$ for an isolated molecule.

From equations (33.17) and (33.18) we find

$$\Delta E_{RM} = -\frac{2\omega_0^2}{3\pi\hbar c^3} |d_{12}|^2 \left\{ (R^2 - M^2 + R)\left(2K + \hbar\omega_0 \ln\left|\frac{K - h\omega_0}{K + h\omega_0}\right|\right) + M\hbar\omega_0 \ln\frac{|K - h\omega_0|\,(K + h\omega_0)}{(\hbar\omega_0)^2} \right\}. \tag{33.19}$$

For $K \gg \hbar\omega_0$ (i.e. for a system with dimensions much less than a wavelength) (33.19) can be written in the form

$$\Delta E_{RM} = -\left(\frac{4\omega_0^2}{3\pi\hbar c^3}\right)|d_{12}|^2 (R^2 - M^2 + R)K.$$

The energy of the RM level is now of the form $E_{RM} + \Delta E_{RM}$. For the frequency shift of the transition $M \to M - 1$ we obtain the expression (for $M \gg 1$)

$$\Delta\omega = \frac{\Delta E_{RM} - \Delta E_{RM-1}}{\hbar} = \left(\frac{4\omega_m\omega_0^2}{3\pi c^3\hbar}\right)|d_{12}|^2\, 2M. \tag{33.20}$$

The similarity of expressions (33.15) and (33.20) is obvious; they are not identical since they refer to different cases: (33.15) gives the line shift of a magnetic moment precessing in a magnetic field; here the magnetic dipole moment has circular polarization. Expression (33.20) refers to the case of an electric dipole transition and linear polarization. It is interesting to compare the magnitude of the frequency shift with the line width. The ratio of these quantities is of the order of $\omega_m/\omega_0 \sim \lambda/l$, i.e. the line shift is λ/l times greater than the natural line width, where λ is the wavelength of the emitted radiation.

34. Radiation from a System whose Dimensions are much Larger than the Wavelength

Up to now we have been discussing spontaneous emission from a system of molecules whose linear dimensions are much less than the wavelength of the emitted radiation. The question arises of whether there can be coherence under conditions where the dimensions of the system are comparable with, or much greater than, the wavelength. Here we have in mind coherence between the emissions from individual molecules. The answer is that if the distances between individual molecules are much less than, or comparable with, the wavelength, then, even if the dimensions of the system are much greater than a wavelength, coherent spontaneous emission is possible, i.e. in this case the emission intensity is not, generally speaking, equal to the sum of the intensities from individual molecules. We have obtained in this chapter a number of relations (in particular the emission intensity) which are valid for any distribution of the system of molecules provided that its dimensions are much less than the wavelength. Only such an integral characteristic as the total dipole moment of the system is important. In the case under discussion at present (the dimensions of the system are comparable with, or greater than, the wavelength) the emission from the system depends essentially on the distribution of the molecular dipole moments†. Therefore to illustrate coherence for a system of molecules whose dimensions are much greater than a wavelength we shall consider the actual distribution of the dipole moments. We shall introduce the dipole moment per unit volume (the polarization vector, or the magnetization in the magnetic case). To take a definite example, we shall discuss the magnetic case, i.e. the case when the molecules have magnetic dipole moments (and the electric dipole moments are zero). Then (19.28), which gives the emission field from any molecular distribution, can be written in the form‡

$$\ddot{q}_\nu + \omega_\nu^2 q_\nu = -\omega_\nu \int (\boldsymbol{H}_\nu(\boldsymbol{r}) \cdot \boldsymbol{M}(\boldsymbol{r}, t))\, dV, \tag{34.1}$$

where $\boldsymbol{M}(\boldsymbol{x}, t)$ is the magnetization of the system as a function of the coordinates and time. Equation (34.1) gives the mean values of the field q_ν and the magnetization $\boldsymbol{M}(\boldsymbol{x}, t)$.

The emission intensity and the emission field found by means of this equation are therefore the regular part of the emission (see section 29). The general solution of equation (34.1) for the quantities q_ν and $p_\nu = \dot{q}_\nu$, giving the radiated magnetic and electric fields, can be written in the form

$$q_\nu = q_\nu^{(0)}(t) - \int\int_0^t (\boldsymbol{H}_\nu(\boldsymbol{r}) \cdot \boldsymbol{M}(\boldsymbol{r}, \tau)) \sin \omega_\nu(t - \tau)\, d\tau\, dV, \tag{34.2}$$

$$p_\nu = p_\nu^{(0)}(t) - \omega_\nu \int\int_0^t (\boldsymbol{H}_\nu(\boldsymbol{r}) \cdot \boldsymbol{M}(\boldsymbol{r}, \tau)) \cos \omega_\nu(t - \tau)\, d\tau\, dV. \tag{34.3}$$

Here $q_\nu^{(0)}(t)$ and $p_\nu^{(0)}(t)$ are the general solutions of the equations when the right-hand side of (34.1) is zero. Expressions (34.2) and (34.3) in the general form give the solution of the problem of the regular part of the field from given distributions of magnetization. Analogous expressions can be obtained for an arbitrary system of two-level molecules and a given energy spin distribution. We shall now examine the special case when the distribution of magnetization is in the form of a standing wave

$$\boldsymbol{M}(\boldsymbol{r}, t) = \boldsymbol{M}_0 \cos k_0 z \cos \omega_0 t,$$

where $k_0 = \omega_0/c$. Equation (34.1) then has the form

$$\ddot{q}_\nu + \omega_\nu^2 q_\nu = b_\nu \cos \omega_0 t, \tag{34.4}$$

† We recall that the whole time we are assuming the validity of the dipole approximation for an individual molecule (its dimensions are much less than a wavelength).

‡ It is obvious that in this form the equation is also valid for systems which cannot be described by the energy spin. The introduction of the energy spin is contained in the equation (19.27). Equation (34.1) is essentially the Maxwell equation which includes the current distribution.

where

$$b_\nu = -\omega_\nu \left(\boldsymbol{M}_0 \cdot \int_V \boldsymbol{H}_\nu(\boldsymbol{r}) \cos k_0 z \, dV \right). \tag{34.5}$$

In order to determine the spontaneous emission intensity we shall find the derivative of the energy of the νth emission mode

$$H_\nu = \frac{1}{2} (\dot{q}_\nu^2 + \omega_\nu^2 q_\nu^2), \quad \frac{dH_\nu}{dt} = \dot{q}_\nu (\ddot{q}_\nu + \omega_\nu^2 q_\nu).$$

Then by using equation (34.4) we obtain

$$\frac{dH_\nu}{dt} = b_\nu \dot{q}_\nu \cos \omega_0 t. \tag{34.6}$$

From equation (34.3), as in (28.6), we obtain the solution when $p_\nu^{(1)} = \dot{q}_\nu^{(1)}$, which gives the spontaneous emission ($\omega_0 t \gg 1$)

$$\dot{q}_\nu^{(1)} \approx b_\nu \frac{\sin (\omega_0 - \omega_\nu) t}{2(\omega_0 - \omega_\nu)} \cos \omega_\nu t. \tag{34.7}$$

In the case of a continuous spectrum of radiation (as might occur in free space) we obtain from (34.6) and (34.7)

$$\frac{dH_\nu}{dt} = \frac{\pi}{4} b_\nu^2 \delta(\omega_0 - \omega_\nu)$$

and, changing to summation over the frequencies, for the spontaneous emission intensity in the range of solid angle $d\Omega$ we have

$$dI = \frac{\pi}{4} \frac{\omega_0^2}{(2\pi c)^3} b_\nu^2 L^3 d\Omega. \tag{34.8}$$

Here we used expression (3.55) for the number of modes in free space within a given range $d\omega_\nu$. The eigenfunctions $\boldsymbol{H}_\nu$, describing the oscillations in free space, are of the form (see Chapter I)

$$\boldsymbol{H}_{\nu_1}^\alpha = K \boldsymbol{e}_\alpha^\nu \cos (\boldsymbol{k}_\nu \cdot \boldsymbol{r}) \quad \text{and} \quad \boldsymbol{H}_{\nu_2}^\alpha = K \boldsymbol{e}_\alpha^\nu \sin (\boldsymbol{k}_\nu \cdot \boldsymbol{r}),$$

where $K = \sqrt{8\pi/L^3}$ is a normalization constant and α has two values corresponding to the two possible independent polarization vectors $\boldsymbol{e}_\alpha^\nu$.

Let us now examine the case when the dimensions A, B and C (along the x, y and z axes) of the molecular system are much less than the wavelength λ. Then we can neglect the integral $\int \boldsymbol{H}_{\nu_2} \cos K_0 z \, dV$. In this approximation it can easily be found that

$$b_\nu^2 = \frac{1}{4}\,\omega_0^2(\boldsymbol{e}_\alpha^\nu \cdot \boldsymbol{M}_0)^2 K^2 \left(\frac{\sin\dfrac{k_x A}{2}}{\dfrac{k_x A}{2}}\right)^2 \left(\frac{\sin\dfrac{k_y B}{2}}{\dfrac{k_y B}{2}}\right)^2$$

$$\times \left(\frac{\sin\dfrac{(k_z - k_0)}{2}\,C}{\dfrac{(k_z - k_0)}{2}\,C}\right)^2 A^2 B^2 C^2. \tag{34.9}$$

These expressions determine the angular distribution of the emission from the system. In the case when $A \sim B \sim C$ the radiation is confined within an angle of the order of

$$\Delta\theta \sim \frac{\lambda}{A}. \tag{34.10}$$

Let us find the total intensity emitted. To do this it is necessary to integrate (34.8) over all solid angles $d\Omega$†. In this case it is sufficient to take only a small range of angles $d\Omega$ about the z-axis

$$d\Omega = \frac{dk_x\,dk_y}{k_0^2}, \tag{34.11}$$

and replace k_z by k_0. Because b_ν^2 is a well-defined function of the angle, integration with respect to k_x and k_y can extend to infinity. Then the total emission intensity is

$$I = 4\frac{\pi}{4}\frac{\omega_0^2}{(2\pi c)^3}\frac{L^3}{k_0^2}\int_0^\infty dk_x \int_0^\infty dk_y\,|b_\nu|^2 = \frac{\omega_0^4}{16\pi c^3}(\boldsymbol{M}_0\cdot\boldsymbol{e})^2 AB\lambda^2 C^2. \tag{34.12}$$

If it is remembered that the quantity M_0ABC is proportional to the number of particles in the system, then it is clear that (34.12) describes coherent emission which, generally speaking, is not equal to the sum of the intensities from the individual molecules (and is not proportional to the number of particles).

In conclusion, we should point out that all the results of this section are equally applicable to radiation in a waveguide, which also has a continuous spectrum, just as in free space. (Of course, the form of a number of the expressions must be changed slightly since the spectrum of radiation in a waveguide differs from the spectrum in free space.)

† It should be stressed that since we are discussing the excitation of standing (and not travelling) waves the integration with respect to $d\Omega$ must be carried out over half the solid angle (see section 3.8).

CHAPTER VIII

Emission in a Resonator

THIS chapter discusses radiation from quantum systems in the case when there is interaction with only one mode of the field in the resonator. A number of simplifying assumptions are made in the theoretical approach to this problem. One of the purposes of this chapter is to set up a simple model as a basis for a quantum treatment of the field in maser amplifiers an dos-cillators, which are considered fully in the second volume. Here, however, we shall not be concerned with details, and in particular we shall discuss the interaction between excited molecules and the field in a resonator in different operating modes (as an amplifier or an oscillator), without considering how negative temperatures can be achieved.

35. The Fundamental Equations

In order to investigate the behaviour of a system of molecules in a resonator we shall make a number of simplifying assumptions which will make it possible to avoid complex mathematical calculations and at the same time help to concentrate our attention on the elucidation of the basic features of the behaviour of the field and the molecules. We shall first assume that there is significant interaction between the system of molecules and only one mode of the resonator, and that the frequency ω_ν of this mode is close to the molecular transition frequency ω_0. We shall further assume that the dimensions of the system of molecules is much less than a wavelength in the resonator or, to be more precise, that the molecules are located in a small enough section of the resonator for any change in the field $A_\nu(x)$ in this section to be neglected (the dipole approximation). In this approximation the behaviour of the system of molecules is entirely determined by the energy spin vector (R_1, R_2, R_3), and the Hamiltonian of a system of two-level molecules interacting with a radiation field can be written in the form (see (19.23))

$$\begin{aligned}\hat{E} &= \hbar\omega_0\hat{R}_3 + \tfrac{1}{2}(\hat{p}_\nu^2 + \omega_\nu^2\hat{q}_\nu^2) - (e_1 \cdot A_\nu)\,\hat{R}_1(\hat{q}_\nu + q_{\nu 0}\sin(\omega t + \theta)) \\ &= \hbar\omega_0\hat{R}_3 + \tfrac{1}{2}\hbar\omega_\nu(\hat{a}^+\hat{a} + \hat{a}\hat{a}^+) - (e_1 \cdot A_\nu)\,\hat{R}_1(\hat{q}_\nu + q_{\nu 0}\sin(\omega t + \theta)). \quad (35.1)\end{aligned}$$

Here, for the sake of simplicity, we have taken $\boldsymbol{e}_2 = 0$ (the molecules are linearly polarized) and introduced the annihilation and creation operators (see (3.28)). A_ν denotes the value of the eigenfunction $A_\nu(\boldsymbol{x})$ at the centre of the system of molecules, $q_{\nu 0}$ is the amplitude of the external electric field, which is a given function of time, and

$$E(t) = -\frac{1}{c}\, A_\nu \dot{q}_{\nu 0}(t) = -\frac{\omega}{c}\, A_\nu q_{\nu 0} \cos(\omega t + \theta). \tag{35.2}$$

We are thus defining the external field classically, which is permissible if its amplitude is large enough, so that we can neglect the quantum fluctuations; we shall assume that this is the case in the following calculations.

Relaxation processes are not taken into consideration in the Hamiltonian (35.1). These processes may be a result of energy loss in the walls of the resonator, or dissipation in the molecular system itself. We shall use the formalism developed in Chapter II to take account of these processes. It follows from (8.18a) that the equations which give the change with time of the mean value of the quantity $\hat{O}$ and include the relaxation processes are†

$$\begin{aligned}\frac{d\langle O\rangle}{dt} = \frac{i}{\hbar}\langle[\hat{E}, \hat{O}]\rangle &+ \Phi^{+}\langle \hat{a}[\hat{O}, \hat{a}^{+}] + [\hat{a}, \hat{O}]\,\hat{a}^{+}\rangle \\ &+ \Phi^{-}\langle \hat{a}^{+}[\hat{O}, \hat{a}] + [\hat{a}^{+}, \hat{O}]\,\hat{a}\rangle \\ &+ \Phi^{+-}\sum_i \langle \hat{r}_i^{-}[\hat{O}, \hat{r}_i^{+}] + [\hat{r}_i^{-}, \hat{O}]\,\hat{r}_i^{+}\rangle \\ &+ \Phi^{-+}\sum_i \langle \hat{r}_i^{+}[\hat{O}, \hat{r}_i^{-}] + [\hat{r}_i^{+}, \hat{O}]\,\hat{r}_i^{-}\rangle \\ &+ \Phi^{00}\sum_i \langle \hat{r}_{3i}[\hat{O}, \hat{r}_{3i}] + [\hat{r}_{3i}, \hat{O}]\,\hat{r}_{3i}\rangle,\end{aligned} \tag{35.3}$$

where $\hat{r}_i^{\pm} = \hat{r}_{1i} \pm i\hat{r}_{2i}$; $\hat{r}_{3i}$ are the energy spin operators of the individual molecules;

Φ^{+-}, Φ^{-+} and Φ^{00} are coefficients which denote the relaxation of an individual spin (see (19.35) and (19.36));

Φ^{+} are coefficients which denote the dissipation of the field in the resonator (see (11.4)).

We assume that individual spins relax independently. In other words, we consider that the coefficients $\Phi_{ii'}^{rs}$ (see section 19) are of the form

$$\Phi_{ii'}^{rs} = \Phi^{rs}\delta_{ii'}.$$

† In the equations (35.3) we are neglecting the frequency shift due to relaxation processes.

In the case when the dissipative system (in particular the walls of the resonator, and the crystal lattice when there is spin–lattice relaxation) is in thermal equilibrium there is the following relation between the coefficients Φ:

$$\Phi^{+} = \Phi^{-} e^{-\hbar\omega_{\nu}/kT}, \quad \Phi^{+-} = \Phi^{-+} e^{-\hbar\omega_{0}/kT}. \tag{35.4}$$

We require the equations for the mean values of R_1, R_2, R_3, a^+, a and their quadratic combinations, which describe the fluctuations. Substituting the operators $\hat{R}_1$, $\hat{R}_2$, $\hat{a}^+$ and $\hat{a}$ for $\hat{O}$ in (35.3) and using the commutation relations for these operators we find

$$\left.\begin{aligned} \dot{a} &= -i\omega_{\nu} a + i\alpha R_1 - \gamma a, \\ \dot{a}^+ &= i\omega_{\nu} a^+ - i\alpha R_1 - \gamma a^+, \end{aligned}\right\} \tag{35.5}$$

$$\dot{R}_1 = -\omega_0 R_2 - \frac{R_1}{T_2}, \quad \dot{R}_2 = \omega_0 R_1 - \frac{R_2}{T_2} + \alpha\langle \hat{R}_3(\hat{a}^+ + \hat{a})\rangle$$
$$+ \alpha R_3(a_1^+ + a_1)\sin(\omega t + \theta), \tag{35.6}$$

where a, a^+, R_1, R_2, R_3 are the mean values of the corresponding operators.
We shall use the following notation

$$a_1^+ + a_1 = \sqrt{\frac{2\omega_{\nu}}{\hbar}}\, q_{\nu 0}, \quad \alpha = \frac{(e_1 \cdot A_{\nu})}{\sqrt{2\hbar\omega_{\nu}}},$$
$$T_2^{-1} = \Phi^{00} + \Phi^{+-} + \Phi^{-+}, \quad \gamma = \Phi^{-} - \Phi^{+}. \tag{35.7}$$

If we can neglect the quantum fluctuations of R_3 (or a^+ and a), then in the right-hand side of the second equation (35.6) $\langle \hat{R}_3(\hat{a}^+ + \hat{a})\rangle$ can be replaced by $R_3(a^+ + a)$. In practice R_3 has a very large value (of the order of the number of molecules) so this substitution is amply justified. If R_3 is a fixed quantity, then bearing in mind the last remark (35.5) and (35.6) make up a closed set of equations for finding the mean fields (a, a^+) and the mean dipole moment R_1, R_2. Generally, however, the variation of R_3 depends upon a, a^+, R_1 and R_2, so the set (35.5) and (35.6) must be supplemented by equations defining R_3. These, as we shall see, are the equations for the quadratic quantities $\langle \hat{a}^+\hat{a}\rangle$, $\langle \hat{R}^+\hat{a}\rangle$, etc. Unlike the quantities a, a^+, R_1 and R_2, which change rapidly with the frequency ω_0 (or $\omega_{\nu} \approx \omega_0$) R_3 consists of two parts, one of which changes little in a time $2\pi/\omega_0$, whilst the other part changes at a frequency $2\omega_0$ and at higher frequencies. It is not difficult to show that the contribution from the high-frequency part of R_3 can be neglected in other equations (35.5) and (35.6) if the parameters α, γ and T_2^{-1} are sufficiently small when compared with ω_0, as is generally the case. In future, therefore, we shall be interested in a value of R_3 averaged over a time greater than $2\pi/\omega_0$, and also in the values of the two quadratic quantities averaged over this time. After simple transformations we find from (35.1) and (35.3)

$$\dot{R}_3 = -\alpha\overline{\langle \hat{R}_2(\hat{a} + \hat{a}^+)\rangle} - \alpha(a_1 + a_1^+)\overline{R_2 \sin(\omega t + \theta)} - \frac{R_3 - R_3^0}{T_1}, \tag{35.8}$$

where the bar denotes time-averaging,

$$T_1^{-1} = 2(\Phi^{+-} + \Phi^{-+}), \quad R_3^0 = \frac{N}{2}\frac{\Phi^{+-} - \Phi^{-+}}{\Phi^{+-} + \Phi^{-+}} \tag{35.9}$$

and N is the total number of molecules.

Let us introduce new variables

$$R^{\pm} = R_1 \pm iR_2, \quad R_1 = \tfrac{1}{2}(R^+ + R^-),$$

$$R_2 = \frac{1}{2i}(R^+ - R^-)$$

and in terms of these variables write the expression

$$\langle \hat{R}_2(\hat{a} + \hat{a}^+)\rangle = \frac{1}{2i}\langle \hat{R}^+\hat{a} - \hat{R}^-\hat{a}^+ + \hat{R}^+\hat{a}^+ - \hat{R}^-\hat{a}\rangle.$$

In the approximation used we shall neglect the mean values of quantities of the $\langle R^+a^+\rangle$, $\langle R^-a\rangle, \ldots,$ type when compared with the quantities $\langle R^+a\rangle$, $\langle R^-a\rangle, \ldots$. After a series of transformations, putting $\omega_0 = \omega_\nu$ for the sake of simplicity, we obtain the system of equations (Fain, 1963c)

$$\dot{R}_3 = U - \frac{1}{T_1}(R_3 - R_3^0) - \alpha(a_1 + a_1^+)\overline{R_2 \sin(\omega t + \theta)}, \tag{35.10}$$

$$\dot{n} = -U - 2\gamma(n - n_0), \tag{35.11}$$

$$\dot{U} = -\frac{\alpha^2}{2}\{\overline{R_1^2} + \overline{R_2^2} + R_3 + 2\langle nR_3\rangle\} - \frac{\alpha^2}{2}(a_1^+ + a_1)\overline{\langle R_3(a^+ + a)\rangle \sin(\omega t + \theta)} - \left(\gamma + \frac{1}{T_2}\right)U, \tag{35.12}$$

$$\frac{d(\overline{R_1^2} + \overline{R_2^2})}{dt} = -\langle R_3U + UR_3\rangle + 2\alpha(a_1 + a_1^+)\langle R_3R_2 + R_2R_3\rangle \sin(\omega t + \theta) - \frac{2}{T_3}\left\{\overline{R_1^2} + \overline{R_2^2} - \frac{N}{2}\right\}, \tag{35.13}$$

where $n = \langle a^+ a \rangle$ is the mean number of photons (the mean energy of the radiation field is $(n + \frac{1}{2})\hbar\omega_0$); n_0 is the equilibrium value of n and U denotes the quantity

$$U = \frac{i\alpha}{2} \langle \overline{R^+ a} - \overline{R^- a^+} \rangle. \tag{35.14}$$

The equations (35.5), (35.6), (35.10)–(35.13) are the fundamental equations relevant to in all the following discussions a system of molecules and a field in a resonator. This set of equations makes it possible to determine not only the mean values of the field R_1, R_2 but also their quadratic fluctuations. We notice at once that when the field and the molecules are in a state of thermal equilibrium at a temperature T

$$n_0 = \frac{1}{e^{\hbar\omega_0/kT} - 1}, \qquad R_3^0 = -\frac{N}{2}\frac{e^{\hbar\omega_0/kT} - 1}{e^{\hbar\omega_0/kT} + 1},$$

$$\frac{N}{2} + R_3^0(2n_0 + 1) = 0. \tag{35.15}$$

These values of n and R_3 give the stationary solution of equations (35.10)–(35.13) (for $a_1 + a_1^+ = 0$) if in addition

$$\overline{R_1^2} + \overline{R_2^2} = \frac{N}{2}, \quad U = 0. \tag{35.16}$$

It is easy to see that in a state of thermal equilibrium the first equation of (35.16) must be satisfied. In fact, in a state of thermal equilibrium the states of the individual molecules are not correlated with each other† and therefore

$$\left(\langle \hat{r}_{1i}\hat{r}_{1j} \rangle = \langle \hat{r}_{2i}\hat{r}_{2j} \rangle = \frac{\delta_{ij}}{4}\right),$$

$$\begin{aligned}\langle R_1^2 + R_2^2 \rangle &= \langle (\sum_i \hat{r}_{1i})^2 + (\sum \hat{r}_{2i}^2)^2 \rangle \\ &= \sum_i r_{1i}^2 + r_{2i}^2 + \sum_{i \neq j} \langle \hat{r}_{1i}\hat{r}_{1j} + \hat{r}_{2i}\hat{r}_{2j} \rangle \\ &= \sum_i r_{1i}^2 + r_{2i}^2 = \frac{N}{2}.\end{aligned}$$

† The density matrix of the whole system of interacting particles is

$$\varrho = C \exp\left(-[\hbar\omega_0/kT] \sum_i \hat{r}_{3i}\right),$$

where C is a normalizing factor. Therefore $\hat{\varrho} = C \prod_i \exp\left(-[\hbar\omega_0/kT]\, \hat{r}_{3i}\right)$, which shows that the individual molecules are statistically independent.

36. Free Motion (with no External Field)

36.1. The Behaviour of a Molecule + Field System in a Resonator in a Time $t \ll T_1, T_2$

We now investigate various solutions of the initial equations (35.5), (35.6) and (35.10)–(35.13). First of all we shall find the free solution of these equations (when there is no external force) for $T_1 = \infty$, $T_2 = \infty$. This means that the solutions obtained will be valid for $t \ll T_1, T_2$. We shall further assume that we can neglect the quantum fluctuations of a, a^+, R^+ and R^-. As has already been pointed out, this can be done if all their values are large enough. The quantum fluctuations of these quantities in a resonator will be specially discussed. Using the assumption that the parameter α is small we shall look for the solution of equations (35.5), (35.6) and (35.10) in the form

$$a = a_0\, e^{-i\omega t}, \quad a^+ = a_0^+\, e^{i\omega t}, \quad R^{\pm} = r_{\pm}\, e^{\pm i\omega t}, \tag{36.1}$$

where $a_0, a_0^+, r_{\pm}$ are functions of time varying slowly compared with $e^{i\omega t}$ and $\omega = \omega_0 = \omega_\nu$. Substituting (36.1) in (35.5), (35.6) and (35.10), and using definition (35.14) we arrive at the following equations:

$$\dot{r}_+ = i\alpha R_3 a_0^+, \quad \dot{r}_- = -i\alpha R_3 a_0, \tag{36.2}$$

$$\dot{a}_0 = -\gamma a + \frac{i\alpha}{2} r_-, \quad \dot{a}_0^+ = -\gamma a - \frac{ia}{2} r_+, \tag{36.3}$$

$$\dot{R}_3 = \frac{i\alpha}{2}(r_+ a_0 - r_- a_0^+). \tag{36.4}$$

We shall select a phase of the field p_ν, q_ν such that $a_0 = a_0^+$. It is easy to see that this does not invalidate the generality of the discussion. We differentiate the right and left-hand sides of the equation (36.4) and use equations (36.2) and (36.3). As a result we obtain

$$\ddot{R}_3 + \frac{\alpha^2 a_0^2}{2} R_3 - \frac{\dot{a}_0}{a_0}\dot{R}_3 = 0. \tag{36.5}$$

The general solution of this equation, as can be checked by direct substitution, is of the form

$$R_3 = R\cos\theta, \quad \theta = \frac{\alpha}{\sqrt{2}}\int_0^t a_0(t_1)\, dt_1 + \theta_0. \tag{36.6}$$

From equations (35.10) and (35.11) we obtain the relation

$$\dot{R}_3 + \dot{n} + 2\gamma(n - n_0) = 0, \tag{36.7}$$

which expresses the energy balance of the molecule + field system. For the sake of simplicity let us look at the case when $n \gg n_0$ (the equilibrium value of the number of photons). Then from (36.6) and (36.7) we obtain an equation for the angular variable θ (Fain, 1959 a):

$$\ddot{\theta} - \frac{\alpha^2}{4} R \sin\theta + 2\gamma\dot{\theta} = 0, \quad a_0 = \frac{\sqrt{2}}{\alpha}\dot{\theta}. \tag{36.8}$$

This is the equation for a pendulum with viscous damping. (To reduce it to the usual form θ must be replaced by $\theta_1 + \pi$.) If initially the molecules are in an excited state ($\theta \neq \pi$) they then move towards a state of equilibrium ($\theta = \pi$) and excite a field in the resonator; the field in its turn excites the molecules, and so on until damping stops the process†. If a strong enough field is excited at the start in the resonator the vector $\boldsymbol{R}$, apart from precessing around direction 3, will make a rotation determined by the angle θ (Fain, 1957 b). Expressions (36.6) and (36.8) provide the complete solution to the problem of the behaviour of a system of molecules interacting with radiation in a time $t \ll T_1, T_2$, provided that the molecular transition frequency is the same as the frequency of the radiation.

36.2. Stationary Solutions. Decay Processes

We shall now move on to discuss stationary states and the decay processes into these states for a molecule + field system in a resonator, in the absence of an external force. We obtain from equations (35.5) and (35.6) (neglecting the fluctuations of R_3), changing to the variables a_0, $\overset{+}{a}_0$, r_+ and r_-, the following set of equations, which are similar to (36.2) and (36.3):

$$\begin{aligned} \dot{r}_+ &= -\frac{1}{T_2} r_+ + i\alpha R_3 \overset{+}{a}_0, \quad \dot{r}_- = -\frac{1}{T_2} r_- - i\alpha R_3 a_0, \\ \dot{\overset{+}{a}}_0 &= -\gamma \overset{+}{a}_0 - \frac{i\alpha}{2} r_+, \quad \dot{a}_0 = -\gamma a_0 + \frac{i\alpha}{2} r_-. \end{aligned} \tag{36.9}$$

The change of R with time is given by the set of equations (35.10)–(35.13). In a stationary state the field amplitudes $\overset{+}{a}_0$, a, the energy spin amplitudes $r_\pm$, R_3, n, U and the other quantities are time-independent. In this case (36.9)

† The equilibrium value of the angle θ would be changed if the term $2\gamma n_0$, which was neglected in going from (36.7) to (36.8), were taken into consideration.

and (35.10)–(35.13) change into sets of algebraic equations. In particular (36.9) changes into the homogeneous linear set of equations

$$-\frac{1}{T_2} r_+ + i\alpha R_3 a_0^+ = 0, \quad -\frac{1}{T_2} r_- - i\alpha R_3 a_0 = 0,$$
$$-\frac{i\alpha}{2} r_+ - \gamma a_0^+ = 0, \quad \frac{i\alpha r_-}{2} - \gamma a_0 = 0. \tag{36.10}$$

The determinant of each of these sets is

$$\frac{\gamma}{T_2} - \frac{\alpha^2 R_3}{2}.$$

If this determinant is finite all the amplitudes a_0^+, a_0, $r_\pm$ must be zero:

$$a_0^+ = a_0 = 0, \quad r_+ = r_- = 0. \tag{36.11}$$

If, however, the determinant is equal to zero the amplitudes a_0, a_0^+, $r_\pm$ may be finite. In this case

$$R_3 = \frac{2\gamma}{\alpha^2 T_2}. \tag{36.12}$$

Therefore stationary states are possible with both zero and finite amplitudes of the dipole moment and field. Let us examine the characteristics of these states in greater detail. We shall first solve the problem of the stability of a state with zero amplitudes. For this we shall examine small oscillations of the field and dipole moment (energy spin) about the state given by (36.11). In this case we can neglect the change in R_3, since in accordance with (35.10) and (35.14) the change in R_3 is a quantity of the second order of smallness in $r_\pm$ or a_0^+, a_0. Eliminating the quantities $r_\pm$ from the system (36.9) and keeping R_3 constant we arrive at the linear second-order equations for a_0 and a_0^+:

$$\ddot{a}_0 + \left(\gamma + \frac{1}{T_2}\right)\dot{a}_0 + \left(\frac{\gamma}{T_2} - \frac{\alpha^2}{2} R_3\right) a_0 = 0,$$
$$\ddot{a}_0^+ + \left(\gamma + \frac{1}{T_2}\right)\dot{a}_0^+ + \left(\frac{\gamma}{T_2} - \frac{\alpha^2}{2} R_3\right) a_0^+ = 0. \tag{36.13}$$

The general solution of these equations is of the form

$$a_0 = C_1\, e^{\beta_1 t} + C_2\, e^{\beta_2 t}, \quad a_0^+ = C_1^+\, e^{\beta_1 t} + C_2^+\, e^{\beta_2 t}, \tag{36.14}$$

where

$$\beta_1 = -\frac{1}{2}\left(\gamma + \frac{1}{T_2}\right) + \sqrt{\frac{1}{4}\left(\gamma + \frac{1}{T_2}\right)^2 - \left(\frac{\gamma}{T_2} - \frac{\alpha^2}{2} R_3\right)};$$

$$\beta_2 = -\frac{1}{2}\left(\gamma + \frac{1}{T_2}\right) - \sqrt{\frac{1}{4}\left(\gamma + \frac{1}{T_2}\right)^2 - \left(\frac{\gamma}{T_2} - \frac{\alpha^2}{2} R_3\right)}.$$

It is now easy to see that a state in which the amplitudes of the field and the dipole moment are zero is stable (equations (36.9) have no exponentially increasing solutions) for all R_3 which satisfy the condition

$$\eta = \frac{\alpha^2 R_3 T_2}{2\gamma} < 1. \tag{36.15}$$

On the other hand, when the self-excitation condition $\eta > 1$ is satisfied the system reaches a steady state which has finite amplitudes a_0 and a_0^+. This is a state of continuous oscillation†. In this state R_3 takes up the stationary value (36.12). At first glance it appears that there is a contradiction between this statement and the self-excitation condition which requires that η should be greater than unity, whilst in the state (36.12) η is just equal to unity. However, in order for self-excitation of the system to occur R_3 must initially be such that η is greater than unity, but when equilibrium is reached R_3 takes up a value (36.12) corresponding to $\eta = 1$. Let us now determine the stationary values of the field amplitudes a_0, a_0^+, and of the dipole moment $r_\pm$ in the oscillatory state. To do this we use (35.10), (35.14) and (36.12). It follows from these equations that in the stationary state (36.12) (when $\dot{R}_3 = 0$)

$$U = \frac{i\alpha}{2} \langle \overline{R^+ a} - \overline{R^- a^+} \rangle = \frac{1}{T_1} \frac{R_3^0}{\eta_0} (1 - \eta_0), \tag{36.16}$$

where

$$\eta_0 = \frac{\alpha^2 R_3^0 T_2}{2\gamma} > 1; \quad R_3 = \frac{1}{\eta_0} R_3^0. \tag{36.17}$$

Strictly speaking (36.16) alone does not give the stationary values of the field and dipole moment amplitudes. In order to determine these amplitudes it is necessary to find a relation between U, the mean of a quadratic combination of the field and the dipole moment, and the actual amplitudes $r_\pm$, a_0 and a_0^+. This can be done if in the calculation of the stationary values of the amplitudes we neglect quadratic fluctuations of the $\langle R^+ a \rangle - \langle R^+ \rangle \langle a \rangle$ type. In this case, using (36.10) we obtain

$$a_0^+ a_0 = \frac{\eta_0}{2\gamma T_2 R_3^0} r_+ r_- = \frac{\eta_0 - 1}{\eta_0} \frac{R_3^0}{2\gamma T_1}. \tag{36.18}$$

Choosing the oscillation phase arbitrarily we can put

$$a_0^+ = a_0.$$

† Maser amplifiers and oscillators are discussed in detail in the second volume.

Then the expressions for the steady-state oscillation amplitudes

$$a_0 = \sqrt{\frac{\eta_0 - 1}{\eta_0} \frac{R_3^0}{2\gamma T_1}}, \quad r_+ = i\sqrt{\frac{T_2}{T_1}(\eta_0 - 1)\frac{R_3^0}{\eta_0}} \tag{36.19}$$

follow from (36.18).

36.3. The Stationary Values of the Quadratic Quantities

We notice that condition (36.15) is always satisfied if $R_3 < 0$ (this holds in a state of thermal equilibrium). If, however, $R_3 > 0$ (non-equilibrium systems, negative temperatures), then η may be less than unity; in this case, as we shall see below, the system of molecules in the resonator may amplify an external signal (amplification mode). If $\eta < 1$, then β_1 and $\beta_2 < 0$ and after sufficient time ($t \gg -1/\beta_1$) the amplitudes of the radiation and the dipole moment become zero. However, the quadratic quantities associated with the field and the molecules (in particular the field energy), which give the fluctuations of the field and the dipole moment, remain finite. Let us find their values for the amplification mode and the oscillation mode. For this we use the set of equations (35.10)–(35.13). In the stationary state (for $a_1^+ + a_1 = 0$)

$$\dot{R}_3 = 0, \quad \dot{n} = 0, \quad \dot{U} = 0 \quad \text{and} \quad \frac{d}{dt}(\overline{R_1^2} + \overline{R_2^2}) = 0$$

and (35.10)–(35.13) becomes a set of algebraic equations. This system of equations is not closed since the mean values $\langle nR_3 \rangle$ and $\langle R_3U + UR_3 \rangle$ cannot, generally speaking, be expressed in terms of the other quantities $R_3 n, \ldots$ However, by using the smallness of the quantity $\alpha^2 T_2/2\gamma\eta_0 = 1/R_3^0$ (see (36.17)) we can approximately use the system of equations (35.10)–(35.13) neglecting correlations of the $\langle \Delta n\, \Delta R_3 \rangle$ type.

Let us first examine the amplification mode. In this case the solution can be found in the form

$$\left.\begin{aligned} R_3 &= \varrho R_3^{(0)} + \varrho_0 + \cdots, \\ n &= \nu + \cdots, \\ U &= u + \cdots, \\ \overline{R_1^2} + \overline{R_2^2} &= rR_3^0 + r + \cdots, \end{aligned}\right\} \tag{36.20}$$

where the dots denote terms which vanish as $R_3^0 \to \infty$. Substituting (36.20) in (35.10)–(35.13) (in the steady state case), replacing $\alpha^2/2$ by $(\gamma\eta_0/T_2)(1/R_3^0)$, comparing terms of first and zero order in R_3^0 and dropping the rest of the terms we obtain a solution in the form† (r_0 here is still undetermined)

$$\varrho = 1; \quad \varrho_0 = T_1 u; \quad \nu = n_0 - \frac{u}{2\gamma}$$

$$r = \frac{N}{2R_3^0} - T_2 u; \quad u = -\frac{\eta_0 \gamma \left(\frac{N}{2R_3^0} + 2n_0 + 1\right)}{(1-\eta_0)(1+\gamma T_2)}; \quad \eta_0 < 1.$$

From (36.20) we obtain in the amplification mode

$$n = \eta_0 + \frac{\eta_0(n_0 + \frac{1}{2} + \frac{1}{4}N/R_3^0)}{(1-\eta_0)(1+\gamma T_2)}, \tag{36.21}$$

$$R_3 = R_3^0 - \frac{2\eta_0 \gamma T_1(n_0 + \frac{1}{2} + \frac{1}{4}N/R_3^0)}{(1-\eta_0)(1+\gamma T_2)}, \tag{36.22}$$

$$\overline{R_1^2} + \overline{R_2^2} = \frac{N}{2} + \frac{2\eta_0 \gamma T_2 R_3^0(n_0 + \frac{1}{2} + \frac{1}{4}N/R_3^0)}{(1-\eta_0)(1+\gamma T_2)}, \tag{36.23}$$

$$U = -\frac{2\eta_0 \gamma(n_0 + \frac{1}{2} + \frac{1}{4}N/R_3^0)}{(1-\eta_0)(1+\gamma T_2)}. \tag{36.24}$$

The solution (36.21)–(36.24) is valid for all $\eta_0 < 1$ with the exception of the very small region $1 - \eta_0$ corresponding to the condition $T_1 u \gtrsim R_3^0$. In this region the solution of U cannot be given in the form (36.20).

In the oscillation mode the solution can be found in the form

$$\left.\begin{aligned} R_2 &= \varrho R_3^0 + \cdots, \\ n &= \nu R_3^0 + \cdots, \\ U &= u R_3^0 + \cdots, \\ \overline{R_1^2} + \overline{R_2^2} &= r R_3^{02} + \cdots. \end{aligned}\right\} \tag{36.25}$$

The dots here denote terms having a lower power of R_3 which must be dropped to be able to neglect correlations of the $\langle \Delta n \, \Delta R_3^0 \rangle$ type. In this approximation the solutions are of the form

$$n = \frac{(\eta_0 - 1)}{2\gamma T_1} \frac{R_3^0}{\eta_0}, \tag{36.26}$$

$$R_3 = \frac{R_3^0}{\eta_0}, \tag{36.27}$$

† We take here into account the fact that N is of the same order of magnitude as R_3^0.

$$\overline{R_1^2} + \overline{R_2^2} = \frac{T_2}{T_1}(\eta_0 - 1)\left(\frac{R_3^0}{\eta_0}\right)^2, \tag{36.28}$$

$$U = -\frac{1-\eta_0}{\eta_0 T_1} R_3^0. \tag{36.29}$$

Here we must also exclude the above-mentioned $1 - \eta_0$ region since the assumption (36.25) is not satisfied in this region. We note that the expressions for U and R_3 in the oscillation mode have already been obtained from the condition for the existence of non-zero amplitudes of the field and dipole moment (see (36.12), (36.16) and (36.17)). The expressions obtained in this way agree with (36.27) and (36.29). This agreement means that in the approximation used we have in essence neglected the fluctuations when treating the oscillation mode. In actual fact it follows from the solutions that the quantity $n - \langle a^+\rangle\langle a\rangle = \langle a^+ a\rangle - \langle a^+\rangle\langle a\rangle$ and the quantities similar to it, which are fluctuations, vanish.

Let us briefly discuss expressions (36.21)–(36.24) which apply when $\eta_0 < 1$. If the radiation and the molecules are at the same temperature the terms proportional to η_0 disappear in accordance with (35.15) and R_3, n, $\overline{R_1^2} + \overline{R_2^2}$ take up their equilibrium values. If, however, $R_3^0 > 0$ (the molecules have a negative temperature), whilst the radiation has a positive temperature, and $0 < \eta_0 < 1$ (the system can amplify an external signal), then the expressions (36.21)–(36.24) describe the non-equilibrium fluctuations of the system†.

Expression (36.21) can be used in particular for estimating the noise of an amplifier. The energy of the radiation field in a resonator is of the form

$$H_v = \left(n + \frac{1}{2}\right)\hbar\omega_0 = \frac{1}{2}\hbar\omega_0 + n_0\hbar\omega_0 + \frac{\eta_0}{(1-\eta_0)(1+\gamma T_2)} \times \left(n_0\hbar\omega_0 + \frac{\hbar\omega_0}{2} + \frac{N}{4R_3^0}\hbar\omega_0\right), \tag{36.30}$$

where $n_0 = 1/(e^{\hbar\omega_0/kT} - 1)$.

The first term in (36.30) is the zero-point vacuum fluctuations in a resonator, the second term is the thermal fluctuations of the radiation field and the third term is the energy of the fluctuations of the spontaneous emission from the molecules and of the field stimulated by the thermal fluctuations (a term proportional to n_0).

† We recall that the mean square dispersion of $\hat{O}$ is $\langle \Delta O\rangle^2 = \langle \hat{O}^2\rangle - \langle O\rangle^2$ and in the case when $\langle \hat{O}\rangle = 0$ (which is the case for the magnitude of the field and the dipole moment in the amplification mode) the mean squares describe the fluctuations themselves.

37. Stimulated and Spontaneous Emission in a Resonator

37.1. The Behaviour of a System of Molecules in an External Field

Let us first examine the behaviour of a system of molecules (for $T_1 = T_2 = \infty$) in an externally applied field. In this case we can use equation (19.27), or the equivalent equations (35.6) and (35.10) (in which we neglect the eigen-field). Changing to the variables r_+ and r_- and discarding small terms containing harmonics of $\omega_0 \approx \omega$, we arrive at

$$\dot{R}_3 = \frac{1}{2\hbar}\left(V_{21} r_+ \, \mathrm{e}^{i(\delta t - \theta)} + V_{12} r_- \, \mathrm{e}^{-i(\delta t - \theta)}\right), \tag{37.1}$$

$$\dot{r}_+ = -\frac{V_{12}}{\hbar} R_3 \, \mathrm{e}^{-i(\delta t - \theta)}, \quad \dot{r}_- = -\frac{V_{21}}{\hbar} R_3 \, \mathrm{e}^{i(\delta t - \theta)}, \tag{37.2}$$

where

$$V_{12}(t) = V_{12} \sin(\omega t + \theta), \quad V_{12} = -\tfrac{1}{2}(e_1 \cdot A_\nu)\, q_{\nu 0} = V_{21} \tag{37.3}$$

is the energy of the interaction of one molecule with the field† and $\delta = \omega_0 - \omega$. Eliminating r_+ and r_- from (37.1) and (37.2) we obtain

$$\dddot{R}_3 + \Omega_0^2 \dot{R}_3 = 0, \quad \Omega_0^2 = \frac{|V_{12}|^2}{\hbar^2} + \delta^2 . \tag{37.4}$$

The general solution of this equation is of the form

$$R_3 = A \cos \Omega_0 t + B \sin \Omega_0 t + C. \tag{37.5}$$

In order to determine the constants A, B, C the initial conditions must be given for $R_3(0)$, $\dot{R}_3(0)$ and $\ddot{R}_3(0)$. The initial conditions for the derivatives $\dot{R}_3(0)$ and $\ddot{R}_3(0)$ can be found from (37.1) and (37.2)

$$\dot{R}_3(0) = \frac{1}{2\hbar}\left(V_{21} r_+(0)\, \mathrm{e}^{-i\theta} + V_{12} r_-(0)\, \mathrm{e}^{i\theta}\right), \tag{37.6}$$

$$\ddot{R}_3(0) = -\frac{|V_{12}|^2}{\hbar^2} R_3(0) + \frac{i\delta}{2\hbar}\left(V_{21} r_+(0)\, \mathrm{e}^{-i\theta} - V_{12} r_-(0)\, \mathrm{e}^{i\theta}\right). \tag{37.7}$$

† The interaction energy is, in accordance with (35.1), of the form $-(e_1 \cdot A_\nu) R_1 q_\nu(t)$. For a single molecule $R_1 = \frac{1}{2}\begin{pmatrix} 0 & 1 \\ 1 & 0 \end{pmatrix}$ from which (37.3) follows. The quantities $V_{12} = V_{21}$ must be introduced for comparison with the solution to the problem of the behaviour of a single molecule in a monochromatic field (see the end of this sub-section). In the general case $V_{12} = V_{21}^*$.

Using these values we find

$$A = -\ddot{R}_3(0)\,\Omega_0^{-2}, \quad B = \dot{R}_3(0)\,\Omega_0^{-1}, \quad C = R_3(0) + \ddot{R}_3(0)\,\Omega^{-2}. \tag{37.8}$$

In particular if the system of molecules is initially in a definite energy level (or the density matrix of the molecular system is diagonal in the energy representation), then $r_+(0) = r_-(0) = 0$ and

$$R_3 = R_3(0)\,[|V_{12}|^2\,\hbar^{-2}\cos\Omega_0(t - t_0) + \delta^2]\,\Omega_0^{-2}. \tag{37.9}$$

Therefore an electromagnetic field whose frequency is close to the molecular transition frequency causes the population difference of the system of molecules (the quantity $2R_3$) to vary harmonically with a frequency Ω_0 which depends on the amplitude of the field.

In the case when the electromagnetic field is in resonance with the molecular system but the amplitude is changing slowly (compared with ω_0)

$$q_\nu(t) = q_{\nu 0}(t)\cos\omega_0 t\,,$$

the equation for the quantity R_3 becomes (it is assumed that the amplitude of the interaction energy is $V_{12} = V_{21} = V_0$)

$$\ddot{R}_3 - \frac{\dot{V}_0}{V_0}\,\dot{R}_3 + \frac{V_0^2}{\hbar}\,R_3 = 0. \tag{37.10}$$

This has the solution

$$R_3 = R(0)\cos\theta, \quad \theta = \int_0^t \frac{V_0}{\hbar}\,dt. \tag{37.11}$$

We notice that the results of this section could be obtained by examining the behaviour of a single two-level molecule in an external field. In a field each molecule behaves independently of all the other molecules. On the other hand allowing for the eigenfield may lead to a connection (coherence) between the individual molecules. This shows that questions of coherence are important when discussing spontaneous emission from a system of molecules (see Chapter VII). A solution to the problem of the behaviour of a single molecule in the presence of a monochromatic signal is given in the book by Landau and Lifshitz (1963) (the problem in section 40). This solution shows that the probability of finding a molecule at a time t in the same state as it occupied at a time $t = t_0$ is

$$w(t) = \left\{\frac{|V_{12}|^2}{\hbar^2}\cos\Omega_0(t - t_0) + \delta^2\right\}\Omega_0^{-2}.$$

It is not difficult to obtain (37.9) from this.

37.2. Stimulated Emission in a Resonator

Let us now examine the case when both the external and the eigenfields of the radiation are significant. Taking R_3 as given we find from (35.5) and (35.6), when $\omega_\nu = \omega_0$,

$$\dot{r}_+ + \frac{1}{T_2} r_+ - i\alpha R_3 a_0^+ = \frac{\alpha}{2} R_3(a_1^+ + a_1)\, e^{i\theta},$$
$$\dot{r}_- + \frac{1}{T_2} r_- + i\alpha R_3 a_0 = \frac{\alpha}{2} R_3(a_1^+ + a_1)\, e^{-i\theta}; \tag{37.12}$$

$$\dot{a}_0 + \gamma a_0 - i\frac{\alpha}{2} r_- = 0,$$
$$\dot{a}_0^+ + \gamma a_0^+ + i\frac{\alpha}{2} r_+ = 0. \tag{37.13}$$

Eliminating r_+ and r_- from these equations we obtain for the amplitudes a_0 and a_0^+

$$\ddot{a}_0 + \left(\gamma + \frac{1}{T_2}\right)\dot{a}_0 + \left(\frac{\gamma}{T_2} - \frac{\alpha^2}{2} R_3\right) a_0^+ = \frac{i\alpha^2 R_3}{4} (a_1^+ + a_1)\, e^{-i\theta},$$

$$\ddot{a}_0^+ + \left(\gamma + \frac{1}{T_2}\right)\dot{a}_0^+ + \left(\frac{\gamma}{T_2} - \frac{\alpha^2}{2} R_3\right) a_0 = -\frac{i\alpha^2 R_3}{4} (a_1^+ + a_1)\, e^{i\theta}. \tag{37.14}$$

By using (36.14) and (36.15) we can find the general solution for the inhomogeneous equations (37.14) in the form

$$a_0 = C_1\, e^{\beta_1 t} + C_2\, e^{\beta_2 t} + \frac{i}{2}\frac{\eta}{1-\eta}(a_1^+ + a_1)\, e^{-i\theta},$$

$$a_0^+ = C_1^+\, e^{\beta_1 t} + C_2^+\, e^{\beta_2 t} - \frac{i}{2}\frac{\eta}{1-\eta}(a_1^+ + a_1)\, e^{i\theta}. \tag{37.15}$$

We shall now discuss the case when $\eta < 1$. In this case $\beta_1, \beta_2 < 0$ and after a time $t \gg -1/\beta_1, -1/\beta_2$ the exponents $e^{\beta_1 t}$, $e^{\beta_2 t}$ can be neglected. Therefore the steady-state amplitude of the field in the resonator will be

$$a_0 = \frac{i}{2}\frac{\eta(a_1^+ + a)}{1-\eta}\,\mathrm{e}^{-i\theta}, \quad a_0^+ = -\frac{i}{2}\frac{\eta}{1-\eta}(a_1^+ + a)\,\mathrm{e}^{i\theta}. \tag{37.16}$$

The total field (eigen and external) in the steady state becomes

$$q_\nu + q_{\nu 0}\sin(\omega t + \theta) = q_{\nu 0}\frac{1}{1-\eta}\sin(\omega t + \theta), \tag{37.17}$$

when $\eta < 0$ ($R_3 < 0$) the total field decreases (absorption occurs); when $1 > \eta > 0$ the total field increases and amplification occurs. We notice that the amplified field has the same phase as the external field†. The amplification factor for the field in the resonator is of the form

$$k = \frac{1}{1-\eta}. \tag{37.18}$$

Let us now examine the development of the radiation field in a resonator. Let there be no eigenfield at $t = 0$, and let the amplitudes of the dipole moment r_+, r_- be zero. Then by using (37.12) and (37.13) we have the following initial conditions:

$$a_0(0) = 0, \quad \dot{a}_0(0) = 0, \quad a_0^+(0) = 0, \quad \dot{a}_0^+(0) = 0. \tag{37.19}$$

The solution of (37.15) that satisfies these initial conditions is of the form

$$\begin{aligned} a_0 &= \frac{i}{2}\frac{\eta}{1-\eta}\left\{1 + \frac{-\beta_2\,\mathrm{e}^{\beta_1 t} + \beta_1\,\mathrm{e}^{\beta_2 t}}{\beta_2 - \beta_1}\right\}\mathrm{e}^{i\theta}(a_1^+ + a_1), \\ a_0^+ &= -\frac{i}{2}\frac{\eta}{1-\eta}\left\{1 + \frac{-\beta_2\,\mathrm{e}^{\beta_1 t} + \beta_1\,\mathrm{e}^{\beta_2 t}}{\beta_2 - \beta_1}\right\}\mathrm{e}^{i\theta}(a_1^+ + a_1). \end{aligned} \tag{37.20}$$

For $t \ll -1/\beta_2;\ -1/\beta_1$ these expressions become

$$a_0 = \frac{i}{2}\frac{\gamma\eta}{T_2}\,t^2\,\mathrm{e}^{-i\theta}(a_1^+ + a_1) = \frac{i}{4}\alpha^2 R_3 t^2\,\mathrm{e}^{-i\theta}(a_1^+ + a_1),$$

$$a_0^+ = -\frac{i}{4}\alpha^2 R_3 t^2\,\mathrm{e}^{i\theta}(a_1^+ + a_1). \tag{37.21}$$

Therefore with $R_3 > 0$ we are dealing with ordinary stimulated emission (see (28.7)).

† This occurs only when the frequency of the emitted radiation is the same as the frequency of the external field.

It has already been pointed out that the distinction between spontaneous and stimulated emission is generally unambiguous for only a short time in the perturbation theory approximation. Often, however, a reasonable division can be made into spontaneous and stimulated emission over a long time in a steady-state mode and not in the perturbation theory approximation. The radiation given by (37.20), for example, can be called stimulated since it is proportional to the amplitude of the external field, and becomes zero when the external field is zero. The rest of the radiation (which is still present when the external field is zero) can be called spontaneous. It should, however, be remembered that if a long time is involved, and in particular in a stationary state, spontaneous emission defined in this way includes stimulated emission due to the eigenfield (which originally appeared spontaneously). The mean spontaneous emission field for the particular initial conditions in (37.19) is zero. The energy of the spontaneous emission changes in accordance with equations (35.10)–(35.13).

37.3. The Intensity of Spontaneous and Stimulated Emission in a Resonator with no External Field

Let us now examine the intensities of the spontaneous and stimulated emission fields in a resonator with no external field. It follows from (35.11) (for $a_1 + a_1^+ = 0$) that the change in the energy of the field in a resonator is given by

$$\frac{dH_\nu}{dt} = -U\hbar\omega_0 - 2\gamma\hbar\omega_0(n - n_0). \tag{37.22}$$

The second term on the right-hand side represents the losses (in the walls of the resonator, or those due to any output from the resonator); the first term

$$I = -U\hbar\omega_0 \tag{37.23}$$

is the intensity of the radiation emitted from the molecules inside the resonator, which can be divided into spontaneous and stimulated parts. Let us take various time intervals after initiating the interaction with the field in the resonator. Let $t \ll 1/\gamma, T_1, T_2$; then from (35.12) we find

$$I = \frac{\alpha^2\hbar\omega_0}{2}\{\overline{R_1^2} + \overline{R_2^2} + R_3 + 2nR_3\}\, t. \tag{37.24}$$

From this, in exact agreement with (28.16) and (28.17), we can determine the intensity of the spontaneous emission:

$$I^s = \frac{\alpha^2\hbar\omega_0}{2}(\overline{R_1^2} + \overline{R_2^2} + R_3)\, t \tag{37.25}$$

and of the stimulated (induced) emission:

$$I^i = \frac{\alpha^2 \hbar\omega_0}{2} 2nR_3 t. \tag{37.26}$$

In (37.24)–(37.26) the values of $\overline{R_1^2} + \overline{R_2^2}$, R_3 and n are taken at $t = 0$. Let us now examine the steady-state behaviour of the system

$$t \gg \frac{1}{-\beta_1}, \quad \frac{1}{-\beta_2}, \quad T_1.$$

In this case, in accordance with (37.23) and (35.12) the total intensity of the emitted radiation is

$$I = \frac{\hbar\omega_0\alpha^2}{2\left(\gamma + \frac{1}{T_2}\right)} \{\overline{R_1^2} + \overline{R_2^2} + R_3 + 2nR_3\}. \tag{37.27}$$

As in (37.25) and (37.26) we can separate the intensities of the stimulated and spontaneous emission:

$$I^s = \frac{\hbar\omega_0\alpha^2}{2\left(\gamma + \frac{1}{T_2}\right)} \{\overline{R_1^2} + \overline{R_2^2} + R_3\}, \tag{37.28}$$

$$I^i = \frac{\hbar\omega_0\alpha^2}{2\left(\gamma + \frac{1}{T_2}\right)} 2nR_3. \tag{37.29}$$

Here we take the stimulated emission to be that part which has an intensity proportional to the number of photons in the resonator, and the rest as spontaneous emission. The quantities R_1^2, R_2^2, R_3 and n in (37.28) and (37.29) are defined in (36.21)–(36.24) and (36.26)–(36.29).

Let us examine separately the cases when $\eta_0 < 1$ (amplification mode) and when $\eta_0 > 1$ (oscillation mode). When $\eta_0 < 1$, by using (36.23) and neglecting the difference

$$R_3 + \frac{N}{2} = N_+,$$

where N_+ is the number of molecules in the upper level, we obtain the spontaneous emission intensity in the form

$$I^s = \frac{\hbar\omega_0\alpha^2}{2\left(\gamma + \dfrac{1}{T_2}\right)} \left\{ N_+ + \frac{2\eta_0\gamma T_2 R_3^0 \left(n_0 + \dfrac{1}{2} + \dfrac{N}{4R_3^0}\right)}{(1-\eta_0)(1+\gamma T_2)} \right\}. \tag{37.30}$$

The presence of the second term shows that the spontaneous emission intensity is different from the sum of the intensities from the individual molecules. Allowing for the field reaction in the higher approximations of perturbation theory† leads to a certain degree of coherence between the emissions from the individual molecules. (In a state of thermal equilibrium this coherence disappears.)

Let us now examine the intensities of spontaneous and stimulated emissions in the oscillation mode, when $\eta_0 > 1$. By using (36.26)–(36.28) we can, in accordance with (37.28) and (37.29), write the intensities of the spontaneous and stimulated emissions in the oscillation mode in the following form:

$$I^s = \frac{\hbar\omega_0\alpha^2}{2\left(\gamma + \dfrac{1}{T_2}\right)} \left\{ \frac{T_2}{T_1} (\eta_0 - 1) \left(\frac{R_3^0}{\eta_0}\right)^2 + \cdots \right\}, \tag{37.31}$$

$$I^i = \frac{\hbar\omega_0\alpha^2}{2\left(\gamma + \dfrac{1}{T_2}\right)} \left\{ \frac{(\eta_0 - 1)}{\gamma T_1} \left(\frac{R_3^0}{\eta_0}\right)^2 + \cdots \right\}, \tag{37.32}$$

where the dots denote terms of second or lower order in R_3^0, which are connected with taking the fluctuations into consideration.

Therefore the total emission intensity in a maser oscillator can be divided into the part which is stimulated and that which is spontaneous. Generally speaking these two parts have the same order of magnitude. With certain values of the parameters, however, one part may predominate over the other. For example for $\gamma T_2 \ll 1$ stimulated emission is predominant. Both stimulated and spontaneous emission contain a "coherent" part proportional simply to the number of particles††.

† In first order perturbation theory for the parameter α^2 the second term in (37.30) may be neglected since η_0 is proportional to α^2.

†† We have used quotation marks since the part that is proportional to R_3^0 has a certain degree of coherence. The purely incoherent part is proportional to N_+.

37.4. The Probability of Spontaneous Emission in a Resonator (Fain, 1959, 1963; Bloembergen and Pound, 1954; Bunkin and Oraevskii, 1959)

It follows from (37.27) that the probability per unit time of spontaneous emission from a single molecule ($n = 0$, $\overline{R_1^2} + \overline{R_2^2} + R_3 = 1$) is

$$w_s = \frac{\alpha^2}{2\left(\gamma + \dfrac{1}{T_2}\right)}. \tag{37.33}$$

As we can see, w_s is not time-dependent: this is so provided that

$$t \gg \frac{1}{\gamma},\ T_2. \tag{37.34}$$

It may be said that in our case the part of the quantity $\delta\omega \sim \omega^*$ characterizing the continuous spectrum (see section 27, Chapter II and Appendix I) is played by γ or $1/T_2$.

Expression (37.33) has been obtained rigorously as the solution of the problem of radiation from a molecule in a resonator. We shall now carry out another derivation of this expression which is not rigorous, but which has a certain clarity. From section 5 of Chapter I the transition probability is

$$w = \frac{2\pi}{\hbar^2} |V|^2\, \delta(\omega_\nu - \omega_0), \tag{37.35}$$

where ω_0 is the transition frequency: ω_ν is the frequency of the radiation; $|V$ is the modulus of the matrix element of the transition. The energy of interaction of a single molecule with a field in a resonator is

$$\hat{V} = -\frac{1}{c}(\dot{\hat{d}} \cdot A_\nu)\hat{q}_\nu = -\hat{r}_1(e_1 \cdot A_\nu)\,\hat{q}_\nu = -\hat{r}_1\hat{q}_\nu\alpha\sqrt{2\hbar\omega_\nu}.$$

Here we have used the definition of α from (35.7). The matrix element V corresponding to a spontaneous transition with the emission of a photon is

$$V = -\frac{i}{c}\,\omega_{12}\,(d_{12} \cdot A_\nu)\sqrt{\frac{\hbar}{2\omega_\nu}} = -\frac{1}{2}\,\hbar\alpha. \tag{37.36}$$

The total probability of spontaneous emission can be obtained, having integrated (37.35) over the spectrum of the field,

$$w_s = \frac{2\pi}{\hbar^2} |V|^2 \varrho(\omega_0), \tag{37.37}$$

where $\varrho(\omega_0)$ is the density of states at the frequency ω_0.

Strictly speaking we are dealing with a radiation in a single mode, so we cannot use the density of states. If we remember, however, that the spectrum of the resonator is broadened by attenuation, the density of states can be introduced if a little latitude is allowed. A comparison with the exact expression (37.33) shows that this density must be†

$$\varrho(\omega_0) = \frac{1}{\pi} \frac{1}{\gamma + \dfrac{1}{T_2}}. \tag{37.38}$$

If (37.38) and (37.36) are substituted in (37.37) we obtain (37.33)

$$w_s = \frac{\alpha^2}{2\left(\gamma + \dfrac{1}{T_2}\right)} = \frac{|(\boldsymbol{d}_{12} \cdot \boldsymbol{A}_\nu)|^2 \omega_0}{\hbar c^2 \left(\gamma + \dfrac{1}{T_2}\right)}. \tag{37.39}$$

Here $\boldsymbol{d}_{12}$ is a matrix element of the dipole moment, $\boldsymbol{A}_\nu$ is the value of $\boldsymbol{A}_\nu(\boldsymbol{r})$ at the molecule.

† This expression can be understood as follows. Let for example $\gamma \gg 1/T_2$. In this case it may be considered that one mode of the resonator (one state) has a frequency width of the order of γ. Then a distribution that satisfies the condition $\int_{-\infty}^{\infty} \varrho(\omega)\, d\omega = 1$ is of the form $\varrho(\omega) = (1/\pi)\, \gamma/[(\omega - \omega_0)^2 + \gamma^2]$.

CHAPTER IX

Non-linear Effects in Optics

IN SUFFICIENTLY intense electric and magnetic fields the properties of a medium (the dielectric constant and the magnetic permeability) become field dependent (see section 22). A number of classical effects in optics are caused by this change in the properties of solids in constant electric and magnetic fields: the linear electro-optical effect, the Kerr effect, the Faraday effect, the Cotton–Mouton effect (see, e.g., Landau and Lifshitz, 1957). In this chapter we shall discuss non-linear effects in alternating electromagnetic fields. The development of lasers (see Chapter XII in Volume 2), which can generate intense electromagnetic fields with an extremely small spectral width, has made it possible to observe and use these effects in optics. In the radio region the non-linear properties of a medium are widely used in the design of oscillators based on both classical and quantum effects. Estimates of the fields in which the non-linear properties are significant are given in sub-section 22.5. We should like simply to state here that non-linear effects in a plasma appear with far smaller fields than in a solid (see, e.g., Ginzburg, 1960).

The non-linear optical effects which we shall study in this chapter can be divided into two classes: the parametric interaction of electromagnetic waves discussed by Khokhlov (1961), Akhmanov and Khokhlov (1962), and Armstrong, Bloembergen, Ducuing and Pershan (1962), and the Raman effect and other two-quantum processes. We shall discuss these two phenomena in detail, but for the present it is interesting to see how these effects are connected with the expansion of the polarization in powers of the field (see section 22)

$$P_a(t) = \chi_{ab}(\omega_l)\, E_b(\omega_l)\, e^{-i\omega_l t} + \chi_{abc}\,(\omega_s, \omega_l)\, E_b(\omega_s)\, E_c(\omega_l)\, e^{-i(\omega_s+\omega_l)\,t}$$
$$+ \chi_{abcd}(\omega_s, \omega_l, \omega_r)\, E_b(\omega_s)\, E_c(\omega_l)\, E_d(\omega_r)\, e^{-i(\omega_s+\omega_l+\omega_r)\,t} + \ldots \quad (22.25)$$

To do this we write the change in the energy of the field in the νth mode due to the work done by this field on the polarization (22.25). In accordance with (17.7) the derivative of the energy of the field in the νth mode H_ν is (with the opposite sign) equal to the change in the internal energy of the system and is of the form†

$$\frac{dH_\nu}{dt} = \int P_a \frac{d}{dt} [E_a(\omega_\nu)\, e^{-i\omega_\nu t} + E_a(-\omega_\nu)\, e^{i\omega_\nu t}]\, dV. \qquad \text{(IX.1)}$$

By substituting the various terms of the expansion (22.25) in this expression we can separate out the different wave interaction processes. Before discussing the results of this substitution we shall pick out the steady component of the rate of change of energy. The derivative $\dot{H}_\nu$ contains rapidly alternating terms such as $e^{i(\omega_s+\omega_l)t}$ $(\omega_s + \omega_l \neq 0)$ and terms with zero frequency; the latter give the systematic change in the energy‡. Therefore for such a change to occur each term of (22.25) must satisfy a definite relation between the frequencies—the time synchronism or frequency condition. We then take the case when matter fills the whole of the resonator (or, in the case of free space, it fills the whole of space). In the latter case we shall assume that the amplitudes are

$$E_a(\omega_\nu) \sim e^{i(k_\nu r)},$$

i.e. we are dealing with travelling waves. In this case the space integrals on the right-hand side are finite only for a definite relation between the wave vectors—the spatial synchronism or phase matching condition.

Substituting the first term of (22.25) in (IX.1) and separating out the systematic part of the energy change we obtain the frequency and phase matching conditions:

$$\omega_\nu = \pm\omega_l; \quad \boldsymbol{k}_\nu = \pm\boldsymbol{k}_l.$$

Thus the first term gives the work done by the field on the polarization created by a field of the same frequency and with the same wave vector. The change in the field energy is proportional to the integral of the square of the field, i.e. the total field energy or the number of photons in the field. Therefore the first term in the expansion (IX.1)–(22.5) is the usual stimulated emission (or absorption). The rest of the terms are connected with non-linear effects. The frequency and phase matching conditions for the second term of the expansion (IX.1)–(22.25) are

$$\omega_s + \omega_l \pm \omega_\nu = 0, \quad \boldsymbol{k}_s + \boldsymbol{k}_l \pm \boldsymbol{k}_\nu = 0.$$

(The latter condition is easy to obtain if we remember that $\int e^{i((k_s+k_l\pm k_\nu)r)} dV$ is finite only in the case when the exponent is zero.) The change in the energy

† Here there is no summation over ν. There is summation over all the repeated suffixes.

‡ These terms can be obtained after averaging for a time much greater than ω_ν^{-1}.

to this order of perturbation theory is proportional to the product of the three amplitudes of the field at the different frequencies. This term depends essentially on the phase relations between these fields, unlike the stimulated emission. (It is interesting to compare this term with the "stimulated" emission which depends on the relative phases of the oscillator and the field; see (28.11).) This term, which contains the three amplitudes, cannot be taken as proportional to the numbers of photons. Therefore there is no such term in ordinary perturbation theory (see, e.g., Heitler, 1954), where it is assumed that definite numbers of photons exist initially. According to the uncertainty relation between the number of photons and the phase (see Chapter I)

$$\Delta n \Delta\varphi \gtrsim 1$$

in a state with definite numbers n the phases are completely uncertain. This explains why parametric interaction† cannot be handled by the usual methods of perturbation theory.

Let us move on now to the third term of the expansion (IX.1)–(22.25). The frequency and phase matching conditions for this term are of the form

$$\omega_l + \omega_s + \omega_r \pm \omega_v = 0, \quad \boldsymbol{k}_l + \boldsymbol{k}_s + \boldsymbol{k}_r \pm \boldsymbol{k}_v = 0.$$

These conditions are satisfied by terms which have a different physical meaning. In the first place, these are terms with completely different amplitudes and frequencies which give the parametric interaction of four waves. They depend essentially on the phase relations between fields at different frequencies. Secondly, in the same order of perturbation theory we can find terms with pairs of equal amplitudes:

$$E_a(\omega_v) = E_b^*(-\omega_s), \quad E_c(\omega_l) = E_d^*(-\omega_r), \quad (a = b; c = d),$$

$$\omega_v = -\omega_s, \quad \omega_l = -\omega_r.$$

This part of the derivative $\dot{H}_v$ is proportional to the product of the numbers of photons of frequencies ω_v and ω_l:

$$n_v n_l \propto |E(\omega_v)|^2 \, |E(\omega_l)|^2$$

and does not depend on the phase relations between these fields. In the usual perturbation theory approach these terms correspond to two-quantum transitions and, in particular, to the stimulated Raman effect. We could continue the discussion by taking further terms in the expansion. It is quite clear, however, that the change in the energy in any order of perturbation theory can be connected either with parametric interaction, which depends

† As will be clear from what follows parametric interaction is connected with this term.

on the phase relation between the fields†, or with Raman interaction, which depends upon the squares of the amplitudes (the numbers of photons) and does not depend on the phases.

Sections 38 and 40 are devoted to two-quantum processes and the Raman effect and section 39 to parametric interaction. In section 40 we also compare parametric and Raman systems.

38. Two-quantum Processes. The Raman Effect, Stimulated and Spontaneous Emission (Fain and Yashchin, 1964)

38.1. Elementary Two-quantum Processes

In this sub-section we shall discuss in detail the third term in the expansion of the polarization (22.25) in powers of the field. As we have already pointed out, two-quantum processes and the parametric interaction of four fields are connected with this term (the tensor $\chi_{abcd}(\omega_s, \omega_l, \omega_r)$); we shall not deal with the latter effect. We shall be able to explain the essential features of parametric interaction by taking three fields, as discussed below (section 39). For the present we shall deal with a number of topics connected with two-quantum processes. We notice at once that two-quantum processes include in particular the process of spontaneous Raman emission or Raman scattering—an effect first observed by Landsberg and Mandelstam (1928) and Raman and Krishnan (1928).

The fundamental processes connected with the emission and absorption of two photons can be classified as follows. Let E_1 and E_2 be two levels of a substance ($E_2 > E_1$) and let there be initially n_α quanta of energy $\hbar\omega_\alpha$, and n_β quanta of energy $\hbar\omega_\beta$ in the radiation field. Then in a system consisting of a sample of the substance interacting with radiation, two-quantum transitions of the following kinds are possible.

1. The substance makes a transition from the upper state E_2 to the lower one E_1 and two new photons appear, with energies $\hbar\omega_\alpha$ and $\hbar\omega_\beta$: $n'_\alpha = n_\alpha + 1$, $n'_\beta = n_\beta + 1$. To satisfy the law of the conservation of energy for the whole system it is necessary that

$$\hbar\omega_\alpha + \hbar\omega_\beta = E_2 - E_1. \tag{38.1}$$

This process is called two-photon emission (see Fig. IX.1, *a*).

2. If the substance was originally in the lower state E_1 the action of n_α photons of energy $\hbar\omega_\alpha$ and n_β photons of energy $\hbar\omega_\beta$ can cause a transition

† In calculations in which the numbers of photons are given phase-averaging is actually carried out and these terms drop out even in the classical limit when the number of photons is large.

into the upper state with the absorption of two photons with energies $\hbar\omega_\alpha$ and $\hbar\omega_\beta$: $n'_\alpha = n_\alpha - 1$, $n'_\beta = n_\beta - 1$. In this case the same condition (38.1) follows from the law of the conservation of energy. A process of this kind is called two-photon absorption (see Fig. IX.1, *b*)

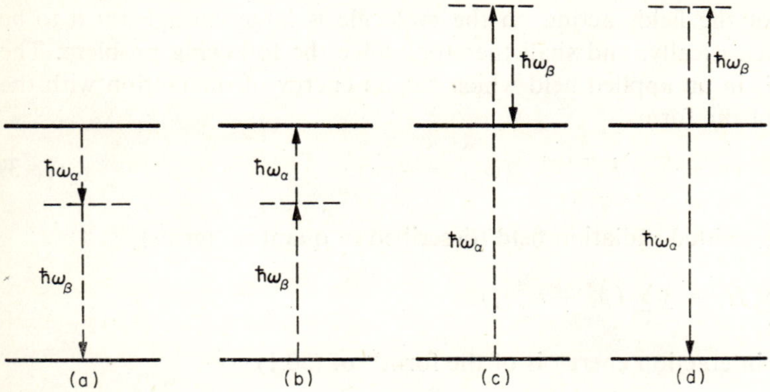

FIG. IX.1. Transitions in fundamental two-quantum processes.

The dotted line with the arrow indicates the direction of the transition: (*a*) emission of two photons $\hbar\omega_\alpha$ and $\hbar\omega_\beta$; (*b*) absorption of two photons $\hbar\omega_\alpha$ and $\hbar\omega_\beta$; (*c*) emission of a photon of energy $\hbar\omega_\beta$ and absorption of a photon with an energy ω_α; (*d*) emission of a photon with an energy $\hbar\omega_\alpha$ and absorption of a photon with an energy $\hbar\omega_\beta$. There are no energy levels at the places indicated by horizontal dotted lines.

3. If the energy of one of the photons is greater than the energy difference $E_2 - E_1$, then Raman scattering processes and stimulated Raman emission and absorption are possible. Let $\hbar\omega_\alpha > E_2 - E_1$; then if the substance is originally at the level E_1, then $n'_\alpha = n_\alpha - 1$, $n'_\beta = n_\beta + 1$ and the law of the conservation of energy gives the frequency condition

$$\hbar(\omega_\alpha - \omega_\beta) = E_2 - E_1. \tag{38.2}$$

Processes of spontaneous and stimulated Raman emission occur at a frequency ω_β (see Fig. IX.1, *c*) (the so-called Stokes component of the radiation). Absorption of a photon occurs at a frequency ω_α.

4. If the substance was originally in the upper level, then $n'_\alpha = n_\alpha + 1$, $n'_\beta = n_\beta - 1$, and stimulated and spontaneous Raman emission occur at a frequency ω_α (the anti-Stokes component) (see Fig. IX.1, *d*). At the same time there is absorption of a quantum $\hbar\omega_\beta$.

38.2. The Intensity of Two-quantum Emission

If the energy levels E_2 and E_1 of the substance (in future we shall refer to a molecule) are discrete and the radiation has a continuous spectrum we can introduce the transition probabilities for the processes listed above. These probabilities have been calculated (Heitler, 1954; Göppert-Mayer, 1931; Placzek, 1935); we shall not repeat the calculations here, but we shall find the intensities of two-photon emission and absorption in the general case without making any assumptions about the discreteness or continuity of the energy spectra of the molecule and the field†. We shall consider, however, that one of the fields acting on the molecule is large enough for it to be described classically, and shall therefore solve the following problem. The molecule is in an applied field which has an energy of interaction with the molecule of the form

$$\hat{V}_1(t) = \sum_l \hat{V}^l e^{-i\omega_l t} \quad (\omega_{-l} = -\omega_l), \tag{38.3}$$

and in the emitted radiation field (described in quantum terms)

$$\sum_\nu \hat{H}_\nu \equiv \tfrac{1}{2} \sum_\nu (\hat{p}_\nu^2 + \omega_\nu^2 \hat{q}_\nu^2),$$

where the interaction energy is of the form‡ of (30.1)

$$\hat{V}_2 = -\sum_\nu \hat{B}_\nu \hat{q}_\nu, \quad \hat{B}_\nu = \sum_{s=1}^{n} \frac{e_s}{m_s c} (\hat{p}_s \cdot A_\nu(s)). \tag{38.4}$$

The complete Hamiltonian of the system can be written as

$$\hat{H} = \mathscr{H}_0 + \tfrac{1}{2} \sum_\nu (\hat{p}_\nu^2 + \omega_\nu^2 \hat{q}_\nu^2) - \sum_\nu \hat{B}_\nu \hat{q}_\nu, \tag{38.5}$$

where

$$\mathscr{H}_0 = \hat{H}_0 + \hat{V}_1(t) \tag{38.6}$$

is the Hamiltonian of the molecule in the applied field. We shall discuss spontaneous and stimulated emission from a system with a Hamiltonian $\mathscr{H}_0$. Just as in section 30 we find the operator of the energy derivative of the νth mode

$$\frac{d\hat{H}_\nu}{dt} = \hat{B}_\nu \hat{p}_\nu.$$

We then find the mean value of this quantity (i.e. we find the emission intensity from the system described by the Hamiltonian (38.6)). To do

† If it is assumed that the energy spectrum is discrete the line width of the emitted radiation is equal to the natural width of spontaneous emission. In the general case discussed below the intensities of the corresponding processes may include line widths not connected with radiation, i.e. radiationless transitions are also taken into consideration.

‡ Here we are neglecting the term A_ν^2 for the same reasons that it was neglected when discussing ordinary spontaneous and stimulated emission. At the same time we are not taking into consideration coherent scattering with an unchanged frequency (see Heitler, 1954).

this we use the equation for the density matrix in the interaction representation $\hat{V}_2$

$$i\hbar \frac{\partial \hat{\varrho}}{\partial t} = [\hat{V}_2, \hat{\varrho}].$$

(According to the results of section 5 this equation also holds in the case when the Hamiltonian of an unperturbed system is explicitly time-dependent.) Then, using the same approximations as in section 30, we arrive at the expression

$$\frac{d\bar{H}_\nu}{dt} = \overline{B_\nu(t)}\,\overline{p_\nu(t)} + \frac{i}{\hbar}\sum_\nu \int_0^t \{\overline{B_\nu(t)\,B_\lambda(t_1)}\,\overline{p_\nu(t) q_\lambda(t_1)}$$
$$- \overline{B_\lambda(t_1)\,B_\nu(t)}\,\overline{q_\lambda(t_1)\,p_\nu(t)}\}\,dt_1. \tag{38.7}$$

This expression differs from (30.9) in that in (38.7) the time function $\hat{B}_\nu(t)$ appears because of the presence of the external field $\hat{V}_1(t)$. We shall consider the second term in (38.7); this term gives the emission that does not depend on the phase relation between $B_\nu(t)$ and the field described by the p_ν and q_ν. After averaging over a time much longer than the period of the radiation, using the conditions (30.10) we obtain the intensity of the emitted radiation in the form†

$$I_\nu = \frac{1}{2}\int_0^t \cos\omega_\nu(t - t_1)\,\{\overline{B_\nu(t)\,B_\nu(t_1)} + \overline{B_\nu(t_1)\,B_\nu(t)}\}\,dt_1$$
$$- i\left(\bar{n}_\nu + \frac{1}{2}\right)\int_0^t \sin\omega_\nu(t - t_1)\,\overline{[B_\nu(t), B_\nu(t_1)]}\,dt_1. \tag{38.8}$$

where $\overline{B_\nu(t)\,B_\nu(t_1)}$ is the part of this correlation which depends only upon $t - t_1$, and $\bar{n}_\nu$ is the mean number of photons of the νth mode. We now have only to find $\hat{B}_\nu(t)$ to a first approximation in $\hat{V}_1(t)$. We assume that the field is sufficiently small, so we can use perturbation theory with respect to $\hat{V}_1(t)$. In accordance with the non-stationary perturbation theory of section 5

$$\hat{B}_\nu(t) = \hat{S}^{-1}\hat{B}_\nu\hat{S},$$

where $\hat{B}_\nu$ is an operator in the Schrödinger representation and the unitary transformation operators $\hat{S}^{-1}$ and $\hat{S}$ obey the equations

$$\frac{d\hat{S}^{-1}}{dt} = \frac{i}{\hbar}\hat{S}^{-1}(\hat{H}_0 + \hat{V}_1(t)), \quad \frac{d\hat{S}}{dt} = -\frac{i}{\hbar}(\hat{H}_0 + \hat{V}_1(t))\,\hat{S}. \tag{38.9}$$

† Time-averaging here leaves only the stationary part of the correlation $\overline{B_\nu(t)\,B_\nu(t_1)}$ and $\overline{p_\nu(t)\,p_\nu(t_1)}$, i.e. the part depending only on the difference $t - t_1$.

Expanding $\hat{S}^{-1}$ and $\hat{S}$ into a series in powers of $\hat{V}_1$

$$\hat{S}^{-1} = \hat{S}_0^{-1} + \hat{S}_1^{-1} + \dots, \quad \hat{S} = \hat{S}_0 + \hat{S}_1 + \dots,$$

including terms up to those in $\hat{V}_1$ we find:

$$\frac{d\hat{S}_0}{dt} = -\frac{i}{\hbar}\hat{H}_0\hat{S}_0, \quad \frac{d\hat{S}_1}{dt} = -\frac{i}{\hbar}\hat{V}_1\hat{S}_0 - \frac{i}{\hbar}\hat{H}_0\hat{S}_1,$$

$$\frac{d\hat{S}_0^{-1}}{dt} = \frac{i}{\hbar}\hat{S}_0^{-1}\hat{H}_0, \quad \frac{d\hat{S}_1^{-1}}{dt} = \frac{i}{\hbar}\hat{S}_0^{-1}\hat{V}_1 + \frac{i}{\hbar}\hat{S}_1^{-1}\hat{H}_0. \tag{38.10}$$

It is easy to check by direct substitution that $\hat{S}_1$ and $\hat{S}_1^{-1}$ chosen so that

$$\hat{S}_1 = -\frac{i}{\hbar}\int_{-\infty}^{t} dt_1\, e^{-\frac{i}{\hbar}\hat{H}_0(t-t_1)}\,\hat{V}_1(t_1)\, e^{-\frac{i}{\hbar}\hat{H}_0 t_1},$$

$$\hat{S}_1^{-1} = \frac{i}{\hbar}\int_{-\infty}^{t} dt_1\, e^{\frac{i}{\hbar}\hat{H}_0 t_1}\,\hat{V}_1(t_1)\, e^{\frac{i}{\hbar}\hat{H}_0(t-t_1)}, \tag{38.11}$$

satisfy (38.10)†. The quantity $\hat{B}_\nu(t)$ (just like any other operator described by the Hamiltonian (38.6) to terms in the first power of $\hat{V}_1$) is of the form

$$\hat{B}_\nu(t) = \exp\left[\frac{i}{\hbar}\hat{H}_0 t\right]\hat{B}_\nu \exp\left[-\frac{i}{\hbar}\hat{H}_0 t\right] + \hat{S}_0^{-1}\hat{B}_\nu\hat{S}_1 + \hat{S}_1^{-1}\hat{B}_\nu\hat{S}_0.$$

In (38.8) we must include terms in $\hat{B}_\nu^{(1)}(t)$‡

$$\hat{B}_\nu^{(1)}(t) = \hat{S}_0^{-1}\hat{B}_\nu\hat{S}_1 + \hat{S}_1^{-1}\hat{B}_\nu\hat{S}_0. \tag{38.12}$$

† The general expansion of $\hat{S}$ in powers of $\hat{V}_1(t)$ is of the form

$$\hat{S}(t) = \exp\left[-\frac{i}{\hbar}\hat{H}_0 t\right] - \frac{i}{\hbar}\int_{-\infty}^{t} dt_1 \exp\left[-\frac{i}{\hbar}\hat{H}_0(t-t_1)\right]\hat{V}_1(t_1)$$

$$\times \exp\left[-\frac{i}{\hbar}\hat{H}_0 t_1\right] + \dots + \left(-\frac{i}{\hbar}\right)^n \int_{-\infty}^{t} dt_n \int_{-\infty}^{t_n} dt_{n-1} \dots$$

$$\dots \int_{-\infty}^{t_2} dt_1 \exp\left[-\frac{i}{\hbar}\hat{H}_0(t-t_n)\right]\hat{V}_1(t_n)\exp\left[-\frac{i}{\hbar}\hat{H}_0(t_n-t_{n-1})\right]\dots$$

$$\dots \hat{V}_1(t_1)\exp\left[-\frac{i}{\hbar}\hat{H}_0 t_1\right] + \dots$$

‡ When substituting in (38.8) we are including in the product $\hat{B}_\nu(t)\,\hat{B}_\nu(t_1)$ terms down to the second order in $\hat{V}_1$. If we write the expansion of $\hat{B}_\nu$ in the form $\hat{B}_\nu = \hat{B}_\nu^{(0)} + \hat{B}_\nu^{(1)} + \hat{B}_\nu^{(2)} + \dots$, then in the product $\overline{\hat{B}_\nu(t_1)\,\hat{B}_\nu(t_1)}$ there will be terms of the form

By using (38.11), (38.12) and (38.3) we can find the matrix elements $\hat{B}_\nu^{(1)}$ in the form

$$B^{(1)}_{\nu nm}(t) = \sum_l \mathrm{e}^{i(\omega_{nm}-\omega_l)t} b^l_{\nu nm}, \tag{38.13}$$

where

$$b^l_{\nu nm} = \sum_k \left\{ \frac{B_{\nu nk} V^l_{km}}{\hbar} \zeta(\omega_l - \omega_{km}) - \frac{V^l_{nk} B_{\nu km}}{\hbar} \zeta(\omega_l - \omega_{nk}) \right\}. \tag{38.14}$$

It follows from (38.13) and the Hermitian nature of $\hat{B}_\nu^{(1)}$ that

$$b^l_{\nu nm} = (b^{-l}_{\nu mn})^*. \tag{38.15}$$

We shall use this relation in future. The mean value of the product of the quantities $\hat{B}_\nu(t)$ in the steady state with a diagonal density matrix is of the form

$$\overline{B_\nu^{(1)}(t)\, B_\nu^{(1)}(t_1)} = \sum_{n,k,l,l'} \varrho_n b^l_{\nu nk} b^{l'}_{\nu kn}\, \mathrm{e}^{i\omega_{nk}(t-t_1)}\, \mathrm{e}^{-i\omega_l t - i\omega_{l'} t_1}.$$

The part of this correlation that depends only on $t - t_1$, as can easily be seen, is of the form

$$\overline{B_\nu^{(1)}(t)\, B_\nu^{(1)}(t_1)} = \sum_{nkl} \varrho_n b^l_{\nu kn} b^{-l}_{\nu kn}\, \mathrm{e}^{i(\omega_{nk}-\omega_l)(t-t_1)}. \tag{38.16}$$

Let us briefly examine the part played by terms like $\overline{B_\nu^{(0)}(t)\, B_\nu^{(1)}(t_1)}$ and $\overline{B_\nu^{(0)}(t)\, B_\nu^{(2)}(t_1)}$ (see the last footnote). As in (38.13) we can write

$$B^{(2)}_{\nu mn}(t) = \sum_{ls} \mathrm{e}^{i(\omega_{mn}-\omega_l-\omega_s)t} b^{l,s}_{\nu mn},$$

$$B^{(0)}_{\nu mn}(t) = \mathrm{e}^{i\omega_{mn}t} b_{\nu mn}.$$

(The last equation is obvious.) Using these equations we can write

$$\overline{B_\nu^{(0)}(t)\, B_\nu^{(1)}(t_1)} = \sum_{n,m,l} \varrho_n b_{\nu nm} b^l_{\nu mn}\, \mathrm{e}^{i\omega_{nm}(t-t_1) - i\omega_l t_1},$$

$$\overline{B_\nu^{(0)}(t)\, B_\nu^{(2)}(t_1)} = \sum_{n,m,l,s} \varrho_n b_{\nu nm} b^{l,s}_{\nu mn}\, \mathrm{e}^{i\omega_{nm}(t-t_1) - i\omega_l t_1 - i\omega_s t_1},$$

$\overline{\hat{B}_\nu^{(0)}(t)\, \hat{B}_\nu^{(0)}(t_1)}$, $\overline{\hat{B}_\nu^{(0)}(t)\, \hat{B}_\nu^{(1)}(t_1)}$, $\overline{\hat{B}_\nu^{(0)}(t)\, \hat{B}_\nu^{(2)}(t_1)}$ and $\overline{\hat{B}_\nu^{(1)}(t)\, \hat{B}_\nu^{(1)}(t_1)}$.

Terms of the form of $\overline{\hat{B}_\nu^{(0)}(t)\, \hat{B}_\nu^{(0)}(t_1)}$ make a contribution to the stimulated and spontaneous emission when there is no field $\hat{V}_1$; terms of the form of $\overline{\hat{B}_\nu^{(0)}(t)\, \hat{B}_\nu^{(2)}(t)}$ and $\overline{B_\nu^{(0)}(t)\, B_\nu^{(1)}(t_1)}$, as will be shown below, make no contribution to the time-averaged intensity of the Raman and two-quantum emission (38.8). Therefore only terms of the form of $\overline{\hat{B}_\nu^{(1)}(t)\, \hat{B}_\nu^{(1)}(t_1)}$ remain.

when $\omega_l \neq 0$ there is a part depending only on $t - t_1$ in $\overline{B_\nu^{(0)}(t)\, B_\nu^{(2)}(t_1)}$ if $\omega_l = -\omega_s$ (when l, $s = \pm 1$, $\omega_l = \omega_1$ and $\omega_s = -\omega_1$ or vice versa). Substituting the stationary part

$$\overline{B_\nu^{(0)}(t)\, B_1^{(2)}(t_1)} = \sum_{n,m,l} \varrho_n b_{\nu nm} b_{\nu mn}^{l,-l}\, e^{i\omega_{nm}(t-t_1)}$$

in (38.8) we can check that the emitted intensity is finite when

$$\omega_\nu = \omega_{nm}.$$

In other words, the correlation $\overline{B_\nu^{(0)}(t)\, B_\nu^{(2)}(t_1)}$ modifies the emission (absorption) intensity at one of the resonance frequencies of the system. The effect of the external field is to change the emission intensity. This change is small compared with the basic emission (absorption) effect which occurs even when there is no external field. In future, therefore, we shall not take terms of the type of $\overline{B_\nu^{(0)}(t)\, B_\nu^{(2)}(t_1)}$ into consideration. Substituting (38.16) in (38.8) and dealing with time intervals much greater than the characteristic time $\tau_c = 1/\delta\omega$ (where $\delta\omega$ characterizes the spectrum of the molecule or the field), we obtain an expression similar to (30.11)†

$$I_\nu = \pi \sum_{n,k,l} |b_{\nu nk}^l|^2\, \delta(\omega_{n\kappa} + \omega_\nu + \omega_l)\, \{\varrho_\kappa + \bar{n}_\nu(\varrho_\kappa - \varrho_n)\}. \tag{38.17}$$

(We have used relation (38.15).)

Let us examine the special case when a field of frequency ω_1 acts on a molecule. In this case

$$\hat{V}_1(t_1) = \hat{V}_1^1\, e^{-i\omega_1 t} + \hat{V}_1^{-1}\, e^{i\omega_1 t} \tag{38.18}$$

and l takes up the values ± 1. We find from (38.17) that the emitted frequency ω_ν can have the values

$$\omega_\nu = \omega_{kn} \pm \omega_1. \tag{38.19}$$

The minus sign corresponds to two-photon emission and absorption (cases 1 and 2 at the beginning of the section; see (38.1)), and the plus sign corresponds to the Raman effect (cases 3 and 4, see (38.2)). The probabilities of the corresponding transitions are obtained by dividing I_ν by $\hbar\omega_\nu$. The total emission intensity can be obtained from (38.17) by summation over all the modes ν.

† See Appendix I for the change for large t to the δ-function.

38.3. A General Discussion of Absorption in a System Acted upon by an External Field

It follows from (38.17) that if a field with a frequency ω_1 acts on a molecule there is additional absorption (positive or negative). We shall now show that this is a general property of all non-linear systems possessing resonance frequencies. In other words we shall show that if there is resonance absorption in an arbitrary non-linear system at a frequency ω_0 and the system is acted upon by an external force of frequency $\omega_1 > \omega_0$, then at a frequency $\omega_2 = \omega_1 - \omega_0$ negative absorption appears (if the system was not originally excited) and at a frequency ω_1 positive absorption appears (in the presence of a given force of frequency ω_2). If, however, $\omega_1 + \omega_2 = \omega_0$, then positive absorption appears at both the frequencies ω_1 and ω_2 (if the system is initially excited, so that there is negative absorption at the frequency ω_0, then at the frequencies ω_1 and ω_2 negative absorption occurs). We shall prove these properties of non-linear systems in the most general case independently of the physical nature of the systems under discussion (molecules, *LCR* circuits in electronics, a plasma, mechanical systems, etc.) and independently of the nature of the active forces (electrical forces, sound, mechanical forces, etc.).

Let external forces with frequencies ω_l act on an arbitrary system which can be described by a Hamiltonian $\hat{H}$ which is explicitly time-independent. The energy of the interaction of the system with these forces can be written in the form (38.3) (where l in particular takes up the values ± 1 and ± 2; the frequencies $\pm\omega_1$ and $\pm\omega_2$). We wish to discover whether there is absorption at the frequency ω_2 in the presence of the external force. Absorption at the frequency ω_2 is measured by the imaginary part of the susceptibility at the frequency ω_2. Without loss of generality it can be stated that the interaction energy with a weak field of frequency ω_2 has the form of (14.1)

$$\hat{V}_2 = -\sum_a f_a^{(2)}\hat{x}_a\,\mathrm{e}^{-i\omega_2 t} + f_a^{(-2)}\hat{x}_a\,\mathrm{e}^{i\omega_2 t}. \tag{38.20}$$

In order to find the susceptibility at the frequency ω_2 we must find the mean value of x_a at the frequency ω_2 in the presence of a "strong" field $\hat{V}_1$ and a weak field $\hat{V}_2$†. It is not difficult to see, taking (22.25) as an example, that an effect at the frequency ω_2 due to the action of forces of frequencies ω_1 and ω_2 appears only in the third approximation in the interaction energy

$$\hat{V} = \hat{V}_1 + \hat{V}_2 = \sum_l \hat{V}^l\,\mathrm{e}^{-i\omega_l t}.$$

† Of course $\langle x_a\rangle$ contains Fourier components with frequencies other than ω_2, but it is sufficient to know its value at ω_2 in order to calculate the absorption at ω_2.

In this case there may be satisfaction of the condition

$$\omega_s + \omega_l + \omega_r = \pm\omega_2, \tag{38.21}$$

which should hold for non-zero frequencies $\omega_s = \pm\omega_1; \pm\omega_2$. To find the third approximation correction to $\langle x_a \rangle$ at the frequency ω_2 we make use of the density matrix $\varrho_{mn}^{(3)}(t)$ (see (22.7))

$$\begin{aligned}\langle x_a \rangle = \sum_{nm} \varrho_{mn}^{(3)} x_{anm}\, e^{i\omega_{nm}t} = \frac{1}{\hbar^3} \sum_{r,l,s,m,p,k,n} e^{-i(\omega_r+\omega_l+\omega_s)t} \\ \times \zeta(\omega_r + \omega_l + \omega_s - \omega_{mn}) \{V_{mp}^r V_{pk}^s V_{kn}^l [(\varrho_n - \varrho_k)\, \zeta(\omega_l - \omega_{kn}) \\ - (\varrho_k - \varrho_p)\, \zeta(\omega_s - \omega_{pk})]\, \zeta(\omega_l + \omega_s - \omega_{pn}) \\ - V_{mk}^s V_{kp}^l V_{pn}^r\, [(\varrho_p - \varrho_k)\, \zeta(\omega_l - \omega_{kp}) \\ - (\varrho_k - \varrho_m)\, \zeta(\omega_s - \omega_{mk})]\, \zeta(\omega_l + \omega_s - \omega_{mp})\}\, x_{anm}.\end{aligned} \tag{38.22}$$

A contribution to the required susceptibility is made by terms that satisfy condition (38.21). After changing the summation suffixes these terms can be written in the form†

$$\begin{aligned}\langle x_a \rangle = \frac{1}{\hbar^3} \sum_{\substack{r,l,s\\1,2,n,k}} e^{-i\omega_2 t} \{\zeta(\omega_2 - \omega_{n1})\, V_{n2}^r V_{2k}^s V_{k1}^l x_{a1n}\, [(\varrho_1 - \varrho_k]\, \zeta(\omega_l - \omega_{k1}) \\ - (\varrho_k - \varrho_2)\, \zeta(\omega_s - \omega_{2k})] - \zeta(\omega_2 - \omega_{2n})\, V_{2k}^s V_{k1}^l V_{1n}^r x_{an2} \\ \times [(\varrho_1 - \varrho_k)\, \zeta(\omega_l - \omega_{k1}) - (\varrho_k - \varrho_2)\, \zeta(\omega_s - \omega_{2k})]\} \\ \times \zeta(\omega_l + \omega_s - \omega_{21}) + \text{c.c.}\end{aligned} \tag{38.23}$$

Let us further examine the contribution to $\langle x_a \rangle$ connected with the δ-function $\delta(\omega_l + \omega_s - \omega_{21})$ (we recall that $\zeta(x) = (P/x) - i\pi\delta(x)$). This means that we are assuming the resonance condition‡

$$\omega_l + \omega_s = \omega_{21}, \tag{38.24}$$

which, in particular, reduces to (38.1) and (38.2). By carrying out simple transformations and remembering that l, $s = \pm 1, \pm 2$, and making the force $f_a^{(\pm 2)}$ small (leaving only the linear terms in $f_a^{(\pm 2)}$) we derive the following

† We have specially denoted the two summation suffixes by the numbers 1 and 2 since the frequency $|\omega_{21}|$ is the resonance frequency ω_0 mentioned above. Naturally summation over the suffixes 1 and 2 proceeds in the usual way (these suffixes must not be confused with $r, l, s = \pm 1, \pm 2$).

‡ Condition (38.24) can be satisfied approximately within the limits of the line width of ω_{21}, but in this case the additional assumption must be made that there are no other resonances in the system, i.e. $\omega_l - \omega_{nk} \neq 0$ and $\zeta(x) = 1/x$.

expression for the diagonal element of the imaginary part of the susceptibility at the frequency ω_2^*†

$$\chi''_{aa}(\omega_2) = -\frac{\pi}{\hbar} \sum_{\substack{1,2 \\ l=\pm 1}} (\varrho_1 - \varrho_2)\,|A^l_{12}|^2\, \delta(\omega_l + \omega_2 - \omega_{12}), \quad (38.25)$$

where (see (38.15))

$$A^l_{12} = (A^{-l}_{21})^* = \sum_k \left\{ \frac{x_{a1k} V^l_{k2}}{\hbar} \zeta(\omega_l - \omega_{k2}) - \frac{V^l_{1k} x_{ak2}}{\hbar} \zeta(\omega_l - \omega_{1k}) \right\}, \quad (38.26)$$

and V^l_{mn} is a matrix element of the interaction energy with an applied field of frequency ω_1.

It follows from (38.25) that $\chi''(\omega_2)$ is connected with the fourth-rank cross-susceptibility tensor $\chi_{abcd}(\omega_s, \omega_l, \omega_r)$ (see section 22) in the special case when conditions (38.21) and (38.24) are satisfied. Expression (38.25) refers specifically to two-quantum processes. When $l = -1$ the susceptibility at a frequency ω_2, as can easily be seen from (38.25), will be negative if $\omega_2 < \omega_1$ and the molecular system is not excited initially ($\omega_{12} < 0$ and $\varrho_2 < \varrho_1$). If, however, $\omega_2 > \omega_1$, then $\chi''_{aa}(\omega_2) > 0$ ($\omega_{12} > 0$, $\varrho_1 < \varrho_2$). In the case of an excited system, on the other hand, the susceptibility at the lower frequency ($\omega_2 < \omega_1$) is positive and at the higher one ($\omega_2 > \omega_1$) is negative. It is clear that the first case is the Stokes component of the stimulated Raman emission and the second case (excited systems) is the anti-Stokes component. The sign of the susceptibility for two-photon ($l = 1$) emission (absorption) does not depend on the relation between ω_1 and ω_2; in the case of an excited molecular system ($\omega_{12} > 0$ and $\varrho_2 < \varrho_1$) the susceptibility is negative (stimulated two-photon emission). Two-photon absorption occurs in the case of a non-excited molecular system. It is easy to check that the conservation laws (38.1) and (38.2) are satisfied in all the cases listed.

The intensity of two-quantum processes, as can be seen from (38.25), is essentially determined by the matrix elements x_{a1k} and x_{ak2} and is not determined by the matrix element of the transition between the initial and final states. For molecular systems with spherical symmetry this means in the case of electric dipole transitions that only levels having the same parity (levels between which there can be no transition) take part in the two-quantum transition. The intermediate levels (k) are levels which have a parity different from that of the initial and final states (the so-called alternative forbiddenness; see Placzek, 1935).

† We recall that in accordance with the definitions of section 14

$$\langle x_a(t) \rangle = [\chi_{aa}(\omega_2) f_a^{(2)} e^{-i\omega_2 t} + \chi^*_{aa}(\omega_2) f_a^{(-2)} e^{i\omega_2 t}].$$

38.4. The Anharmonic Oscillator

We should stress once again the very general nature of the above conclusions drawn from (38.25). We have not taken into consideration any of the actual properties of the system under discussion or of the forces acting on it. At the same time it should be stressed that these conclusions must be valid both for quantum systems and for purely classical systems. As an example let us take a classical anharmonic oscillator with a frequency ω_0 on which an external force of a frequency $\omega_1 > \omega_0$ is acting. Then negative absorption should appear at the frequency $\omega_2 = \omega_1 - \omega_0$. (In other words, instability appears in the system with respect to a disturbance at the frequency ω_2.) The equation of motion of a damped anharmonic oscillator acted upon by forces at frequencies ω_1 and ω_2 is of the form

$$\ddot{x} + \omega_0^2 x + \alpha\dot{x} - \lambda x^2 = f_1 \cos \omega_1 t + f_2 \cos(\omega_2 t + \theta). \qquad (38.27)$$

We shall try to find the solution of this equation in the form

$$x = \sum_{m,n} x^{(m,n)} \, e^{i(m\omega_1 t + n\omega_2 t)}.$$

We substitute this expression in (38.27), equate the coefficients of identical exponential terms and assume that the forces f_1 and f_2 are small (we keep terms of the second order in f_1 and of the first order in f_2.) As a result we obtain

$$[-(\omega_1 - \omega_2)^2 + \omega_0^2 + i(\omega_1 - \omega_2)\alpha]\, x^{(1,-1)} - 2\lambda x^{(0,-1)} x^{(1,0)} = 0$$

(we discard terms of the type $x^{(1,1)}$ as being non-resonance terms, and take $\omega_1 - \omega_2 \approx \omega_0$). To a first approximation

$$x^{(-1,0)} = x^{(1,0)} = \frac{f_1}{2(\omega_0^2 - \omega_1^2)} \qquad (38.28)$$

and

$$x^{(1,-1)} = \frac{\lambda f_1 x^{(0,-1)}}{[\omega_0^2 - (\omega_1 - \omega_2)^2 + i(\omega_1 - \omega_2)\alpha]\,(\omega_0^2 - \omega_1^2)}. \qquad (38.29)$$

Equating the terms in $e^{-i\omega_2 t}$ we find

$$[-\omega_2^2 + \omega_0^2]\, x^{(0,-1)} - 2\lambda x^{(-1,0)} x^{(1,-1)} = \tfrac{1}{2} f_2 \, e^{-i\theta}.$$

Substituting (38.28) and (38.29) in this and retaining the small terms of the first order in f_2 we obtain

$$x^{(0,-1)} = \frac{f_2 \, e^{-i\theta}}{2(\omega_0^2 - \omega_2^2)} + \frac{\lambda^2 f_1^2 f_2 \, e^{-i\theta}}{2(\omega_0^2 - \omega_1^2)^2 (\omega_0^2 - \omega_2^2)^2 [\omega_0^2 - (\omega_1 - \omega_2)^2 + i(\omega_1 - \omega_2)\alpha]}.$$

From this, by using the definition of the susceptibility (14.9) and (14.10), we find (putting $\omega_0 \approx \omega_1 - \omega_2$)

$$\chi(\omega_2) = \frac{1}{\omega_0^2 - \omega_2^2} + \frac{\lambda^2 f_1^2 \left[\omega_0 - (\omega_1 - \omega_2) - \dfrac{i\alpha}{2}\right]}{2(\omega_0^2 - \omega_1^2)^2 (\omega_0^2 - \omega_2^2)^2 \omega_0 \left\{[\omega_0 - (\omega_1 - \omega_2)]^2 + \dfrac{\alpha^2}{4}\right\}}. \quad (38.30)$$

As was found from the general theory, the imaginary part of the susceptibility is negative†.

38.5. The Fluctuation–Dissipation Theorem

By using (38.25) we can derive a number of general relations. In particular we can find the analogue of the fluctuation–dissipation theorem. We shall relate the susceptibility (38.25) in the presence of a force of frequency ω_1 to the equilibrium fluctuations in the system in the presence of this force. The steady-state fluctuation spectrum of the quantities $\hat{x}_a(t)$ and $\hat{x}_b(t)$ is given by (17.3)

$$\tfrac{1}{2}\langle \hat{x}_{a\omega}\hat{x}_{b\omega'} + \hat{x}_{b\omega'}\hat{x}_{a\omega}\rangle = (x_a x_b)_{-\omega}\, \delta(\omega + \omega'). \quad (38.31)$$

In the presence of an external force the mean value of the left-hand side will contain terms which do not include $\delta(\omega + \omega')$. These terms correspond to the non-stationary part of the fluctuations of the quantities x_a and x_b; they disappear when time-averaged for $t \gg 2\pi/\omega$. In future we shall neglect the fluctuations which do not disappear in the absence of an external force. A simple analysis shows that in this case the first-approximation corrections $x_a^{(1)}$ and $x_b^{(1)}$ (to the energy of interaction with the applied field $\hat{V}^l$) must be substituted in (38.31). As in (38.13) we find

$$x_{anm}^{(1)}(t) = \sum_l e^{i(\omega_{nm} - \omega_l)t} A_{nm}^l, \quad (38.32)$$

where

$$A_{nm}^l = \sum_k \left\{ \frac{x_{ank} V_{km}^l}{\hbar}\, \zeta(\omega_l - \omega_{km}) - \frac{V_{nk}^l x_{akm}}{\hbar}\, \zeta(\omega_l - \omega_{nk}) \right\}. \quad (38.33)$$

† For positive α, which corresponds to an unexcited system. Negative α would mean the presence of negative absorption at ω_0 and in this case, in accordance with the general theory, $\chi''(\omega_2) > 0$.

From this we can find the matrix elements

$$(x_{a\omega}^{(1)})_{nm} = \frac{1}{2\pi}\int_{-\infty}^{\infty} x_{anm}^{(1)}(t)\, e^{i\omega t}\, dt = \sum_l A_{nm}^l \delta(\omega_{nm} - \omega_l + \omega).$$

The left-hand side of (38.31) becomes

$$\sum_{m,n} \tfrac{1}{2} \varrho_n[(x_{a\omega}^{(1)})_{nm}(x_{b\omega'}^{(1)})_{mn} + (x_{b\omega'}^{(1)})_{nm}(x_{a\omega}^{(1)})_{mn}]$$
$$= \tfrac{1}{2}\sum_{nml} (\varrho_n + \varrho_m)\, A_{nm}^l B_{mn}^{-l}\, \delta(\omega_{nm} - \omega_l + \omega)\, \delta(\omega + \omega')$$

(here we have discarded the terms which do not contain $\delta(\omega + \omega')$). The quantity B_{mn}^l differs from A_{mn}^l by the substitution of x_a by x_b. In particular for the spectral density of the fluctuations of x_a we find

$$(x_a^{(1)2})_\omega = \tfrac{1}{2}\sum_{nml} (\varrho_n + \varrho_m)\, A_{nm}^l A_{mn}^{-l}\, \delta(\omega_{nm} + \omega_l + \omega). \tag{38.34}$$

We are now in a position to derive the fluctuation–dissipation theorem which relates the stationary part of the fluctuations in the presence of an external field of frequency ω_l to the susceptibility (38.25) in the presence of the same force. Here we must assume that the unperturbed system was in thermal equilibrium. Replacing the suffixes n and m in (38.34) by 2 and 1 and remembering that in thermal equilibrium

$$\frac{\varrho_1 - \varrho_2}{\varrho_1 + \varrho_2} = \frac{1 - e^{\frac{\hbar\omega_{12}}{kT}}}{1 + e^{\frac{\hbar\omega_{12}}{kT}}}$$

as in (17.12), we obtain

$$(x_a^{(1)2})_{\omega_2} = \frac{\hbar}{2\pi}\,\frac{1 + e^{-\hbar(\omega_l+\omega_2)/kT}}{1 - e^{-\hbar(\omega_l+\omega_2)/kT}}\,\chi_{aa}''(\omega_2), \tag{38.35}$$

where Raman emission corresponds to $l = -1$; $\omega_l = -\omega_1$ and two-photon absorption to $l = 1$; $\omega_l = \omega_1$. Relation (38.35) can easily be generalized to the case when the tensor χ characterizing the properties of the medium in the presence of external forces has the off-diagonal components

$$(x_a^{(1)}\, x_b^{(1)})_{\omega_2} = \frac{i\hbar}{4\pi}\,\frac{1 + e^{-\hbar(\omega_l+\omega_2)/kT}}{1 - e^{-\hbar(\omega_l+\omega_2)/kT}}\,[\chi_{ab}^*(\omega_2) - \chi_{ba}(\omega_2)]. \tag{38.36}$$

It can be seen from (38.35) that, when $\omega_1 > \omega_2$ and $l = -1$, the positive value of $(x_a^{(1)2})_{\omega_2}$ means that $\chi''(\omega_2)$ is negative, as should be the case in a state of thermal equilibrium.

We should point out another important relation that follows from (38.25). For identical amplitudes f_1 and f_2 at the frequencies ω_1 and ω_2 ($\omega_1 - \omega_2 = \omega_0$)

$$\chi''(\omega_2) = -\chi''(\omega_1), \tag{38.37}$$

where ω_0 is the resonance frequency of the unperturbed system.

From this follows another relation

$$\frac{Q_1}{\omega_1} + \frac{Q_2}{\omega_2} = 0, \tag{38.38}$$

where $Q_1 = (\omega_1/2)\,\chi''(\omega_1)\,|f^{(1)}|^2$ and $Q_2 = (\omega_2/2)\,\chi''(\omega_2)\,|f^{(1)}|^2$ are the mean values of the power absorbed at the frequencies ω_1 and ω_2 in the presence of fields of these frequencies, if there are no other causes of absorption.

38.6. The Connection between the Intensity of Two-quantum Emission and the Imaginary Part of the Susceptibility

Let us look at the connection between fluctuations (38.34) and the intensity of two-quantum emission (38.17). In the dipole approximation, which we consider for simplicity, the quantity $\hat{B}_\nu$ in (38.4) can be written in the form

$$\hat{B}_\nu = \frac{1}{c}(\dot{d} \cdot A_\nu) = \frac{1}{c}\,\dot{d}_a A_\nu, \tag{38.39}$$

where $\dot{d}_a$ is the projection of the derivative of the dipole moment of the system onto the direction of A_ν. (Here A_ν is evaluated at the centre of gravity of the molecule.)

The emission intensity into the νth mode, using (38.17), can be written in the form

$$I_\nu = \pi \sum \delta(\omega_{nk} + \omega_\nu + \omega_l)\,|b^l_{\nu nk}|^2\,\{\tfrac{1}{2}(\varrho_n + \varrho_k) + (\varrho_k - \varrho_n)(\bar{n}_\nu + \tfrac{1}{2})\}.$$

Then, having taken the quantity d_a as x_a in (38.34) and (38.25), and using (38.39) and (38.26) we obtain

$$I_\nu = \frac{A_\nu^2\omega_\nu^2}{c^2}\left\{\pi(d_a^{(1)2})_{\omega_\nu} - \left(\bar{n}_\nu + \frac{1}{2}\right)\hbar\chi''(\omega_\nu)\right\}. \tag{38.40}$$

Just as before (section 30), we find that the spontaneous emission intensity ($\bar{n}_\nu = 0$) consists of two parts: one part is connected with the fluctuations of the dipole moment, the other part with the absorption (or stimulated emission) due to the fluctuations of the vacuum field. Then by using (38.35) we obtain (in the case of spontaneous Raman emission)

$$I_\nu^s = \frac{A_\nu^2\omega_\nu^2}{c^2}\,\hbar\chi''(\omega_\nu)\,\frac{1}{e^{-\hbar(\omega_1-\omega_\nu)/kT} - 1}. \tag{38.41}$$

A similar relation can be written for two-photon emission. Therefore $\chi''(\omega_2)$ provides information concerning not only absorption but also spontaneous emission in a state of thermal equilibrium.

39. The Propagation of Parametrically Coupled Electromagnetic Waves

39.1. The Wave Equation in a Non-linear Medium. Phase Matching

In this section we shall discuss the interaction of electromagnetic waves in a non-linear medium, defined by the relation between the polarization and the electric field (see (22.25)). Let us examine the non-linearity due to the second term in (22.25)†. The Maxwell equations describing the interaction of electromagnetic waves can be written in the form

$$\text{curl}\, \boldsymbol{E} = -\frac{1}{c}\frac{\partial \boldsymbol{H}}{\partial t}, \tag{39.1}$$

$$\text{curl}\, \boldsymbol{H} = \frac{1}{c}\frac{\partial \boldsymbol{D}}{\partial t} = \frac{1}{c}\frac{\partial \varepsilon \cdot \boldsymbol{E}}{\partial t} + \frac{4\pi}{c}\frac{\partial \boldsymbol{P}^{NL}}{\partial t}, \tag{39.2}$$

where $\boldsymbol{P}^{NL}$ denotes the non-linear part of the polarization; the linear part of the polarization is allowed for in the definition of the dielectric constant tensor ε (the dot in (39.2) denotes the inner product of ε and $\boldsymbol{E}$). In (39.1) and (39.2) we have put the magnetic permeability equal to unity. After eliminating the magnetic field from (39.1) and (39.2) we obtain the wave equation in the form

$$\text{curl curl}\, \boldsymbol{E} + \frac{1}{c^2}\frac{\partial^2(\varepsilon \cdot \boldsymbol{E})}{\partial t^2} = -\frac{4\pi}{c^2}\frac{\partial^2 \boldsymbol{P}^{NL}}{\partial t^2}. \tag{39.3}$$

Let us first examine the interaction between stationary waves that are harmonically time-dependent. Two such waves with frequencies ω_1 and ω_2 propagated along the z-direction are of the form

$$\begin{aligned} \boldsymbol{E}_j &= \boldsymbol{E}^{(j)}(z)\, e^{-i\omega_j t} + \boldsymbol{E}^{(-j)}(z)\, e^{i\omega_j t} = \text{Re}\, A_j(z)\, \boldsymbol{e}_j\, e^{i(k_j z - \omega_j t)} \\ &= \boldsymbol{e}_j m_j(z) \cos(k_j z - \omega_j t + \varphi_j(z)), \end{aligned} \tag{39.4}$$

where $\boldsymbol{e}_j$ is a unit vector in the direction of $\boldsymbol{E}_j$, and $m_j(z) = |A_j(z)|$.

These waves produce polarization at a frequency

$$\omega_3 = \omega_1 + \omega_2, \tag{39.5}$$

† For the sake of simplicity we shall draw no distinction between the macroscopic and actual fields.

which in its turn generates a wave with a frequency ω_3. The electric field of this wave can do work on the polarization at ω_3. The sign of this work depends on the phase difference between the field and the polarization at ω_3. If the work is negative the energy in the wave $\boldsymbol{E}_3$ rises at the expense of the other fields. Therefore the possibility of wave generation exists in principle. This generation depends essentially on the phase relations between the different fields, and, since all the fields are related by the wave equation (39.3), the possibility of such a process depends finally on the boundary conditions which determine the phases. It is also obvious that the work done by the field on the polarization will be large enough only when there is phase matching, i.e. in the case when the wave vectors of the field and the polarization coincide. The dependence of the polarization at the frequency ω_3 on the coordinates is determined by the wave vector $\boldsymbol{k}_1 + \boldsymbol{k}_2$ (and also by $\boldsymbol{k}_1 - \boldsymbol{k}_2$). The dependence of the field on the coordinates is given by the wave vector $\boldsymbol{k}_3(\omega_3)$. Because the dispersion of the medium is non-linear, we have generally

$$\boldsymbol{k}_3(\omega_3 = \omega_1 + \omega_2) \neq \boldsymbol{k}_1(\omega_1) + \boldsymbol{k}_2(\omega_2).$$

In certain cases we can by a special selection of the parameters of the system achieve phase matching for a definite group of waves. For example, in experiments (Giordmaine, 1962; Maker, Terhune, Nisenoff and Savage, 1962; Franken and Ward, 1963) on harmonic generation in the optical range phase matching was achieved for ordinary and extraordinary waves in an anisotropic crystal. These experiments made use of the fact that in certain anisotropic crystals a direction of propagation can be chosen in which the

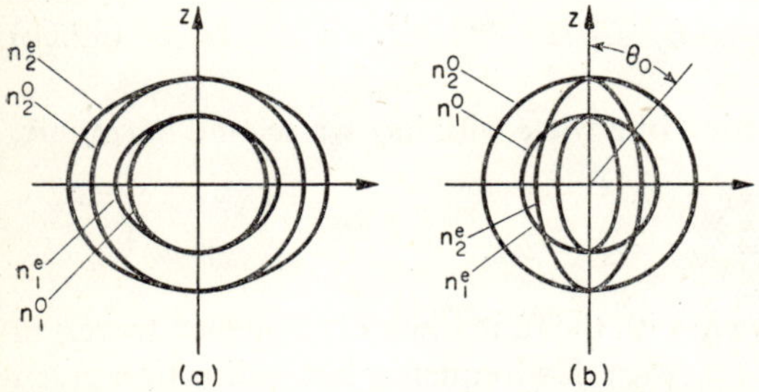

FIG. IX.2. Refractive index surfaces for quartz (*a*) and KDP (*b*). The suffixes 1 and 2 relate to radiation at the fundamental frequency and at the second harmonic respectively. The indices *o* and *e* relate to ordinary and extraordinary rays respectively. Not drawn to scale.

refractive index of an ordinary wave at one frequency is the same as the refractive index of an extraordinary wave at another frequency. Figure IX.2 shows diagrammatically the cross-section of the wave surfaces of a uniaxial crystal of KDP (potassium dihydrogen phosphate). This crystal was used in experiments on the frequency doubling of an incident light wave (see below).

A similar cross-section in a quartz crystal, in which it is impossible to achieve phase matching, is shown for comparison.

Let us continue the qualitative discussion of the interaction of electromagnetic waves in a crystal. When solving the wave equation (39.3), we cannot generally limit ourselves to a discussion of the waves with frequencies ω_1, ω_2 and ω_3 that satisfy (39.5). In fact, polarizations (and correspondingly waves) appear at frequencies $\omega_3 - \omega_1 = \omega_2$ and $\omega_3 + \omega_1 = \omega_4$ because of the presence of waves with frequencies ω_3 and ω_1; the waves at ω_4 and ω_3 generate waves at frequencies $\omega_1 = \omega_4 - \omega_3$ and $\omega_5 = \omega_4 + \omega_3$ and so on. However, we shall discuss the case when phase matching holds for one group of waves and does not hold for the rest. In other words, the case when

$$\Delta \boldsymbol{k} = \boldsymbol{k}_1 + \boldsymbol{k}_2 - \boldsymbol{k}_3$$

is much less than the corresponding differences for the other groups of waves. Of course this assumption can be made only for a medium in which dispersion is significant. In a medium without dispersion we naturally cannot limit ourselves to a single group of waves and all the combination frequencies must be considered.

After these remarks we can proceed to the solution of the wave equation.

39.2. The Interaction of Waves with Frequencies ω and 2ω

The non-linear part $\boldsymbol{P}^{NL}$ of the polarization corresponding to the parametric interaction of three waves is, in accordance with (22.25), of the form

$$P_i^{NL} = \sum_{i,j,l,s,r} \chi_{ijl}(\omega_s, \omega_r)\, E_j^{(s)} E_l^{(r)}\, \mathrm{e}^{-i(\omega_s+\omega_r)t}. \tag{39.6a}$$

Let us now examine the interaction of a wave with its own second harmonic. In this case

$$\boldsymbol{k}_1 = \boldsymbol{k}_2, \quad \omega_1 = \omega_2 = \omega, \quad \omega_3 = 2\omega.$$

Let us take the case when the wave with the fundamental frequency travels in the direction b and the wave with twice this frequency has a direction given by the suffix a. Expression (39.6a) becomes†

$$\begin{aligned} P_\alpha^{NL}(2\omega) &= \chi_{abb}(\omega, \omega)\, E_b^{(1)2}\, \mathrm{e}^{-i2\omega t}, \\ P_b^{NL}(-\omega) &= 2\chi_{bab}(-2\omega, \omega)\, E_a^{(-2)} E_b^{(1)}\, \mathrm{e}^{i\omega t}. \end{aligned} \tag{39.6b}$$

For the time being we shall limit ourselves to the isotropic case. (Anisotropy will be considered below.) Here, from equations (39.3) and (39.6b), after

† We recall that in the summation in (39.6) the terms $\chi_{ijl}\ \chi(\omega_s, \omega_r)\, E_j^{(s)} E_l^{(r)}\, \mathrm{e}^{-i(\omega_s+\omega_r)t}$ occur twice if (s, j) and (r, l) are different.

substituting the first and second harmonic waves in the form given in (39.4) we obtain for the amplitudes of the fundamental and the second harmonic

$$\frac{\partial^2 E_b^{(-1)}}{\partial z^2} + k_1^2 E_b^{(-1)} + \frac{8\pi}{c^2}\,\omega^2 \chi_{bab}(-2\omega, \omega)\, E_a^{(-2)} E_b^{(1)} = 0, \quad (39.7)$$

$$\frac{\partial^2 E_a^{(2)}}{\partial z^2} + k_2^2 E_a^{(2)} + \frac{16\pi}{c^2}\,\omega^2 \chi_{abb}(\omega, \omega)\, E_b^{(1)^2} = 0. \quad (39.8)$$

The amplitudes $A_j(z)$ (which do not depend on z unless non-linearity is present) change so slowly that we can take

$$\frac{\partial A}{\partial z} \ll kA \quad (39.9)$$

because of the small effect of the non-linearity. Now, by using (39.4) and (39.9) we can obtain the complex amplitudes A in the form

$$\frac{\partial A_1^*}{\partial z} = -i\,\frac{2\omega^2 K}{k_1}\,A_2^* A_1\, e^{i(2k_1-k_2)z}, \quad (39.10)$$

$$\frac{\partial A_2}{\partial z} = i\,\frac{4\omega^2 K}{k_2}\,A_1^2\, e^{i(2k_1-k_2)z}, \quad (39.11)$$

where (see (22.38))

$$K = \frac{2\pi}{c^2}\,\chi_{abb}(\omega, \omega) = \frac{2\pi}{c^2}\,\chi_{bab}(-2\omega, \omega). \quad (39.12)$$

Before moving on to the solution of (39.7), (39.8), (39.10) and (39.11) we should point out the obvious analogy between these equations and the equations that describe harmonically coupled point (undistributed) systems. Vitt and Gorelik (1933) have discussed systems of this kind taking as an example an elastic pendulum, i.e. a weight hanging on a spring whose upper end is fixed. Slightly simplified equations describing systems of this kind can be written in the form

$$\ddot{x} + 4\omega^2 x + \alpha y^2 = 0, \quad (39.13)$$

$$\ddot{y} + \omega^2 y + \beta x y = 0. \quad (39.14)$$

In order to understand how an interaction is achieved between systems x and y let us examine the following special cases.

1. Let x be large enough for us to be able to consider that for a certain length of time it is given by

$$x = x_0 \cos 2\omega t;$$

then for y we obtain a Mathieu type equation

$$\ddot{y} + (\omega^2 + \beta x_0 \cos 2\omega t)\, y = 0. \tag{39.15}$$

As is well known (see, e.g., Landau and Lifshitz, 1958), this equation describes parametric resonance: pumping of a quantity y with a frequency ω takes place under the action of a force of frequency 2ω. A characteristic feature of parametric resonance is that when the initial values of y and $\ddot{y}$ are strictly zero they remain zero subsequently, unlike the case of ordinary resonance in which there is a rise in the displacement with time (proportional to t) even when the initial value is zero.

2. Let y be given by

$$y = y_0 \cos \omega t;$$

then we obtain for x the equation of an oscillator acted upon by an external force with a component at a frequency 2ω:

$$\ddot{x} + 4\omega^2 x = -\frac{\alpha y_0^2}{2}(1 + \cos 2\omega t). \tag{39.16}$$

Therefore at the frequency 2ω ordinary resonance appears which leads to pumping of the oscillations of x.

It is clear that equations (39.15) and (39.16) can be used only for a short time while the displacement of y (or of x in the second case) is small enough. In these equations no allowance is made for the inverse action of the "pumped" oscillator on the "pumping" one. That such an action exists follows directly from the fact that the system as a whole is conservative: the energy of the "pumped" oscillation can rise only at the expense of the "pumping" energy.

The interaction of the waves we are discussing is essentially a space-time analogue of the parametric interaction of point systems. In future we shall therefore speak of parametric wave interaction (Khokhlov, 1961; Akhmanov and Khokhlov, 1962). Equations (39.7) and (39.8) are analogous to (39.14) and (39.13) respectively. The field $E_a^{(-2)}$ is a coefficient, alternating in the coordinate z, of the quantity $E_b^{(-1)}$, whilst the field $(E_b^{(1)})^2$ in equation (39.7) plays the part of the constraining force.

Before solving equations (39.10) and (39.11) we shall take account of the crystal anisotropy. Between the direction of the energy flux (the Poynting vector) and the direction of propagation (the z-axis in our case) there is in an anisotropic medium a finite angle α equal to the angle between $\boldsymbol{E}$ and $\boldsymbol{D}$ (see, e.g., Landau and Lifshitz, 1957). It is easy to show (Armstrong, Bloembergen, Ducuing and Pershan, 1962) that including the anisotropy in (39.10) and (39.11) leads to

$$\frac{\partial A_1^*}{\partial z} = -i\,\frac{2\omega^2 K}{k_1 \cos^2 \alpha_1}\, A_2^* A_1\, \mathrm{e}^{i(2k_1-k_2)z}, \tag{39.10'}$$

$$\frac{\partial A_2}{\partial z} = i\,\frac{4\omega^2 K}{k_2 \cos^2 \alpha_2}\, A_1^2\, \mathrm{e}^{i(2k_1-k_2)z}, \tag{39.11'}$$

where the angles α_1 and α_2 refer to the waves 1 and 2 respectively and the coefficient K takes the form

$$K = \frac{2\pi}{c^2}\, \boldsymbol{e}_2 \cdot \chi(\omega, \omega) : \boldsymbol{e}_1 \boldsymbol{e}_1,$$

where the dot and the colon respectively denote the inner products of the vector $\boldsymbol{e}_2$, the third-rank tensor χ and the second-rank tensor $\boldsymbol{e}_1\boldsymbol{e}_1$. To solve equations (39.10′) and (39.11′) we express A_j in terms of the real amplitudes m_j and phases φ_i in accordance with (39.4)

$$A_j = m_j(z)\, \mathrm{e}^{i\varphi_i(z)} \tag{39.17}$$

and write out separately the real and imaginary parts of (39.10′) and (39.11′). As a result we obtain:

$$\frac{\partial m_1}{\partial z} = (2\omega^2 K / k_1 \cos^2 \alpha_1)\, m_1 m_2 \sin\theta, \tag{39.18}$$

$$\frac{\partial m_2}{\partial z} = -(4\omega^2 K / k_2 \cos^2 \alpha_2)\, m_1^2 \sin\theta, \tag{39.19}$$

$$\frac{\partial \theta}{\partial z} = \Delta k + 4\omega^2 K[(m_2/k_1 \cos^2 \alpha_1) - (m_1^2/m_2 k_2 \cos^2 \alpha_2)] \cos\theta, \tag{39.20}$$

where

$$\theta = 2\varphi_1(z) - \varphi_2(z) + \Delta kz \quad \text{and} \quad \Delta k = 2k_1 - k_2.$$

From the first two equations follows the law of conservation of power flux in an electromagnetic field in the direction of propagation†

$$W = \frac{c^2}{8\pi\omega}\left[k_1 m_1^2 \cos^2 \alpha_1 + \frac{1}{2} k_2 m_2^2 \cos^2 \alpha_2\right], \quad \frac{dW}{dz} = 0. \tag{39.21}$$

† We recall (Landau and Lifshitz, 1957) that the Poynting vector in an anisotropic medium is of the form

$$\boldsymbol{S} = \frac{c^2}{4\pi\omega}\, \{\boldsymbol{k}\overline{E^2} - \overline{(\boldsymbol{k}\cdot\boldsymbol{E})\,\boldsymbol{E}}\},$$

where the bar denotes time-averaging. The projection of this vector on to the vector $\boldsymbol{k}$ is

$$\frac{(\boldsymbol{S}\cdot\boldsymbol{k})}{k} = \frac{c^2}{4\pi\omega k}\, \{k^2\overline{E^2} - \overline{(kE)^2}\} = \frac{c^2 k \cos^2 \alpha}{4\pi\omega}\, \overline{E^2} = \frac{c^2 k \cos^2 \alpha}{8\pi\omega}\, m^2. \tag{39.22}$$

Substituting (39.18) and (39.19) in (39.20) leads to

$$\frac{d\theta}{dz} = \Delta k + \cot\theta \frac{d}{dz} \ln(m_1^2 m_2). \tag{39.23}$$

We shall now introduce new variables:

$$u = [c^2 k_1 \cos^2 \alpha_1 / 8\pi\omega W]^{1/2} m_1,$$

$$v = [c^2 k_2 \cos^2 \alpha_2 / 16\pi\omega W]^{1/2} m_2, \tag{39.24}$$

$$\zeta = \left(\frac{2\omega^2 K}{k_1 \cos^2 \alpha_1}\right) (16\pi\omega W / c^2 k_2 \cos^2 \alpha_2)^{1/2} z.$$

Equations (39.18), (39.19) and (39.23) become

$$\frac{du}{d\zeta} = uv \sin\theta, \quad \frac{dv}{d\zeta} = -u^2 \sin\theta, \quad \frac{d\theta}{d\zeta} = \Delta s + \cot\theta \frac{d}{d\zeta} \ln(u^2 v), \tag{39.25}$$

where

$$\Delta s = \frac{\Delta k}{(2\omega^2 K / k_1 \cos^2 \alpha_1)(16\pi\omega W / c^2 k_2 \cos^2 \alpha_2)^{1/2}}. \tag{39.26}$$

We notice that the conservation of energy flux in the new variables u and v can be written in the form

$$u^2 + v^2 = 1, \tag{39.27}$$

i.e. u^2 and v^2 denote the relative energy fluxes in the fundamental and the second harmonic. Let us now move on to discuss various cases†.

A. Exact phase matching: $\Delta k = \Delta s = 0$, $2k_1 = k_2$. In this case the third equation in (39.25) can be integrated at once

$$u^2 v \cos\theta = \Gamma = \left(\frac{2k_1}{k_2}\right) (c^2 k_2 / 16\pi\omega W)^{3/2} m_1^2(0) m_2(0)$$
$$\times \cos^2 \alpha_1 \cos \alpha_2 \cos[\varphi_2(0) - 2\varphi_1(0)], \tag{39.28}$$

where $m_1(0)$, $m_2(0)$, $\varphi_1(0)$ and $\varphi_2(0)$ are the boundary values of the amplitudes and phases of the fundamental and second harmonic.

From (39.25), (39.26) and (39.28) we obtain

$$\frac{d}{d\zeta}(v^2) = \pm 2[v^2(1 - v^2)^2 - \Gamma^2]^{1/2}, \tag{39.29}$$

† Here and later we follow Armstrong, Bloembergen, Ducing and Pershan (1962) for the solution of equations (39.18)–(39.20).

where the "±" sign is determined by the sign of sin θ. The solution of (39.29) can be expressed by the elliptic integral

$$\zeta = \pm\frac{1}{2}\int_{v^2(0)}^{v^2(\zeta)} \frac{d(v^2)}{[v^2(1-v^2)^2 - \Gamma^2]^{1/2}}. \tag{39.30}$$

Since v is real and less than or equal to 1, v^2 lies between the two smallest positive roots of the equation

$$v^2(1-v^2)^2 - \Gamma^2 = 0.$$

We denote these roots by v_a and v_b ($v_a < v_b$); the quantity v oscillates between v_a and v_b with the period

$$\Pi_\zeta = \int_{v_a^2}^{v_b^2} \frac{d(v^2)}{[v^2(1-v^2)^2 - \Gamma^2]^{1/2}}. \tag{39.31}$$

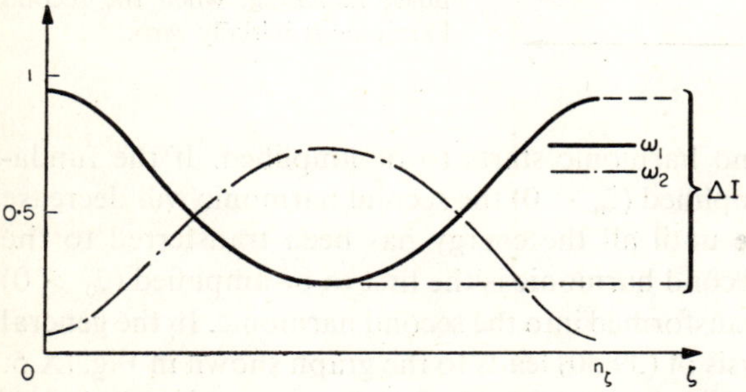

FIG. IX. 3. The change in the power of the fundamental and second harmonic for arbitrary initial conditions. The relative power is plotted as the ordinate.

A typical solution is shown in Fig. IX. 3. The initial condition $\Gamma = 0$ means that $v_a^2 = 0$, $v_b^2 = 1$ and $\Pi_\zeta \to \infty$. In this case the integration of (39.30) gives

$$v_{\Gamma=0}(\zeta) = \pm \tanh(\zeta + \zeta_0), \quad u_{\Gamma=0}(\zeta) = \pm \operatorname{sech}(\zeta + \zeta_0). \tag{39.32}$$

The integration constant is $\zeta_0 = 0$ if the initial value of the amplitude of the second harmonic is zero: $m_2(0) = 0$. Figure IX. 4 shows the solution in this case. The characteristic interaction distance in which about 75 per cent of the energy flux is transferred from the fundamental to the second harmonic is

$$l^{-1} = 2\omega^2 K k_1^{-1} m_1(0). \tag{39.33}$$

If the boundary condition is $m_1(0) = 0$, then $\zeta_0 = -\infty$. Strictly speaking, in this case $m_1 = 0$ is the solution of the basic equations. Just as with ordinary parametric excitation, if the value of the excited quantity (and its derivative) is zero no oscillations are excited. It is sufficient, however, for there to be a small initial displacement for an oscillation to be excited.

The solutions (39.32) are valid when $m_1(0) \neq 0$ and $m_2(0) \neq 0$ if

$$\theta(0) = 2\varphi_1(0) - \varphi_2(0) = \pm\pi/2.$$

In the case of exact matching this relative phase is preserved, i.e. $\theta(z)$ is also equal to $\pm\pi/2$. Depending upon the sign of the initial phase either the

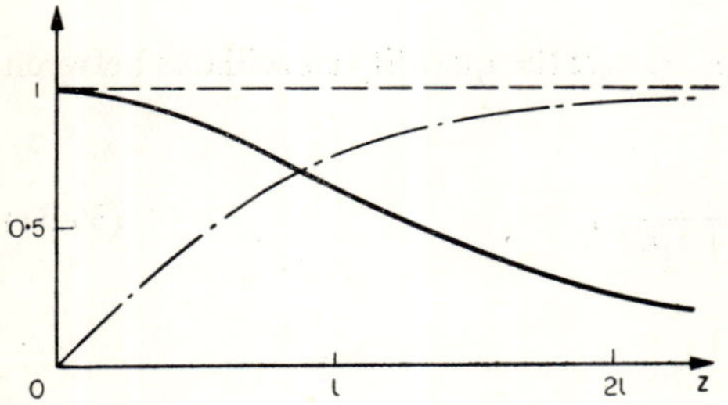

FIG. IX. 4. The rise in the normalized amplitude of the second harmonic and the decrease in amplitude of the fundamental frequency with exact phase matching, when the second harmonic is initially zero.

fundamental or the second harmonic starts to be amplified. If the fundamental is the first to be amplified ($\zeta_0 < 0$) the second harmonic will decrease to zero and then increase until all the energy has been transferred to the second harmonic. If the second harmonic is the first to be amplified ($\zeta_0 > 0$) the energy is completely transformed into the second harmonic. In the general case when $\Gamma \neq 0$ an analysis of (39.30) leads to the graph shown in Fig. IX.5. The area bounded by the curves for u^2 and v^2 is the range of variation of these quantities as functions of the boundary value Γ^2.

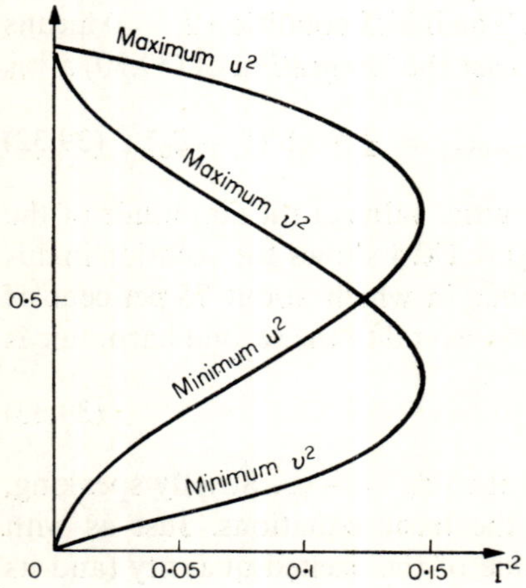

FIG. IX. 5. Range of variation of the power in the field at the fundamental frequency and the second harmonic as a function of the parameter Γ^2 (defined in the text). The lower of the curves gives the minimum of v^2 in the lower part and the maximum of v^2 in the upper part. Likewise the upper curve gives the minimum of u^2 in the lower part and the maximum of u^2 in the upper part.

B. No phase matching: $\Delta k \neq 0$, $2k_1 \neq k_2$. Here we must distinguish two sub-cases. If

$$\Delta s \approx \Delta kl \ll 1, \tag{39.34}$$

where l is defined by (39.33) the situation is approximately the same as in case *A*. If

$$\Delta s \gg 1, \tag{39.35}$$

then we must turn to equations (39.18) and (39.20). The case when m_2 remains small for all z is of practical interest. In this case m_1 may be taken as

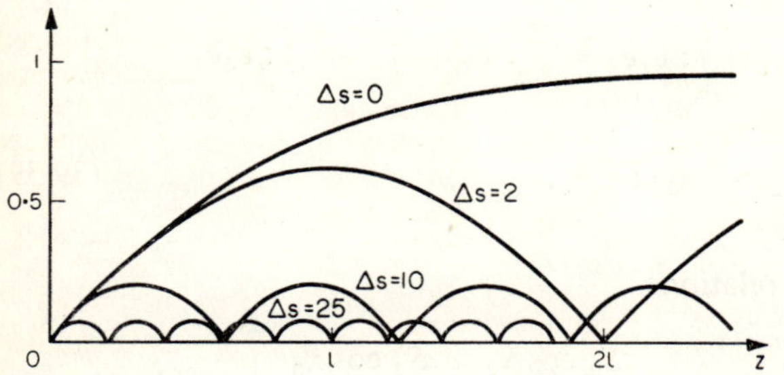

FIG. IX.6. Rise in the amplitude of the second harmonic for different degrees of phase mismatching.

approximately constant. If the boundary value of $m_2(0)$ is zero the approximate solution of (39.19) and (39.20) is

$$\theta = \Delta kz/2 - \pi/2, \tag{39.36}$$

$$m_2 = \frac{8\omega^2 K m_1^2}{k_2 \cos^2 \alpha_2 \Delta k} \sin \frac{\Delta kz}{2} \approx \frac{m_1(0)}{\Delta kl/2} \sin \frac{\Delta kz}{2}, \tag{39.37}$$

where the appropriate equality is written for the condition $\cos^2 \alpha_2 \approx 1$, $2k_1 \approx k_2$. Figure IX.6 shows the solutions for the second harmonic for different values of Δs. (The absolute value of the second harmonic amplitude is $|m_2|$ and when the sign of m_2 changes there is a corresponding change in the phase θ.)

39.3. *The Interaction of three Waves with Frequencies ω_1, ω_2 and ω_3*

We can treat in a similar way the case of the interaction of three waves with the different frequencies ω_1, ω_2 and ω_3 that satisfy (39.5). For the complex amplitudes of these waves being propagated along the z-axis we obtain the set of equations:

$$\frac{dA_1^*}{dz} = -i(\omega_1^2 K/k_1 \cos^2 \alpha_1)\, A_3^* A_2\, e^{i\Delta kz},$$

$$\frac{dA_2^*}{dz} = -\, i(\omega_2^2 K/k_2 \cos^2 \alpha_2)\, A_3^* A_1\, e^{i\Delta kz}, \tag{39.38}$$

$$\frac{dA_3}{dz} = i(\omega_3^2 K/k_3 \cos^2 \alpha_3)\, A_1 A_2\, e^{i\Delta kz},$$

where K in accordance with (22.38) is defined as

$$K = \frac{2\pi}{c^2}\, \boldsymbol{e}_3 \chi(\omega_1, \omega_2) : \boldsymbol{e}_1 \boldsymbol{e}_2 = \frac{2\pi}{c^2}\, \boldsymbol{e}_2 \chi(\omega_3, -\omega_1) : \boldsymbol{e}_3 \boldsymbol{e}_1$$

$$= \frac{2\pi}{c^2}\, \boldsymbol{e}_1 \chi(\omega_3, -\,\omega_2) : \boldsymbol{e}_3 \boldsymbol{e}_2. \tag{39.39}$$

From (39.38) follow the relations

$$\frac{S_1 \cos \alpha_1}{\omega_1} + \frac{S_3 \cos \alpha_3}{\omega_3} = \text{const}, \qquad \frac{S_2 \cos \alpha_2}{\omega_2} + \frac{S_3 \cos \alpha_3}{\omega_3} = \text{const},$$

$$\frac{S_2 \cos \alpha_2}{\omega_2} + \frac{S_3 \cos \alpha_3}{\omega_3} = \text{const}, \tag{39.40}$$

which are similar to the Manley–Rowe (1956) relations for parametric amplifiers. In accordance with (39.22) the quantities $S_i \cos \alpha_i$ denote the energy fluxes in the direction of propagation of the waves. At the same time the law of conservation of the energy flux holds:

$$S_1 \cos \alpha_1 + S_1 \cos \alpha_2 + S_3 \cos \alpha_3 = \text{const}. \tag{39.41}$$

Here we shall not carry out an analysis of the solutions of the equations (39.38) but refer the reader to the paper already quoted (Armstrong, Bloembergen, Ducuing and Pershan, 1962). We shall merely point out that, just as the case of the interaction of the fundamental and second harmonic is analogous to parametrically coupled point systems, so is the interaction of three waves ω_1, ω_2 and ω_3 completely analogous to the parametric interaction of oscillations at the three frequencies; this kind of interaction is dealt with in the theory of parametric amplifiers (see, e.g., Bloom and Chang, 1957). Armstrong, Bloembergen, Ducuing and Pershan (1962) also discuss the parametric interaction of four waves and the generation of the third harmonic. The parametric interaction of four waves is connected with the non-linearity given by the fourth-rank cross-susceptibility tensor $\chi_{abcd}(\omega_s, \omega_l, \omega_r)$ (see (22.21)). A characteristic feature of parametric interaction is the fact that it is essentially dependent on the phase relations between the interacting waves, and in particular, as is clear from the above discussion of the interaction between the fundamental and the second harmonic, on the boundary values of the phases.

39.4. Non-stationary Processes

Up to now we have been discussing processes that are stationary in time. We shall now move on to discuss the non-stationary processes of electromagnetic wave interaction (Karpman, 1963). We turn once more to the basic wave equation (39.3) to do this.

We shall look for solutions of this equation in the form of the sum of the eigen waves $\boldsymbol{E}(\boldsymbol{k})$ and $\boldsymbol{H}(\boldsymbol{k})$ with time-dependent coefficients a_k:

$$\boldsymbol{E} = \sum_k a_k(t)\, \boldsymbol{E}(\boldsymbol{k})\, \mathrm{e}^{-i\omega_k t}, \quad \boldsymbol{H} = \sum_k a_k(t)\, \boldsymbol{H}(\boldsymbol{k})\, \mathrm{e}^{-i\omega_k t}. \tag{39.42}$$

The normalization and the form of the functions $\boldsymbol{E}(\boldsymbol{k})$ and $\boldsymbol{H}(\boldsymbol{k})$ will be defined more exactly later. The suffix $\boldsymbol{k}$ is not necessarily the wave vector and we can consider $\boldsymbol{E}(\boldsymbol{k})$, $\boldsymbol{H}(\boldsymbol{k})$ to be a complete set of eigenfunctions of an equivalent resonator filled with a dielectric $\varepsilon_{\alpha\beta}$. Therefore we shall allow for the linear part of the polarization by an appropriate selection of the eigenfunctions $\boldsymbol{E}(\boldsymbol{k})$ and $\boldsymbol{H}(\boldsymbol{k})$. If we do not allow for the interaction with the non-linear part of the polarization, then the coefficients a_k are not time-dependent, i.e. we assume that $\boldsymbol{E}(\boldsymbol{k})\, \mathrm{e}^{-i\omega_k t}$ and $\boldsymbol{H}(\boldsymbol{k})\, \mathrm{e}^{-i\omega_k t}$ are the solutions of (39.3) when the right-hand side is zero. Using the Maxwell equations (with $\boldsymbol{P}_{NL} = 0$) we find

$$\operatorname{curl} \boldsymbol{H}(\boldsymbol{k}) = -\frac{i\omega_k}{c}\, \varepsilon \cdot \boldsymbol{E}(\boldsymbol{k}), \quad \operatorname{curl} \boldsymbol{E}(\boldsymbol{k}) = \frac{i\omega_k}{c}\, \boldsymbol{H}(\boldsymbol{k}). \tag{39.43}$$

In the case when the expansion of (39.42) is in travelling plane waves it follows from (39.43) that

$$\begin{aligned} &\frac{\omega_k}{c}\, \boldsymbol{H}(\boldsymbol{k}) = [\boldsymbol{k} \wedge \boldsymbol{E}(\boldsymbol{k})], \\ &\frac{\omega_k}{c}\, \varepsilon_{\alpha\beta} E_\beta(\boldsymbol{k}) = -[\boldsymbol{k} \wedge \boldsymbol{H}(\boldsymbol{k})]_\alpha, \\ &H_\alpha(\boldsymbol{k})\, H^*_\alpha(\boldsymbol{k}) = \varepsilon_{\alpha\beta} E^*_\alpha(\boldsymbol{k})\, E_\beta(\boldsymbol{k}). \end{aligned} \tag{39.44}$$

From these equations we obtain

$$\left(k_\alpha k_\beta - k^2 \delta_{\alpha\beta} + \frac{\omega_k^2}{c^2}\, \varepsilon_{\alpha\beta}\right) E_\beta(\boldsymbol{k}) = 0. \tag{39.45}$$

(Here, and in future, the convention for summation over repeated suffixes is assumed.) The condition that the determinant of the system of equations (39.45) is zero

$$\left| k_\alpha k_\beta - k^2 \delta_{\alpha\beta} + \frac{\omega_k^2}{c^2} \varepsilon_{\alpha\beta} \right| = 0 \tag{39.46}$$

gives the Fresnel equation defining the dispersion law, i.e. the relationship between the frequency ω_k and the wave vector $\boldsymbol{k}$. In order to make further use of the wave equation (39.3), in which we shall substitute (39.42) taking into account (39.43), we must define more precisely the law governing the differentiation of quantities such as $\hat{\varepsilon}_{\alpha\beta} f(t)$.

The point here is that, if dispersion is taken into account, the dielectric constant tensor must be looked upon as an operator acting on a function of time. If $f(t) = a_0(t)\, \mathrm{e}^{-i\omega t}$, where $a_0(t)$ is a slowly varying time function compared with $\mathrm{e}^{i\omega t}$, then it is easy to show (see, e.g., Landau and Lifshitz, 1957, p. 323†) that

$$\frac{\partial \hat{\varepsilon}_{\alpha\beta} f(t)}{\partial t} = -\, i\omega \varepsilon_{\alpha\beta}(\omega)\, a_0(t)\, \mathrm{e}^{-i\omega t} + \frac{d(\omega \varepsilon_{\alpha\beta})}{d\omega} \frac{\partial a_0}{\partial t}\, \mathrm{e}^{-i\omega t}$$

and

$$\frac{\partial^2 \hat{\varepsilon}_{\alpha\beta} f(t)}{\partial t^2} = -\, \omega^2 \varepsilon_{\alpha\beta}(\omega)\, a_0(t)\, \mathrm{e}^{-i\omega_0 t} - i\omega \varepsilon_{\alpha\beta} \frac{\partial a_0}{\partial t}\, \mathrm{e}^{-i\omega t} - i\omega \frac{d(\omega \varepsilon_{\alpha\beta})}{d\omega} \frac{\partial a_0}{\partial t}\, \mathrm{e}^{-i\omega t}. \tag{39.47}$$

Using (39.43) and (39.47), after substituting (39.42) in (39.3) and integrating over all space we obtain

$$i\omega_k \left(\varepsilon_{\alpha\beta} + \frac{\partial(\omega_k \varepsilon_{\alpha\beta})}{\partial \omega_k} \right) \int E_\beta(\boldsymbol{k})\, E_\alpha^*(\boldsymbol{k})\, dV \frac{da_k}{dt} = 4\pi \int \frac{d^2 P_\alpha^{NL}}{dt^2} E_\alpha^*(\boldsymbol{k})\, dV\, \mathrm{e}^{i\omega_k t}. \tag{39.48}$$

Here we have assumed orthogonality of the functions $E(k)$. (The integral $\int E_\beta(\boldsymbol{k})\, E_\alpha^*(\boldsymbol{k}_1)\, dV$ is equal to zero if $\boldsymbol{k} \neq \boldsymbol{k}_1$.) Equation (39.48) gives the required relation between the coefficients a_k and the time. The right-hand side of this equation depends on a_k through P_α^{NL} (which depends on the field). In order to determine completely the values of the coefficients a_k we must state more precisely the normalization condition of the functions $E_\alpha(\boldsymbol{k})$. The time-averaged energy in a transparent medium with dispersion can be defined

† P. 255 in the English translation (1960)—Ed.

in the form (Landau and Lifshitz, 1957)

$$W = \frac{1}{8\pi}\sum_k \int dV\left[\frac{\partial(\omega_k \varepsilon_{\alpha\beta})}{\partial\omega_k}\overline{E_\alpha(\omega_k)\,E_\beta(\omega_k)} + \overline{H_\alpha(\omega_k)\,H_\alpha(\omega_k)}\right], \tag{39.49}$$

where the bar denotes time-averaging and $E_\alpha(\omega_k)$ and $H_\alpha(\omega_k)$ denote those parts of the field (39.42) which vary at a frequency ω_k:

$$E_\alpha(\omega_k) = a_k E_\alpha(\boldsymbol{k})\,e^{-i\omega_k t} + a_k^* E_\alpha^*(\boldsymbol{k})\,e^{+i\omega_k t}. \tag{39.50}$$

$H_\alpha(\omega_k)$ can be defined similarly. Substituting (39.50) in (39.49) we obtain

$$W = \frac{1}{4\pi}\sum_k a_k a_k^* \int dV\left[\frac{\partial(\omega_k \varepsilon_{\alpha\beta})}{\partial\omega_k} E_\alpha^*(\boldsymbol{k})\,E_\beta(\boldsymbol{k}) + H_\alpha(\boldsymbol{k})\,H_\alpha^*(\boldsymbol{k})\right]. \tag{39.51}$$

We shall further consider that the expansion is in travelling plane waves. In this case by using the last equation of (39.44) we find

$$W = \frac{1}{4\pi}\sum_k a_k a_k^* \int dV\left[\frac{\partial(\omega_k \varepsilon_{\alpha\beta})}{\partial\omega_k} + \varepsilon_{\alpha\beta}\right] E_\alpha^*(\boldsymbol{k})\,E_\beta(\boldsymbol{k}). \tag{39.52}$$

Here, just as in (39.51), the summation is carried out over the values which correspond to a positive sign for the frequency. In this case $\omega_k = -\omega_{-k}$, $a_k = a_{-k}^*$.

We shall now carry out normalization so that $a_k a_k^*$ denotes the mean number of photons with a given $\boldsymbol{k}$†:

$$\frac{1}{8\pi}\int dV\left[\frac{\partial(\omega_k \varepsilon_{\alpha\beta})}{\partial\omega_k} + \varepsilon_{\alpha\beta}\right] E_\alpha^*(\boldsymbol{k})\,E_\beta(\boldsymbol{k}) = \hbar\omega_k. \tag{39.53}$$

Here the total energy is

$$W = 2\sum_k |a_k|^2\,\hbar\omega_k = \sum_k a_k a_k^*\hbar\omega_k, \tag{39.54}$$

where the second summation is over all values of $\boldsymbol{k}$. Substituting (39.53) in (39.48) we obtain equations for a_k in the form

$$i\omega_k^2\,\frac{da_k}{dt} = \frac{1}{2\hbar}\int \frac{d^2 P_\alpha^{NL}}{dt^2}\,E_\alpha^*(\boldsymbol{k})\,dV\,e^{i\omega_k t}. \tag{39.55}$$

† The whole treatment here, of course, is classical. The quantities a_k and a_k^* correspond to the annihilation and creation operators of a photon with a given $\boldsymbol{k}$.

We can now express the right-hand side of (39.55) in terms of the amplitudes a_k. Using (39.6a) we obtain

$$P_\alpha^{NL} = \sum_{k',k''} \chi_{\alpha\beta\gamma}(\omega_{k'}, \omega_{k''})\, E_\beta(\boldsymbol{k}')\, E_\gamma(\boldsymbol{k}'')\, a_{k'} a_{k''}\, \mathrm{e}^{-i(\omega_{k'}+\omega_{k''})t}.$$

We further remember that for the case of travelling plane waves

$$E_\alpha(\boldsymbol{k}) \sim \mathrm{e}^{i(\boldsymbol{k}\cdot\boldsymbol{r})}$$

the integral in the right-hand side of (39.55) is proportional to

$$\int E_\beta(\boldsymbol{k}')\, E_\gamma(\boldsymbol{k}'')\, E_\alpha^*(\boldsymbol{k})\, dV = V E_\alpha^*(\boldsymbol{k})\, E_\beta(\boldsymbol{k}')\, E_\gamma(\boldsymbol{k}'')\, \delta_{\boldsymbol{k}-\boldsymbol{k}'-\boldsymbol{k}'';0}\,. \tag{39.56}$$

We use this condition and obtain

$$i\frac{da_k}{dt} = -\frac{V}{2\hbar\omega_k^2} \sum_{\boldsymbol{k}'+\boldsymbol{k}''=\boldsymbol{k}} (\omega_k + \omega_{k''})^2\, \chi_{\alpha\beta\gamma}(\omega_{k'}, \omega_{k''})$$

$$\times\, E_\alpha^*(\boldsymbol{k})\, E_\beta(\boldsymbol{k}')\, E_\gamma(\boldsymbol{k}'')\, a_{k'} a_{k''}\, \mathrm{e}^{i(\omega_k-\omega_{k'}-\omega_{k''})t}, \tag{39.57}$$

where V is the normalizing volume. Equations (39.57) are analogous to those discussed above for the field amplitudes as functions of z. Therefore everything that has been said about the parametric interaction of stationary waves can be repeated here. The difference is that the phase matching condition

$$\boldsymbol{k}' + \boldsymbol{k}'' = \boldsymbol{k} \tag{39.58}$$

which replaces condition (39.5), is obligatory here. On the other hand, satisfaction of (39.58) does not mean, on account of the non-linearity of the dispersion, that the frequency condition

$$\omega_k = \omega_{k'} + \omega_{k''}$$

is satisfied, except in special cases. In such cases the infinite set of equations (39.57) is broken up and a set of equations results that relate waves for which the difference

$$\Delta\omega = \omega_k - \omega_{k'} - \omega_{k''}$$

is much less than the corresponding differences for the other waves.

40. Stimulated Raman Emission (Fain and Yashchin, 1964)

40.1. The Raman Effect

In this section we shall discuss an interaction between electromagnetic waves which differs qualitatively from the parametric interaction discussed in the preceding section. As was explained in section 38, if an electromagnetic wave with a frequency $\omega_1 > \omega_0$ is propagated through a medium in which there is resonance absorption at a frequency ω_0, then at the frequency

$$\omega_2 = \omega_1 - \omega_0 \tag{40.1}$$

there is negative absorption. (It is assumed here that the medium was originally in equilibrium.) This means that a medium of this kind is unstable with respect to the appearance of an electromagnetic wave with a frequency ω_2. If a wave of this frequency exists it will be amplified, which in its turn will lead to positive absorption at the frequency ω_1. Therefore a wave with a frequency ω_1 will decrease in amplitude, transforming itself into a wave with a frequency ω_2. Since absorption (at the frequency ω_0) is now being taken into consideration in the medium this kind of transformation will occur with a loss of energy.

Let us study the interaction of travelling plane waves. For the sake of simplicity we shall not take any anisotropy of the medium into consideration, and shall make the z-axis the direction of propagation. Equation (39.3) can be written in the form

$$\text{curl curl } \boldsymbol{E} - \frac{\varepsilon\omega^2}{c^2}\boldsymbol{E} = \frac{4\pi\omega^2}{c^2}\boldsymbol{P}^{\text{NL}},$$

where, as before, we have included the linear part of the polarization in the dielectric constant of the medium. Let the wave of frequency ω_2 be polarized in the x-direction. Then for this wave we obtain

$$\frac{\partial^2 E_x^{(2)}}{\partial z^2} + \frac{\varepsilon\omega_2^2}{c^2}E_x^{(2)} + \frac{4\pi}{c^2}\omega_2^2 P_x^{NL} = 0. \tag{40.2}$$

We shall look for a solution of this equation in the form

$$E_x^{(2)} = \text{Re } A_2(z)\, e^{i(k_2 z - \omega_2 t)}. \tag{40.3}$$

According to section 38 the non-linear part of the susceptibility corresponding to stimulated Raman emission can be written as

$$P_x^{NL} = \text{Re}\,\chi(\omega_2, E_1(z))\, A_2(z)\, e^{i(k_2 z - \omega_2 t)}. \tag{40.4}$$

Further, in accordance with (38.25), if $E_1(z)$ is in the form of a travelling wave, then $\chi''(\omega_2, E_1(z))$ is proportional to the square of the modulus of the amplitude $|A_1(z)|^2$. We shall denote the coefficient of proportionality by $-\xi$

$$\chi''(\omega_2, E(z)) = -\xi\,|A_1|^2. \tag{40.5}$$

We shall further assume that the frequency condition (40.1) is strictly satisfied. In this case it can be shown that the real part of the susceptibility is zero. Substituting (40.3), (40.4), (40.5) in (40.2) and remembering that $A(z)$

changes slowly compared with e^{ikz} we obtain the following equation for the amplitude A_2:

$$\frac{\partial A_2}{\partial z} - \frac{2\pi\omega_2^2}{k_2 c^2}\,\xi\,|A_1|^2 A_2 = 0, \tag{40.6}$$

where $\xi > 0$ (since $\omega_1 > \omega_2$).

By using condition (38.37) we obtain a similar equation for the change in $A_1(z)$ due to the wave A_2:

$$\frac{\partial A_1}{\partial z} + \frac{2\pi\omega_1^2}{k_1 c^2}\,\xi\,|A_2|^2 A_1 = 0. \tag{40.7}$$

With the chosen relation between the frequencies, equations (40.6) and (40.7) describe the increase in the wave A_2 and the decrease of the wave A_1. As in (39.22) we introduce the energy flux densities at ω_1 and ω_2:

$$S_1 = \frac{c^2 k_1}{8\pi\omega_1}|A_1|^2, \quad S_2 = \frac{c^2 k_2}{8\pi\omega_2}|A_2|^2. \tag{40.8}$$

From (40.6) and (40.7) we obtain for these quantities

$$\frac{\partial S_2}{\partial z} - \beta\omega_2 S_1 S_2 = 0, \quad \frac{\partial S_1}{\partial z} + \beta\omega_1 S_1 S_2 = 0, \tag{40.9}$$

where

$$\beta = \frac{32\pi^2\omega_2\omega_1}{k_1 k_2 c^4}\,\xi. \tag{40.10}$$

From (40.9) conservation of

$$I = \frac{S_1}{\omega_1} + \frac{S_2}{\omega_2} = \text{const} = \frac{S_1(0)}{\omega_1} + \frac{S_2(0)}{\omega_2} \tag{40.11}$$

follows when z changes†. Eliminating S_1 from (40.9) and (40.11) we obtain for S_2

$$\frac{\partial S_2}{\partial z} - \beta(\omega_2\omega_1 I - \omega_1 S_2)\,S_2 = 0. \tag{40.12}$$

Integrating this equation we find

$$S_2(z) = \frac{S_2(0)\left[1 + \dfrac{\omega_1}{\omega_2}\dfrac{S_2(0)}{S_1(0)}\right] e^{\beta[\omega_2 S_1(0) + \omega_1 S_2(0)]\,z}}{1 + \dfrac{\omega_1}{\omega_2}\dfrac{S_2(0)}{S_1(0)}\, e^{\beta[\omega_2 S_1(0) + \omega_1 S_2(0)]\,z}}. \tag{40.13}$$

† Compare (40.11) with the general relation (38.38).

Therefore S_2 rises steadily from the initial value $S_2(0)$ to

$$S_2(\infty) = \frac{\omega_2}{\omega_1} S_1(0) + S_2(0). \tag{40.14}$$

Here $S_1(z)$, as (40.11) shows, decreases to zero. It follows from (40.14) that the transformation of the energy from the wave A_1 to the wave A_2 proceeds in the ratio ω_2/ω_1 and the rest of the energy $(\omega_1 - \omega_2)/\omega_1 = \omega_0/\omega_1$ is lost by spontaneous emission at the frequency ω_0. Expression (40.11) can be interpreted as follows. When quanta of frequency ω_1 are propagated they are transformed into quanta $\hbar\omega_2$ with a loss of energy of $\hbar\omega_0 = \hbar\omega_1 - \hbar\omega_2$, so that the total number of quanta of frequencies ω_1 and ω_2 is conserved. Each quantum $\hbar\omega_1$ is transformed into one, and only one, quantum $\hbar\omega_2$. When all the quanta $\hbar\omega_1$ are transformed into quanta $\hbar\omega_2$, energy transformation has occurred in the ratio ω_2/ω_1. The characteristic length in which considerable energy transformation occurs is, from (40.5) and (40.6), given by

$$l^{-1} = \frac{4\pi\omega_2^2}{c} |\chi''(\omega_2, |A_1|^2)| = \frac{4\pi\omega_2^2\xi}{c} |A_1|^2, \tag{40.15}$$

where A_1 is the initial value of the amplitude of the field E_1.

By using the phenomenon of stimulated Raman emission we can design maser amplifiers and oscillators in the optical range (Raman lasers). We shall leave the discussion of this topic to Chapter XII in Volume 2, which deals with lasers.

40.2. A Comparison of the Raman Effect and Parametric Interaction

It is interesting to compare the Raman type of excitation of a system (negative resistance at the frequency ω_2) discussed in this section with parametric excitation. The qualitative features of this comparison can be most easily understood with a discrete model. Let us first examine a typical situation corresponding to parametric pumping (see, e.g., Bloom and Chang, 1957). A pumping field with a frequency ω_1 acts on a non-linear element N (Fig. IX.7, *a*). Two *LCR*-circuits are connected with this non-linear element:

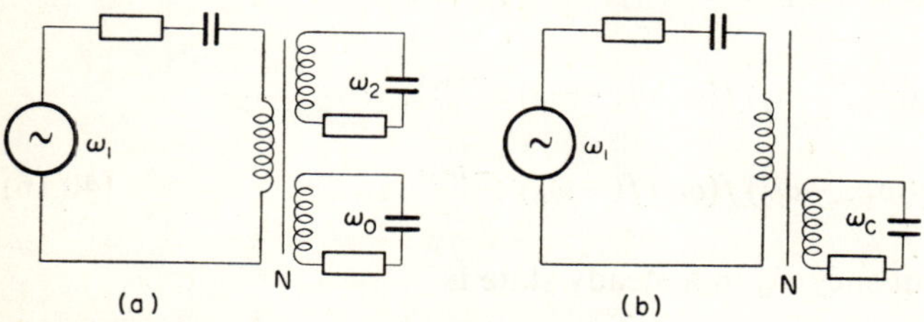

FIG. IX.7. Diagram illustrating parametric pumping at a frequency $\omega_1 = \omega_0 + \omega_2$. A non-linear inductance is taken as the non-linear element. If the circuit and the non-linear element N are looked upon as a single system (*b*), the pumping of this system leads to negative resistance, as follows from section 38, at the frequency $\omega_2 = \omega_1 - \omega_0$.

"signal" and "idler" circuits whose eigenfrequencies ω_0 and ω_2 satisfy the relation

$$\omega_1 = \omega_0 + \omega_2 .$$

Under certain conditions self-excitation occurs (or additional negative resistance appears) in the two circuits. The negative resistance, or the resultant instability, is a consequence of the parametric interaction of the fields in the two circuits in the presence of pumping. The pumping, and the field in one of the circuits, create a field in the other circuit with a phase such that the work done by this field is negative (see the beginning of the chapter). If one of the circuits, e.g. the idler, is taken away from the system the other circuit will not be excited. Therefore the parametric excitation depends not only on the non-linear element but is a property of the electromagnetic system as a whole, i.e. of the complete system of circuits.

Let us now examine the Raman-type excitation of a system. Here a field with a frequency ω_1 also acts on the non-linear element (which has a resonance frequency ω_0). Negative resistance appears in this case at a frequency $\omega_2 = \omega_1 - \omega_0$. However, this negative resistance is the resistance of the non-linear element in the presence of pumping. This resistance can be connected to any circuit with a natural frequency $\sim\omega_2$ and instability will appear in it. Therefore in this type of excitation the negative resistance is a property of the non-linear element itself in the presence of pumping.

It is now easy to establish the connection between parametric and Raman excitation. To do this we take as the non-linear system in the former case a molecular system together with an idler circuit of frequency ω_0 (see Fig. IX.7, *b*). In this kind of system a pumping field with a frequency ω_1 causes the appearance of negative resistance if a test circuit of frequency $\omega_2 = \omega_1 - \omega_0$ is connected. The parametric excitation can be looked upon as stimulated Raman emission in the idler circuit ω_0. The schematic calculation which justifies this argument can be set out as follows. The parametric excitation is connected, as has already been pointed out, with the second term in the expansion (22.25). The responses to the frequencies are respectively of the form

$$x(\omega_0) = \chi(\omega_1, -\omega_2) f(\omega_1) f(-\omega_2)\, e^{-i\omega_0 t},$$

$$x(\omega_2) = \chi(\omega_1, -\omega_0) f(\omega_1) f(-\omega_0)\, e^{-i\omega_2 t}. \tag{40.16}$$

The field at the frequency ω_0 in a steady state is

$$f(-\omega_0) = \alpha x^*(\omega_0) = \alpha\chi^*(\omega_1, -\omega_2) f^*(\omega_1) f^*(-\omega_2)\, e^{i\omega_0 t}, \tag{40.17}$$

where α is a coefficient which depends on the properties of the circuits.

Substituting (40.17) in the second equation of (40.16) we obtain

$$x(\omega_2) = \alpha\chi(\omega_1, -\omega_0)\,\chi^*(\omega_1, -\omega_2)\,|f(\omega_1)|^2 f(\omega_2)\,\mathrm{e}^{-i\omega_2 t},$$

where the quantity $\alpha\chi(\omega_1, -\omega_0)\,\chi^*(\omega_1, -\omega_2)\,|f(\omega_1)|^2$ is the susceptibility at the frequency ω_2 and, in accordance with the general theory, the imaginary part of this quantity should be negative for $\omega_1 > \omega_2$.

Obviously we can also say that Raman excitation is indistinguishable from the parametric case if the "circuit" of the molecular system is taken as the idler circuit. The satisfaction of the relations (39.38) or (40.11), which are possible Manley–Rowe (1956) relations, is comprehensible from this point of view. Relations of this kind can easily be obtained for parametric interaction from the general expression for the change in the field energy given at the beginning of this chapter. The anharmonic oscillator discussed in section 38.4 may be taken as an example of a molecular system circuit.

It should be stressed that if we do not include part of the radiation in the molecular system (as we have just done) and examine the parametric and Raman systems from the standpoint of the interaction of the radiation with the substance, then they are different processes with different interactions between the radiation and the substance. Each of these processes has its own features. Moving on to distributed systems, let us compare the parametric interaction of electromagnetic waves in a non-linear dielectric treated by Khokhlov (1961; Akhmanov and Khokhlov, 1962) and Bloembergen *et al.* (Armstrong, Bloembergen, Ducuing and Pershan, 1962) (see section 39) with the Raman interaction (the present section).

Whilst effective parametric interaction requires that the frequency condition is satisfied, there is no such condition for Raman interaction; negative absorption is created in the medium and there is amplification of a wave at the frequency $\omega_2 = \omega_1 - \omega_0$. In the case of parametric interaction it may also be considered that one wave creates negative or positive resistance for another wave, but this resistance depends essentially on the presence of an idler wave. These are the essential qualitative differences between the two wave interaction mechanisms being discussed.

Let us now compare the order of magnitude of the effects of the parametric and Raman interaction of waves. To make an estimate we can write the explicit expression for the susceptibility in the case when resonance absorption is connected with the two levels of the system. Let the interaction with the dissipative system be spin–lattice in nature. In order to be able to consider ϱ_1 and ϱ_2 in (38.25) as the density matrix of the dynamic sub-system (atom, molecule, etc.) we must make the substitution

$$\zeta(x) \rightarrow \frac{1}{x + i\gamma}, \tag{40.18}$$

where γ is connected with the corresponding relaxation coefficient in the transport equation.

We can take as the basis of the substitution (40.18) the direct calculation of the susceptibility by means of the transport equation (see also section 22.5), which, however, we shall not carry out here but leave to the reader to do for himself. As a result of this calculation (or the substitution of (40.18) in (38.25)) we arrive at the expression

$$\chi''(\omega_2) = \frac{1}{\hbar}\,\frac{(\varrho_2 - \varrho_1)\dfrac{1}{T_2}}{(\omega_1 - \omega_2 - \omega_0)^2 + \dfrac{1}{T_2^2}} \times \left|\sum_k \left(\frac{d_{1k}V_{1k}^{(-1)}}{\hbar(-\omega_1 - \omega_{k2})} - \frac{V_{k2}^{(-1)}d_{k2}}{\hbar(-\omega_1 - \omega_{1k})}\right)\right|^2, \qquad (40.19)$$

where ϱ_1 and ϱ_2 are the populations per unit volume ($\varrho_1 > \varrho_2$) and $T_{2k}^{(-1)}$ is a measure of the width of the absorption line at the frequency ω_0.

We are now in a position to compare the characteristic lengths l_p of the parametric interaction and l_k of the Raman interaction. Under the most favourable conditions when there is exact phase matching the characteristic length of the parametric interaction is given by (39.33). Comparing this with (40.15) and (40.19) when (40.1) holds, and using (22.23), we obtain the following estimate:

$$\frac{l_p}{l_k} = \frac{dm_1}{\hbar T_2^{-1}}, \qquad (40.20)$$

where d is the dipole moment of the molecule or atom and we have assumed that the dipole moment matrix elements contained in (22.23) and (40.19) are quantities of the same order of magnitude as the number of atoms per unit volume. It follows from (40.20) that, although Raman interaction is a higher-order effect than parametric interaction (quadratic in the field), for strong fields (which may be obtained with a laser)

$$m_1 \sim \frac{\hbar T_2^{-1}}{d},$$

the lengths l_p and l_k are comparable.

Let us conclude this section by briefly discussing the degenerate case when $\omega_2 = \omega_1/2$. As follows from (22.19), the response at the frequency $\omega_1/2$ appears in the second application of perturbation theory ($\omega_l + \omega_s = \omega_1$). In this case, however, the susceptibility of the system depends essentially on the phase relations between the pumping field (ω_1) and the signal field (ω_2). In other words, the second term in (22.25) cannot give a negative resistance that is independent of this phase relation. The situation is characteristic of a synchronous parametric amplifier (see section 39.2, which shows that amplification of a fundamental frequency wave may or may not occur depending upon the initial phase). On the other hand, the third term in (22.19) leads to negative resistance (38.25) which does not depend upon the phase relation between the pumping field ω_1 and the signal at the frequency $\omega_1/2$. This difference exists because, during Raman interaction, the idler circuit (the molecular system) and the circuit at the signal frequency are by no means identical, as in the case of a synchronous parametric amplifier.

Appendix I

A.1. The Singular Functions $\delta(x)$, $\zeta(x)$ and P/x

It is well known that singular functions of the type $\delta(x), \zeta(x)$, etc. are widely used in physics; they occur frequently in this book. They are operator quantities which have a meaning only when preceded by an integration sign. The definitions of these functions are as follows:

$$\int_{-a}^{b} \delta(x) f(x)\, dx = f(0), \tag{A.I.1}$$

where the interval $(-a, b)$ includes the point 0;

$$\int_{-a}^{b} \frac{P}{x} f(x)\, dx = \lim_{\varepsilon \to 0} \int_{-a}^{-\varepsilon} \frac{f(x)}{x}\, dx + \int_{+\varepsilon}^{b} \frac{f(x)}{x}\, dx, \tag{A.I.2}$$

where the integral of P/x is taken to be the principal value integral. The function $\zeta(x)$ is defined by

$$\zeta(x) = \frac{P}{x} - i\pi\delta(x). \tag{A.I.3}$$

For reference purposes we give below different forms of these singular functions as limits of continuous functions.

First, however, we shall give an example to illustrate both the meaning of the δ-function and the limits of applicability of expressions using the δ-function.

Let us examine the integral:

$$I = \int_{-a}^{b} \delta_t(\omega - \omega_0) f(\omega)\, d\omega, \quad \delta_t(\omega - \omega_0) = \frac{\sin^2 (\omega - \omega_0)\, t}{\pi t(\omega - \omega_0)^2},$$

where the point ω_0 lies within the interval $(-a, b)$. Introducing a new integration variable $y = (\omega - \omega_0)\, t$ we transform the integral to the form

$$I = \int_{-at-\omega_0}^{bt-\omega_0} \frac{\sin^2 y}{\pi y^2} f\left(\omega_0 + \frac{y}{t}\right) dy.$$

We now let ω^* denote the range of ω near ω_0 in which the function $f(\omega)$ changes only slightly when ω changes by an amount which is small compared with ω^*. In other words, ω^* is the characteristic scale of the variation of the quantity ω near ω_0. We notice further that the function $\sin^2 y/y^2$ is finite in the region $-\pi \leqq y \leqq +\pi$ (i.e. for values $y^2 \sim 1$). By using this fact and the meaning of the quantity ω^* we find that for

$$t \gg \frac{1}{\omega^*}, \quad \frac{1}{b+a} \tag{A.I.4}$$

the integral I becomes

$$I \approx f(\omega_0) \int_{-\infty}^{\infty} \frac{\sin^2 y}{\pi y^2}\, dy = f(\omega_0).$$

Therefore for a value of t sufficiently large to satisfy condition (A.I. 4), the quantity $\delta_t(\omega - \omega_0) = [\sin^2 (\omega - \omega_0)\, t]/[\pi t(\omega - \omega_0)^2]$ is the same as the function $\delta(\omega - \omega_0)$ since

$$\int \delta_t(\omega - \omega_0) f(\omega)\, d\omega \approx f(\omega_0).$$

It is obvious that the criterion for the applicability of the expression $\delta_t(\omega - \omega_0)$ as a δ-function depends essentially on the quantity ω^* characterizing the function $f(\omega)$ to which the δ-function is being applied.

The following are its different forms:

$$\begin{aligned}
\delta(x) &= \frac{1}{\pi} \lim_{K\to\infty} \frac{\sin Kx}{x} = \frac{1}{\pi} \lim_{K\to\infty} \frac{\sin^2 Kx}{Kx^2} \\
&= \frac{1}{\pi} \lim_{K\to\infty} \frac{1 - \cos Kx}{Kx^2} \\
&= \lim_{K\to\infty} \frac{1}{2\pi} \int_{-K}^{K} ds\, e^{ixs} \\
&= \frac{1}{\pi} \lim_{\sigma\to 0} \frac{\sigma}{x^2 + \sigma^2} = \lim \sum_{n=0}^{N} u_n^*(x)\, u_n(0),
\end{aligned} \tag{A.I.5}$$

where $u_n(x)$ is a complete orthonormalized set of functions. The ζ-function and the function P/x are defined by the following relations:

$$\zeta(x) = -i \lim_{K\to\infty} \int_0^K e^{isx}\, ds = \lim_{K\to\infty} \frac{1 - e^{iKx}}{x} = \lim_{\sigma\to 0} \frac{1}{x + i\sigma}, \tag{A.I.6}$$

$$\frac{P}{x} = \lim_{K\to\infty} \frac{1 - \cos Kx}{x} = \lim_{K\to\infty} \int_0^K \sin sx\, ds = \lim_{\sigma\to\infty} \frac{x}{x^2 + \sigma^2}. \tag{A.I.7}$$

In the complex plane the functions P/x, $\zeta(x)$, $\zeta^*(x)$ and $2\pi i\delta(x)$ all behave as $1/x$, provided that the path of integration is chosen carefully for each function. For P/x integration should be carried out along the real axis omitting the region between $-\varepsilon$ and $+\varepsilon$. The functions $\zeta(x)$ and $\zeta^*(x)$ must be integrated along the real axis, taking the integration path near zero above and below zero respectively; the function $2\pi i\delta(x)$ should be integrated anti-clockwise along a closed path round zero.

References

ADAMS, E.N. (1960) *Phys. Rev.* **120,** 675.

AKHIEZER, A.I. and BERESTETSKII, V.B. (1959) *Quantum Electrodynamics* (in Russian), Fizmatgiz. *Elements of Quantum Electrodynamics.* Translated by G.M. Yolkoff, Wiley, New York (1964).

AKHMANOV, S.A. and KHOKHLOV, R.V. (1962) *Zh. eksp. i teoret. fiz.* **43,** 357; (1963) *Soviet Physics JETP*, **16,** 252.

AL'TSHULER, S.A. and KOZYREV, B.M. (1961) *Electron Paramagnetic Resonance* (in Russian), Fizmatgiz. Translated by Scripta Technica, Inc., Academic Press, New York (1964).

ANDREW, E.R. (1955) *Nuclear Magnetic Resonance*, Cambridge University Press.

ARMSTRONG, J.A., BLOEMBERGEN, N., DUCUING, J. and PERSHAN, P.S. (1962) *Phys. Rev.* **127,** 1918.

ASHBY, W.R. (1956) *Introduction to Cybernetics*, Chapman, London.

BASOV, N.G., KROKHIN, O.N. and POPOV, YU.M. (1960) *Uspekhi fiz. nauk*, **72,** 161; (1961) *Soviet Physics Uspekhi*, **3,** 702.

BASOV, N.G. and PROKHOROV, A.M. (1954) *Zh. eksp. i teoret. fiz.* **27,** 431.

BERNARD, W. and CALLEN, H.B. (1959) *Rev. Mod. Phys.* **31,** 1017.

BETHE, H.A. (1947) *Phys. Rev.* **72,** 339.

BLOCH, F. (1946) *Phys. Rev.* **70,** 460.

BLOCH, F. (1957) *Phys. Rev.* **105,** 1206.

BLOEMBERGEN, N. (1956) *Phys. Rev.* **104,** 324.

BLOEMBERGEN, N. and POUND, R.V. (1954) *Phys. Rev.* **95,** 8.

BLOEMBERGEN, N., SHAPIRO, S., PERSHAN, P.S. and ARTMAN, J.O. (1959) *Phys. Rev.* **114,** 445.

BLOOM, S. and CHANG, K.K.N. (1957) *RCA Review*, **18,** 578.

BOGOLYUBOV, N.N. (1947) *Problems of Dynamic Theory in Statistical Physics* (in Russian), Gostekhizdat. Translation in *Studies in Statistical Mechanics*, Vol. I, North-Holland, Amsterdam (1962).

BRILLOUIN, L. (1956) *Science and Information Theory*, Academic Press, New York.

BUNKIN, F.V. (1962) "*Izv. VUZov*", *Radiofiz.* **5,** 687.

BUNKIN, F.V. and ORAEVSKII, A.N. (1959) "*Izv. VUZov*", *Radiofiz.* **2,** 181.

CALLEN, H.B. and WELTON, T.A. (1951) *Phys. Rev.* **83,** 34.

CHANDRASEKHAR, S. (1943) *Rev. Mod. Phys.* **15,** 1.

CLOGSTON, A.M. (1958) *J. Phys. Chem. Solids*, **4,** 271.

DICKE, R.H. (1954) *Phys. Rev.* **93,** 99.

DIRAC, P.A.M. (1926) *Proc. Roy. Soc.* **AIII,** 281.

DIRAC, P.A.M. (1958) *Principles of Quantum Mechanics*, Clarendon Press, Oxford.

EINSTEIN, A. (1917) *Physik. Zeitschr.* **18,** 121.

ELSASSER, W.M. (1937) *Phys. Rev.* **52,** 987.

FABRIKANT, V.A., VUDYNSKII, M.M. and BUTAEVA, F.A. (1951) Patent Cert. No. 148441 (576749/26 of 18 June 1951).

FAIN, V.M. (1957a) *Zh. eksp. i teoret. fiz.* **32,** 607; (1957) *Soviet Physics JETP*, **5,** 501.

FAIN, V.M. (1957b) *Zh. eksp. i teoret. fiz.* **33**, 416; (1958) *Soviet Physics JETP*, **6**, 323.
FAIN, V.M. (1957c) *Zh. eksp. i teoret. fiz.* **33**, 945; (1958) *Soviet Physics JETP*, **6**, 726.
FAIN, V.M. (1958) *Uspekhi fiz. nauk*, **64**, 273.
FAIN, V.M. (1959a) "*Izv. VUZov*", *Radiofizika*, **2**, 167.
FAIN, V.M. (1959b) *Zh. eksp. i teoret. fiz.* **36**, 798; (1959) *Soviet Physics JETP*, **9**, 562.
FAIN, V.M. (1962) *Zh. eksp. i teoret. fiz.* **42**, 1075; (1962) *Soviet Physics JETP*, **15**, 743.
FAIN, V.M. (1963a) *Uspekhi fiz. nauk*, **79**, 641; (1963) *Soviet Physics Uspekhi*, **6**, 294.
FAIN, V.M. (1963b) *Zh. eksp. i teoret. fiz.* **44**, 1915; (1963) *Soviet Physics JETP*, **17**, 1289.
FAIN, V.M. (1963c) "*Izv. VUZov*", *Radiofizika*, **6**, 207.
FAIN, V.M. (1067) *Zh. eksp. i teoret. fiz.* **52**, No.–6; (1967) *Soviet Physics JETP*, **25** No. 6.
FAIN, V.M., KHANIN, YA. I. and YASHCHIN, E.G. (1962) "*Izv. VUZov*", *Radiofizika*, **5**, 697.
FAIN, V.M., KHANIN, YA.I. and YASHCHIN, E.G. (1964) "*Izv. VUZov*", *Radiofizika* **7**, 386.
FAIN, V.M. and YASHCHIN, E.G. (1964) *Zh. eksp. i teoret. fiz.* **46**, 695; (1964) *Soviet Physics JETP*, **19**, 474.
FANO, U. (1957) *Rev. Mod. Phys.* **29**, 74.
FARAGO, P.S. and MARX, G. (1955) *Phys. Rev.* **99**, 1063.
FERMI, E. (1951) *Elementary Particles*, Oxford University Press, London.
FEYNMAN, R.P., VERNON, F.L. and HELLWARTH, R.W. (1957) *J. App. Phys.* **28**, 49.
FOGARASSY, B. (1959) *Acta Phys. Hungarica*, **10**, 305.
FRANKEN, P.A. and WARD, J.F. (1963) *Rev. Mod. Phys.* **35**, 23.
GAPONOV, A.V. (1960) *Zh. eksp. i teoret. fiz.* **39**, 326; (1961) *Soviet Physics JETP*, **12**, 232.
GENKIN, V.N. (1962) *Fiz. tverd. tela*, **4**, 3381; (1963) *Soviet Physics Solid State*, **4**, 2475.
GENKIN, V.N. (1964) *Fiz. tverd. tela*, **6**, 368; (1964) *Soviet Physics Solid State*, **6**, 295.
GENKIN V.N. and MEDNIS, P.M. (1967) "Izv. VUZov", *Radiofizika*, **10**, 192; (1968) *Soviet Radiophysics*, 10.
GINZBURG, V.L. (1939) *Dokl. Akad. Nauk*, **24**, 130.
GINZBURG, V.L. (1940) *Zh. eksp. i teoret. fiz.* **10**, 589, 601.
GINZBURG, V.L. (1943) *Zh. eksp. i teoret. fiz.* **13**, 33.
GINZBURG, V.L. (1947) *Uspekhi fiz. nauk*, **31**, 320.
GINZBURG, V.L. (1952) *Uspekhi fiz. nauk*, **46**, 348.
GINZBURG, V.L. (1959) *Uspekhi fiz. nauk*, **69**, 537; (1960) *Soviet Physics Uspekhi*, **2**, 874.
GINZBURG, V.L. (1960) *Propagation of Electromagnetic Waves in a Plasma* (in Russian). Fizmatgiz. Translated by J.B. Sykes and R.J. Taylor, Pergamon Press, Oxford (1964).
GINZBURG, V.L. and FAIN, V.M. (1957a) *Zh. eksp. i teoret. fiz.* **32**, 162; (1957) *Soviet Physics JETP*, **5**, 123.
GINZBURG, V.L. and FAIN, V.M. (1957b) *Radiotekhn. i elektron.* **2**, 780.
GIORDMAINE, J.A. (1962) *Phys. Rev. Lett.* **8**, 19.
GOLDEN, S. and LONGUET-HIGGINS, H.C. (1960) *J. Chem. Phys.* **33**, 1479.
GÖPPERT-MAYER, M. (1931) *Ann. der Phys.* **9**, 273.
GORDY, W., SMITH, W.V. and TRAMBARULO, R.F. (1953) *Microwave Spectroscopy*, Wiley, New York.
GORDON, J.P., ZEIGER, H.J. and TOWNES, C.H. (1954) *Phys. Rev.* **95**, 282.
GRADSHTEIN, I.S. and RYZHIK, I.M. (1962) *Tables of Series, Integrals and Products* (in Russian), Fizmatgiz. Translated by Scripta Technica, Inc. Academic Press, New York (1965).
GROOT, S.R. DE (1951) *Thermodynamics of Irreversible Processes*, North-Holland, Amsterdam.
HAAR, D. TER (1961) *Elements of Hamiltonian Mechanics*, North-Holland, Amsterdam.
HEITLER, W. (1954) *Quantum Theory of Radiation*, Clarendon Press, Oxford.
HÖHLER, G. (1958) *Zeits. f. Phys.* **152**, 546.

HUBBARD, P.S. (1961) *Rev. Mod. Phys.* **33,** 249.

INGRAM, D. (1955) *Spectroscopy at Radio and Microwave Frequencies*, Butterworth, London.

ITKINA, M.A. and FAIN, V.M. (1958) "*Izv. VUZov*", *Radiofizika*, **1,** 30.

JAVAN, A. (1957) *Phys. Rev.* **107,** 1579.

JAVAN, A., BENNETT, W.R. JR. and HERRIOTT, D.R. (1961) *Phys. Rev. Lett.* **6,** 106.

KARPLUS, R. and SCHWINGER, J. (1948) *Phys. Rev.* **73,** 1020.

KARPMAN, V.I. (1963) *Zh. eksp. i teoret. fiz.* **44,** 1307; (1963) *Soviet Physics JETP*, **17,** 882.

KHALFIN, L.A. (1957) *Zh. eksp. i teoret. fiz.* **33,** 1371; (1958) *Soviet Physics JETP*, **6,** 1053.

KHOKHLOV, R.V. (1961) *Radiotekhn. i elektron.* **6,** 1116.

KIEL, A. (1960) *Phys. Rev.* **120,** 137.

KLEIN, O. (1931) *Zeits. f. Phys.* **72,** 767.

KOGAN, SH.M. (1962) *Zh. eksp. i teoret. phys.* **43,** 304; (1963) *Soviet Physics JETP*, **16,** 217.

KOLOMENSKII, A.A. (1953) *Zh. eksp. i teoret. fiz.* **24,** 167.

KONTOROVICH, V.M. and PROKHOROV, A.M. (1957) *Zh. eksp. i teoret. fiz.* **33,** 1428; (1958) *Soviet Physics JETP*, **6,** 1100.

KOPVILLEM, U. KH. (1960) *Fiz. tverd. tela* **2,** 1829; (1961) *Soviet Physics Solid State*, **2,** 1653.

KRONIG, R. DE L. and BOUWKAMP, C.J. (1938) *Physica*, **5,** 521.

KUBO, R. (1957) *J. Phys. Soc. Japan*, **12,** 570.

LANDAU, L.D. (1927) *Zeits. f. Phys.* **45,** 430; *Collected Papers*, Pergamon and Gordon and Breach, New York (1965), 8.

LANDAU, L.D. and LIFSHITZ, E.M. (1951) *Statistical Physics* (in Russian), Fizmatgiz. Translated by E. Peierls and R.F. Peierls, Pergamon Press, London (1958).

LANDAU, L.D. and LIFSHITZ, E.M. (1957) *Electrodynamics of Continuous Media* (in Russian), Gostekhizdat. Translated by J.B. Sykes and J.S. Bell, Pergamon Press, Oxford (1960).

LANDAU, L.D. and LIFSHITZ, E.M. (1958) *Mechanics* (in Russian), Fizmatgiz. Translated by J.B. Sykes and J.S. Bell, Pergamon Press, Oxford (1960).

LANDAU, L.D. and LIFSHITZ, E.M. (1960) *Classical Theory of Fields* (in Russian), Fizmatgiz. Translated by M. Hamermesh, Pergamon Press, Oxford (1962).

LANDAU, L.D. and LIFSHITZ, E.M. (1963) *Quantum Mechanics* (in Russian), Fizmatgiz. Translated by J.B. Sykes and J.S. Bell, Pergamon Press, London (1958).

LANDSBERG, G. and MANDELSTAM, L. (1928) *Naturwissenschaften*, **16,** 557, 772.

LAX, M. (1960) *Rev. Mod. Phys.* **32,** 25.

LEONTOVICH, M.A. (1941) *J. Phys. (USSR)*, **4,** 499.

MACDONALD, D.K.C. and KOMPFNER, R. (1949) *Proc. IRE*, **37,** 1424.

MAIMAN, T.H. (1960) *Nature (Lond.)*, **187,** 493.

MAKER, P.D., TERHUNE, R.W., NISENOFF, M. and SAVAGE, C.M. (1962) *Phys. Rev. Lett.* **8,** 21.

MANLEY, J.M. and ROWE, H.E. (1956) *Proc. IRE*, **44,** 904.

MUZIKARZH, CH. (1961) *Zh. eksp. i teoret. fiz.* **41,** 1168; (1962) *Soviet Physics JETP*, **14,** 833.

NEUMANN, J. VON (1932) *Mathematische Grundlagen der Quantenmechanik*, Springer, Berlin. Translated by R.T. Beyer, *Mathematical Foundations of Quantum Mechanics*, Princeton University Press, Princeton, N.J. (1955).

NYE, J.F. (1957) *Physical Properties of Crystals*, Clarendon Press, Oxford.

NYQUIST, H. (1928) *Phys. Rev.* **32,** 110.

PLACZEK, G. (1935) *Rayleigh Scattering and the Raman Effect* (in Russian), Gos. nauchn.-tekhn. izd. Ukrainy; *Rayleigh-Streuung und Raman-Effekt*, Leipzig (1934).

PODGORETSKII, M.I. and ROIZEN, I.I. (1960) *Zh. eksp. i teoret. fiz.* **39,** 1473; (1960) *Soviet Physics JETP*, **12,** 1023.

RAMAN, C.V. and KRISHNAN, K.S. (1928) *Nature (Lond.)*, **121,** 501.

References

RYZHOV, YU. A. (1959) "*Izv. VUZov*", *Radiofizika*, **2,** 689.
SCHAWLOW, A.L. and TOWNES, C.H. (1958) *Phys. Rev.* **112,** 1940.
SCHIFF, L.I. (1955) *Quantum Mechanics*, McGraw–Hill, New York.
SCHULMAN, C. (1951) *Phys. Rev.* **82,** 116.
SCOVIL, H.E.D., FEHER, G. and SEIDEL, H. (1957) *Phys. Rev.* **105,** 762.
SENITZKY, I.R. (1954) *Phys. Rev.* **95,** 904.
SENITZKY, I.R. (1955) *Phys. Rev.* **98,** 875.
SENITZKY, I.R. (1956) *Phys. Rev.* **104,** 1486.
SENITZKY, I.R. (1958) *Phys. Rev.* **111,** 3.
SENITZKY, I.R. (1959) *Phys. Rev.* **115,** 227.
SENITZKY, I.R. (1960a) *Phys. Rev.* **119,** 670.
SENITZKY, I.R. (1960b) *Phys. Rev.* **119,** 1807.
SENITZKY, I.R. (1961) *Phys. Rev.* **123,** 1525.
SHTEINSHLEIGER, V.B. (1955) *Wave Interaction Phenomena in Electromagnetic Resonators* (in Russian), Oborongiz.
SILIN, V.P. and RUKHADZE, A.A. (1961) *Electromagnetic Properties of a Plasma and Plasma-Like Media* (in Russian), Gosatomizdat.
SINGER, J.R. (1959) *Masers*, Wiley, New York.
SMITH, L.P. (1946) *Phys. Rev.* **69,** 195.
SOMMERFELD, A.J.W. (1955) *Thermodynamics and Statistical Mechanics* (Russian edition). Translated by J. Kestin, Academic Press, New York (1956).
STRANDBERG, M. (1954) *Microwave Spectroscopy*, Methuen, London.
TOWNES, C.H. and SCHAWLOW, A.L. (1955) *Microwave Spectroscopy*, McGraw–Hill, New York.
TROUP, G.J.F. (1959) *Masers*, Methuen, London.
VAN HOVE, L. (1955) *Physica*, **21,** 517.
VAN VLECK, J. (1948) *Phys. Rev.* **74,** 1229.
VITT, A. and GORELIK, G. (1933) *Zh. tekhr. fiz.* **3,** 294.
VUYLSTEKE, A.A. (1960) *Elements of Maser Theory*, Van Nostrand, Princeton, N.J.
WANGSNESS, R.K. and BLOCH, F. (1953) *Phys. Rev.* **89,** 728.
WEBER, J. (1953a) *IRE Trans. on Electron Devices*, **3,** 1.
WEBER, J. (1953b) *Phys. Rev.* **90,** 977.
WEBER, J. (1954) *Phys. Rev.* **94,** 211, 215.
WEISSKOPF, V. (1935) *Naturwissenschaften*, **23,** 631.
WEISSKOPF, V. and WIGNER, E. (1930a) *Zeits. f. Phys.* **63,** 54.
WEISSKOPF, V. and WIGNER, E. (1930b) *Zeits. f. Phys.* **65,** 18.
WIENER, N. (1948) *Cybernetics*, Wiley, New York.
ZWANZIG, R. (1960) *J. Chem. Phys.* **33,** 1338.

Index

Index

Index